T0360537

Differential Sheaves
and Connections
A Natural Approach to Physical Geometry

SERIES ON CONCRETE AND APPLICABLE MATHEMATICS

ISSN: 1793-1142

Series Editor: Professor George A. Anastassiou
Department of Mathematical Sciences
University of Memphis
Memphis, TN 38152, USA

---

*To view the complete list of the published volumes in the series, please visit:
http://www.worldscientific/series/scaam

Series on Concrete and Applicable Mathematics – Vol.18

Differential Sheaves and Connections

A Natural Approach to Physical Geometry

Anastasios Mallios
Elias Zafiris

National and Kapodistrian University of Athens, Greece

World Scientific

NEW JERSEY · LONDON · SINGAPORE · BEIJING · SHANGHAI · HONG KONG · TAIPEI · CHENNAI · TOKYO

Published by

World Scientific Publishing Co. Pte. Ltd.
5 Toh Tuck Link, Singapore 596224
USA office: 27 Warren Street, Suite 401-402, Hackensack, NJ 07601
UK office: 57 Shelton Street, Covent Garden, London WC2H 9HE

Library of Congress Cataloging-in-Publication Data
Names: Mallios, Anastasios. | Zafiris, Elias, 1970–
Title: Differential sheaves and connections : a natural approach to physical geometry /
    Anastasios Mallios and Elias Zafiris (University of Athens, Greece).
Description: New Jersey : World Scientific, 2015. | Series: Series on concrete and
    applicable mathematics | Includes bibliographical references and index.
Identifiers: LCCN 2015028030 | ISBN 9789814719469 (alk. paper)
Subjects: LCSH: Sheaf theory. | Geometry, Differential. | Algebraic topology.
Classification: LCC QA612.36 .M34 2015 | DDC 514/.224--dc23
LC record available at http://lccn.loc.gov/2015028030

British Library Cataloguing-in-Publication Data
A catalogue record for this book is available from the British Library.

In-house Editors: V. Vishnu Mohan/Kwong Lai Fun

Typeset by Stallion Press
Email: enquiries@stallionpress.com

Printed in Singapore

ἐτεῇ δέ οὐδέν ἤδμεν
ἐν βυθῷ γάρ ἡ ἀλήθεια

Preface

The central nucleus of this treatise targets a perennial issue of Modern Mathematical Physics, or even better, and in alignment with the spirit of this work, of its inversion into *"Physical Mathematics"*, namely the issue of thinking about and modeling or describing *"Physical Geometry"*. The notion of a physical geometry is considered always as the outcome of physical laws pertaining to the regime of a physical theory. The basic question emerging in this context is the following: Does there exist a natural description of physical geometry? By natural we mean a description, which is not based on ad hoc conventions involving a "God-given" geometric coordinate substratum of any local or global form, but it refers to the pertinent physical relations themselves together with their empirical realization in terms of observed events. Of course, the required naturality in the description of a physical geometry should be initially reflected in the way that we set up a sufficient network of concepts and modeling tools in order to capture these relations and become able to distinguish them and eventually to make predictions.

Several issues of an interpretational nature arise together with the requirement of naturality, which force a critical re-examination of the conceptual choices and assumptions initially endorsed from a physical viewpoint. The two most significant ones in relation to a natural approach to physical geometry are the concepts of locality and differentiability. In the opening section of the Prolegomena to the general theory that follows, I will attempt to analyze briefly the related problems, so as to motivate conceptually the focus of the whole treatise and prepare the ground for the proposed resolution provided by the notions of differential sheaves and connections. Technically speaking, the present work constitutes a confluence of the ideas and methods emanating from two sources: The first source is the theory of Abstract (alias Modern) Differential Geometry (ADG), developed by Mallios, using the notions of vector sheaves, connections and sheaf cohomology, and applied in the formulation and interpretation of gauge theories and general relativity, aiming at a background independent understanding of these physical theories. The second source is the theory of topoi, developed by Grothendieck, as generalized localization environments for needs in algebraic geometry and homological algebra going beyond the classical theory of topological spaces. Recently, it has been demonstrated that the theory of Grothendieck topoi offers a precise framework to understand the notion of localization in quantum theories, so it bears a physical significance as well, together with the rich mathematical innovations involved in these developments. These two sources generate a conjoined stream of new concepts and technical tools for a natural approach to physical geometry. The unifying thread is provided by the ubiquitous notion of adjoint functors in category theory, which characterizes the naturality of the purported approach to physical geometry in functorial terms. This is enough to legitimize the title of this work for the time being.

I turn now promptly to the history of the treatise itself, which has been unfortunately non-smooth and disrupted. The treatise has been in the process of scrupulous preparation by Anastasios Mallios as a sole author during the last five years. Sadly, he passed away at the age of 81 years old, when the manuscript has been in the very last stages of its completion. Immediately after being informed about this sad event, George Anastassiou of the University of Memphis, took the initiative to make possible the publication of Mallios's last work and kindly sent an invitation for a final completed version of the initial manuscript to appear in his World Scientific series. So I would like to thank him wholeheartedly for his immediate involvement in preserving Mallios's last writing, his invitation for publication of this work in the context of his series, and for the overall support and encouragement during the final stage of completion of this project.

Next, I feel that I should justify my appearance as a co-author of this book in the present published form. I have been a very close collaborator of Mallios during the last 10 years in the application of ADG in problems of theoretical physics, especially in relation to quantum mechanics and quantum gravity. It was during one of our long regular work meetings in Athens nine years ago that, while talking about a possible topos-theoretic extension of the framework of ADG, I formulated quite clearly the idea that the notion of adjoint functors, and in particular the Hom-Tensor adjunction, should play the role of a connecting bridge between ADG and topos theory. My motivation came from quantum mechanics and more precisely from my previous work on quantum logic, quantum probability, and the problem of quantum measurement, where a topos-theoretic notion of localization in relation to physical representability or observability is precisely rooted on a pair of adjoint functors of the Hom-Tensor universal form. I recall that after making the argument, Mallios was quite critical and hesitant to accept it in the first place. After all, is there any physical meaning underlying such an abstract concept from category theory? We finished that meeting by making the following agreement: I had to write down my argument and provide concrete physical examples, whereas he would think of the suitability of this notion in relation to ADG. Two weeks later we met again at the same place. The first words he told me in his usual humorous style were *"Everything is an adjunction!"*. Since that time our collaboration became more focused and the notion of the Hom-Tensor adjunction has permanently played the role of a compass in searching for a "natural exegesis" of physical geometry. Broadly speaking, somebody may think of the Hom-Tensor adjunction as the universal form of the coordinate free and background independent natural bidirectional relation between an observer and an observandum. The dynamical transgression of this relation, caused for instance by various types of matter sources, leads to a corresponding relation between the connectivity of a physical field and the observable geometric effects following it. From its original inception by Mallios, ADG has demonstrated in a variety of cases, involving general relativity and gauge theories, that the differential geometric mechanism of physical fields is of a homological algebraic origin. Now, the physical dynamical underpinning of the homological Hom-Tensor adjunction, acts as a bridge for transferring ADG to generalized topos-theoretic localization environments, specified by the conditions of observability set up in the regime of different physical theories, in particular quantum theories.

This short preamble also reflects the working title of Mallios's original handwritten manuscript: *"Bohr's Correspondence Principle (: the Commutative Substance of the Quantum), Abstract (: Axiomatic) Quantum Gravity, and Functor Categories"*, which spans over the complete Chapters 1 and 2 of the present book. In view of practical reasons, I took the liberty to shorten the title to the current one. I have not attempted to modify in any way the original conception, structure, formulation and Mallios's precise wording of the material presented in Chapters 1 and 2, which correspond fully and faithfully to the original in this manner. I have merely performed some necessary typographical corrections and also included a few additions and explanatory remarks, which are based on Mallios's handwritten notes. My inspection of these notes in various stages of progress of the manuscript has proved that all the explicated notions have been manufactured carefully and re-worked continuously again and again by Mallios in many consecutive stages until a final formulation has been achieved, which is as clear as possible regarding the targeted physical meaning. I feel really privileged that I took part in the shaping process of some of these notions during all these years of continuous interaction and fruitful joint work with Mallios. My contribution in the writing of this volume spans over Chapter 0, which serves as a kind of extended *"Prolegomena"* to the general theory expounded in Chapter 1 and the six concrete applications of Chapter 2. My main reason for writing these prolegomena is to smooth out somehow the transition to Chapters 1 and 2, as well as to make this book as self-contained as possible, so as to serve both the newcomer and the experienced reader in undertaking the proposed path to a natural and relational approach to *"Physical Geometry"*. In this manner, the present book is situated at the crossroads between the foundations of mathematical analysis with a view toward differential geometry and the foundations of theoretical physics with a view toward quantum mechanics and quantum gravity. The unifying thread is provided by the theory of adjoint functors in category theory and the scrutinization of the concepts of sheaf theory and homological algebra in relation to the description and analysis of dynamically constituted physical geometric spectrums.

The reader should be warned that Mallios's writing style is quite terse and idiosyncratic. Nevertheless, it is precisely these attributes that characterize the uniqueness and originality of his writings. His aim is mainly to engage in an honest dialogue with the reader and make her re-think, re-evaluate and re-constitute her prior knowledge from a new theoretical perspective, which is convincingly demonstrated to be more natural for a geometric description of *"Physis"*. In this way, previous concepts, principles, ideas, schemata, and even calculations, figuring out in modern physical theories are continuously being folded inside out and re-organized coherently from the universal relational viewpoint of naturality or, technically speaking, functoriality. The courageous reader who patiently is going to ascend the ladder to the *"relational panopticon of adjunctions"* will be gratified by the realization that adjoint functors, sheaves and homological algebra break ground for a natural approach to modeling dynamical physical geometry instead of promulgating the "black arts of category and topos theory". This is the reason that the book is basically organized around principles, which are applicable in the whole spectrum of physical theories. Every mathematical notion is carefully scrutinized for the inner or esoteric physical meaning it conveys. This acts as a filtering criterion for the suitability or naturality of these notions in extending a physical theory from one regime to another, for example classical general relativity to quantum gravity.

Coming back to Mallios's idiosyncratic writing style, the text is not organized in the standard form of axioms, propositions, theorems and corollaries but is organized by means of concepts and physical principles. This reflects the change of the center of gravity that a text on *"Physical Mathematics"* should have in comparison to a text on *"Mathematical Physics"* or *"Pure Mathematics"*. I have also followed this way of organizing the material of Chapter 0, where the emphasis is placed on the communicated physical concepts instead of over-stressing the formal mathematical manipulations. Nevertheless, all the necessary mathematical details are presented at least at the level of rigor of a theoretical physics book. Mallios's natural tendency for punctuality and precision, which emanates from his pure mathematical work on the subjects of this treatise, creates in places a "mental strain" in copying with excessive cross-referencing and citation of his former works as well as of all other related sources. This has been actually my main concern for writing the extended *"Prolegomena"*, so as to reduce the necessity of prerequisites to a minimum level. Of course, the interested reader who wishes to delve more deeply in the pure mathematical technicalities of this work, should consult in detail all the cited sources in consecutive stages of study. Therefore, the landscape of the text should be better surveyed by following an *"Ariadne's thread" logical approach.* In the first reading, it would be advisable to go with the natural flow of the arguments originating from the explicated physical principles instead of getting stuck in the technical details and in looking up exhaustively all the cited sources. In subsequent readings, and after figuring out the rudiments of the proposed guiding thread through the Daedalian labyrinth erected by non-natural pre-suppositions or postulates of modern analysis and physics, it would be easier to appreciate the required level of mathematical abstraction as a means to evade the guards of this labyrinth.

At the risk of being momentarily formal in this general preface, if I were to distinguish a single predominant physical principle around which all others amalgamate harmonically, it would be the principle of invariance. More precisely speaking, the principle of *functorial invariance*, and in particular, its manifestation in background-independent differential geometric form via the notion of a *connection*. Taking into consideration the local/global distinction imposed by the constraints of physical observability in terms of appropriate reference frames, the former is realized as invariance with respect to an observable algebra *sheaf* of coefficients. This sheaf is understood simply as a kind of sectional arithmetic utilized to follow and represent some form of dynamical variation induced by a physical field. The representation takes place in terms of tensorial attributes of the corresponding connection, like the curvature or field strength, and is always expressed locally via coefficients taken from the utilized arithmetic. It is precisely this notion of invariance, which generalizes and unifies under a common roof well-known working principles in physics, like the principle of general covariance, the gauge principle and the least action principle. I feel the necessity to emphasize this principle of functorial invariance for physics because it highlights the scope of the whole treatise.

Mallios's original motives for starting to write this exposition, as he has repeatedly emphasized to me, have been mainly conceptual and grounded in physical reasoning. The utterances *"Physical Mathematics"* and *"Physicalization of Geometry"* used repeatedly by him in contradistinction to the standard ones, that is *"Mathematical Physics"* and *"Geometrization of Physics"* always acted as a kind of prevention reminder to the blunder a physicist can commit if blinded by the triumphs of pure mathematics disregards the suitability and ultimate relevance of the involved assumptions in relation to the physical world and the constraints of observability. This is the reason that if there exists a single axiomatic element in the whole treatise, this is condensed in the form of the *axiom of functorial invariance*, whose underpinning is entirely physical, as has been described in concise form previously.

Beside George Anastassiou, I would like to express my gratitude to Aggeliki Petrou and Georgios Petrou for kindly allowing me to access Mallios's handwritten notes. I would also like to express my sincere thanks to Maria Fragoulopoulou, Maria Papatrantafillou and Efstathios Vassiliou for the help they provided. The meeting we've had all together at the Department of Mathematics of the University of Athens in September 2014 has been decisive for the trajectory of this book. Last but not least, I would like to thank Rosa Garderi for the initial typing of Mallios's notes in LaTeX format and George Afendras for the schematic illustrations in these notes.

It is my hope that the theoretical attitude of this work resonates harmoniously with the stance expressed masterfully by Mark Kac in the following excerpt of his essay "Mathematics: Tensions": *"To me and to many of my colleagues mathematics is not just an austere, logical structure of forbidding purity, but also a vital, vibrant instrument for understanding the world, including the workings of our minds, and this aspect of mathematics was all but lost. Complete axiomatization, someone has rightly said, is an obituary of an idea, and Hilbert's great feat was, in a way only a magnificent necrology of geometry. Anyway, there are worse things than being wrong, and being dull and pedantic are surely among them."* I can ascertain with certainty that Mallios's goal in being engaged with this physically motivated treatise during the last five years before passing away, was not to be a necrology of his pure mathematical work on *"Abstract Differential Geometry"*.

Elias Zafiris
Budapest
10 January 2015

Contents

Chapter 0

Prolegomena

0.1 Exordium

The major aim of the application of Analysis and Differential Geometry in Physics is the setting up of a mechanism providing a precise description of the emergence of geometric spectrums due to dynamical interactions, which can be further used for making predictions. In this manner, the notion of a geometric spectrum is considered as the outcome of a physical law or more generally of a dynamical interaction of a particular form. This raises immediately the question if there exists an approach to physical geometric spectrums that is independent of any coordinate point manifold background, in the sense that it refers directly to the physical relations causing the appearance of these spectrums without the intervention of any *ad hoc* coordinate choices. The answer provided to this question in this book is that the theory of *differential vector sheaves*, that is geometric vector sheaves equipped with a connectivity structure and obeying appropriate cohomological conditions, provides the sought after functorial tool for a *universal* and *natural* approach to *physical geometric spectrums*. The major difference of the proposed approach in comparison to the traditional ones based on classical differential calculus and differential geometry of smooth manifolds consists in the realization that a classical analytic technique is susceptible to a natural background-independent generalization if it is *localizable* by sheaf-theoretic means. In this case the technique can be expressed functorially, that is by means of natural transformations of sheaf functors via the machinery of homological algebra. This is of crucial significance for setting up a mechanism describing the emergence of physical geometric spectrums where the notion of background smoothness is inapplicable.

The meticulous development of the concepts of general relativity theory as well as of gauge theories has already taught us an extremely important lesson in this quest. This lesson may be condensed in the following form: Starting from a local description of physical phenomena, one should always seek for a *covariant* or *gauge invariant* formulation of the associated physical laws according to some criterion of *local symmetry* or *local equivalence*. Then, the specification conditions of the corresponding physical geometry, in each particular case, emerge together with the

1

invariant formulation and are independent of the local gauges used in the first place. Several issues of an interpretational nature arise together with the requirement of naturality, which force a critical re-examination of the conceptual choices and assumptions initially endorsed from a physical viewpoint. The two most significant ones in relation to a natural approach to physical geometry are the concepts of *locality* and *differentiability*. We will attempt to comment briefly on the related problems so as to motivate conceptually the focus of the whole treatise.

First, the notion of *locality*, or more broadly *localization*, subsumed in the development of both general relativity and classical gauge theories is based on the notion of *spacetime localization*. In other words, at least a local geometric coordinate spacetime background is initially assumed for the local description of physical phenomena. Not only this, but a further more decisive assumption regarding the global nature of this background as a *smooth manifold* is traditionally imposed, so that all physical quantities are subordinate to a covariant description subject precisely to this underlying global background manifold. Of course, from a gauge-theoretic point of view, the major physical role is played by the *local gauge freedom* in the definition of physical quantities, which is represented or realized by means of internal spaces or fibers of local degrees of freedom over each point of the underlying spacetime manifold. In this manner, it seems unavoidable to have an ontology of dependence on two distinct notions of localization. The first or external one is the localization of spatiotemporal variables being subordinate to the global smooth manifold determination, and the second or internal one is the fiber-wise localization of physical quantities. The advent of quantum theory has posed serious difficulties in interpreting the former notion of external localization. Whereas the internal notion of gauge-theoretic localization *persists upon quantization*, the external one is deprived of its original spatiotemporal semantics. Various attempts at quantizing the classical spacetime manifold geometry have not stood up to the expectations. The success and experimental validation of quantum theory requires, first of all, a careful reconsideration of what we mean by the notion of localization of physical quantities. The pertinent question is if it is possible to have a working notion of localization *independently* of the existence of a geometrical spatiotemporal background. Moreover, since all gauge-theoretic concepts are still physically relevant upon quantization, why should their local formulation be dependent on an underlying smooth manifold base space? Addressing these questions has serious consequences and manifests the significance of a critical remark of Hermann Weyl, who noted: *"While topology has succeeded fairly well in mastering continuity, we do not yet understand the inner meaning of the restriction to differential manifolds. Perhaps one day physics will be able to discard it"*.

Reflecting on this remark, it becomes questionable why the physical notion of gauge-theoretic localization should be implementable exclusively over a smooth manifold. In other words, why should physical quantities be representable as smooth functions up to gauge scaling factor over a smooth manifold? There is *no intrinsic reason* that a local gauge theoretic argument requires the assumption of global smoothness in the functional representation of physical quantities. Notwithstanding this fact, the assumption of continuity seems to be necessary. After all, continuity is a *local* notion and the concept of a topological base space has been precisely formulated in order to express what a continuous function actually is. Thus, from a physical perspective, it would be natural to consider the notion of gauge-theoretic localization for *globally non-smooth* or *distribution-like* continuous physical quantities. Given that these physical quantities form commutative algebras, which can be evaluated in a number field through measurement, the topological representation theorems of Stone and Gelfand provide a concrete *topological realization* of these algebras as algebras of continuous functions over a topological space. Most important, this topological space is neither postulated ad hoc, nor constructed by imposing metrical relations, but is defined by *prime* or indecomposable constituents of the algebras themselves, at least locally, forming in this sense its spectrum.

There are two *caveats* of this modeling philosophy, which should be considered carefully from a physical perspective. The first is that the notion of local gauge theoretic localization *does not* really require, neither the imposition of a global base topological space, nor the assumption that physical quantities should be representable as global continuous functions over such a space. It actually requires much weaker assumptions: First, the notion of a *local scaffolding of control variables*, which may be even determined via operational arguments depending on the experimental procedures for recording events. In a formalization of this type, a global topological space can be only *implicitly* assumed. Put differently, a global topological base space can be only *indirectly* implicated via these local scaffoldings of control parameters, which may be thought of as forming local *covering families* of the former, subject to some well-defined conditions. For example, the open sets of a topological space qualify as local covering families. The most general notion of local covering scaffoldings of this form is provided by the notion of a *Grothendieck topology*. Although the latter notion seems to be very abstract for concrete physical applications, it turns out that precisely the opposite is the case. More concretely, a local scaffolding of control parameters, considered as the locus of gauge localization of a physical quantity, plays the role of a *hole* in a *covering sieve*, viz. an opening through which by specifying the control parameters locally, information can be obtained via measurement. The fact that these openings are localized holes in sieves is of physical significance. This is the case because sieves are *not ad hoc* coordinate objects but they are *functorial*. This is a technical category-theoretic term whose precise meaning in a physical context meets the requirement of *covariance* of local descriptions *independently* of the existence of an underlying global background. Thus, it addresses the need for a natural approach to physical geometry in a background independent manner, and most important, it provides the most appropriate concept of grounding the notion of gauge-theoretic localization of physical quantities. The instantiation of a covering sieve via an efficient physical measurement procedure qualifies the physical quantities as *elements of a presheaf*. The functoriality here means that physical information should be consistent under the operation of *restriction* with respect to the openings of a covering sieve.

Now, there is a technical process called *sheafification* which turns these elements into sections of a sheaf *completing* in this manner the localization procedure. Sheaf functoriality means that physical information should be consistent not only under the operation of restriction, but also under the operation of *extension* with respect to the openings of a covering sieve. There are several consequences of the sheaf-theoretic conceptualization of physical quantities. First, they are not considered as globally defined continuous functions over a global base space, but as *continuous sections* of sheaves, meaning that they are defined locally with respect to the openings employed in each case. Second, physical quantities are not specified by their measured values like globally defined functions do, but they are determined by their *germs*, that is appropriate compatible equivalence classes of sections with respect to these openings. Third, the requirement of compatible amalgamation of local sections under extension with respect to the openings, is expressed through the *gluing* condition characterizing a sheaf of physical quantities. In turn, the gluing condition can be expressed cohomologically in terms of Čech 1-cocycles and their respective *cohomology classes*, which forces a cohomological interpretation of gauge theoretic localization in sheaf-theoretic terms. The latter in conjunction with the above, leads naturally to a *cohomological classification* of state spaces by invariants of a global character. Thus, it paves the way for a coordinate-free description of physical geometry in a background-independent manner. Therefore, the sheaf-theoretic conception of gauge theoretic localization is in accordance to the constraints imposed by quantization.

It is instructive to point out that even in the case of quantum mechanics, all the above aspects become physically relevant. We know that the state of a quantum system is completely determined with respect to some maximal algebra of commuting, and thus, co-measurable observables. Of course, there exist many different commutative algebras of this form, but the effectuation of a measurement procedure always requires the choice of one of them. In this context, the requirement of functoriality is met by considering covering sieves, whose openings see only such local maximal commutative algebras of observables of a quantum system. Moreover, the determination of the state, with respect to a local commutative algebra of observables, is taking place only up to an arbitrary phase, which in turn, specifies the gauge freedom of the state vector of a quantum system. This leads naturally to a cohomological description of quantum state spaces and prepares the ground for a deeper understanding of quantization in relation to observable physical geometries of some form. More precisely, a quantum state space becomes a locally free sheaf of modules of finite rank 1 with respect to a maximal commutative algebra of co-measurable observables, and thus, is described only up to equivalence with respect to isomorphism classes of such sheaves. Thus, the cohomological classification of these sheaves is of utmost physical importance for interpreting the *global aspects of quantization*, for instance Dirac's global quantization condition and the role of the integers.

We may recapitulate the above, by stating the conclusion that the notion of sheaf-theoretic localization of observables and states along the previous lines constitutes a functorial and background manifold independent approach to the localization issue, which is perfectly compatible with the requirements of local gauge freedom and invariance. We stress the fact that according to this approach, the state space *locally* is not only a vector space over some number field but a *free module* over a corresponding local algebra of observables. Thus, the local multiplying coefficients are sections from a commutative algebra of observables and not just numbers from the evaluation field. Still, we can define all the usual notions locally with respect to them, for example we obtain well-defined notions of algebra-valued bilinear forms, inner products and metrics. Notice again that these sections are *not* required to be sections of the sheaf of germs of smooth functions. In contradistinction, the type of sheaf sections should not be chosen ad hoc, but should be specified by other means. Reflecting on this point, there appears the idea that the type of sheaf sections fitting to a particular physical problem should be actually determined by *dynamical means*. Put differently, the notion of invariant sheaf-theoretic localization should not be considered in isolation from the relevant dynamical aspects of a physical theory. After all, a physical theory is a theory of interactions of some kind, and it should be only these interactions that can determine naturally a corresponding physical geometry.

The spectral conception of physical geometry as a result of a dynamical physical law, dictates that locally by duality, the sheaf of observable coefficients should be actually determined as a *local solution* of the dynamical equations of motion. In other words, locally the algebra sheaf of coefficients should be the sole *dynamical variable* determinable as a solution of the equations of motion. It is important to note that this idea is compatible with both the conception of geometry according to Einstein's field equations in the general theory of relativity and the gauge-theoretic principle dictating that the requirement of invariance implies dynamical interactions. Actually, the postulate of invariance, regarding the natural implementation of localization in a physical environment, constitutes the *bridge* between the general mathematical notion of sheaf-theoretic localization and the dynamical constraints imposed by a physical theory. Put concisely, we claim that a process of localization expressed in terms of an algebra sheaf of coefficients in the context of a physical theory should not be chosen ad hoc, but it should be determined as a solution by dynamical or differential means. This claim encapsulates precisely the physical meaning of invariant sheaf-theoretic localization. Notice that we do not distinguish between kinematical and dynamical aspects of a physical theory, since no geometric background substratum is assumed a priori even locally. This implies that even the inertial structure of a physical theory is considered to be of a dynamical character, since it can be formulated by differential means.

The above brings us to the second fundamental issue in relation to a natural approach to physical geometry, viz. the issue of *differentiability*. The pertinent question refers to the problem of formulating what we mean by differentiability in a background manifold independent manner. Clearly, this is the most crucial aspect of a physical theory, which aims to provide predictions about the behavior of a physical system by expressing locally the laws of motion in differential terms. We would dare to say that the *sole purpose* of the background smooth manifold background in physical theories is to afford a *local analytic* mechanism of differentiation, which originates from the locally Euclidean structure. Therefore, the standard differential calculus of Euclidean vector spaces is transferred locally to a smooth manifold, which paves the way for the local analytic formulation of the dynamical laws of motion in terms of smooth functions. Furthermore, the natural physical requirement of independence of geometric objects from the description in a local coordinate system in this context, necessitates the modification of the standard differentiation to the *covariant* one, that is to the notion of the *covariant derivative* along a smooth curve in the manifold. In a nutshell, covariant differentiation provides the means of parallel transport along a smooth curve in the background manifold, and thus the means of *infinitesimal comparison* formulated locally in terms of smooth function coefficients. Despite the immense success of this analytic differentiation concept in theoretical physics, there are serious problems to be addressed: First, is the general notion of covariant differentiation dependent on the smoothness of the underlying curve and the smoothness of the local coefficients used for its formulation? Second, is the above notion of covariant differentiation the natural infinitesimal localization of a general parallel transport process? In other words, does it provide the means of infinitesimal comparison in a local context functorially? Third, the usual description of parallel transport is expressed as a rule of lifting curves from the base manifold to the tangent bundle. From a physical viewpoint, this is problematic because curves on the spacetime manifold are not given a priori, but actually emerge as solutions of the dynamics and are interpreted in this way as smooth histories of observed events.

In the following Section we will explain how all the above issues are addressed from the viewpoint of *Abstract Differential Geometry*, which constitutes the basic conceptual and technical framework of this work. It is instructive at this stage to point out that according to this framework, the general notion of variation or differentiability caused by a dynamical interaction should be expressed intrinsically in terms of a *connection*. The notion of a connection is formulated functorially as a natural transformation from the vector sheaf of states over the observable algebra sheaf of a physical system to the sheaf of vector-valued algebraic differential forms obtained by the sheaf-theoretic localization of the standard algebraic Kähler-de Rham construction. This natural transformation should obey the Leibniz rule, so that in effect can be interpreted physically as a covariant derivative of the sections of the vector sheaf of states according to the pertinent law of variation. From a gauge-theoretic viewpoint, a connection on a vector sheaf of states is interpreted locally as the *potential* of a gauge field, which causes the variation or differentiability of its sections. In this manner, the pair constituted by a vector sheaf of states together with a connection, viz. a differential sheaf, represents the *dynamical variation mechanism* induced by a gauge field, where the connection provides the natural coordinate-free generalization of the notion of a differential equation. The integrability properties of a connection under extension of this mechanism from the local to the global level are derived by means of *sheaf cohomology* theory and not by the application of the usual analytic techniques of the differential geometry of smooth manifolds. The *curvature* of a connection is a tensorial homological observable quantity, which is measured by the gauge field *strength* of the associated local potentials in a *covariant way* with respect to the corresponding observable algebra sheaf. Most important, the formulation of the dynamical variation mechanism should be *functorial* with respect to any utilized observable algebra sheaf. This is expressed as the principle of *invariance* of the differential geometric mechanism with respect to cohomologically suitable algebra sheaves of coefficients. The requirement of functoriality is the crux of a natural approach to physical geometry determined by dynamical means, in the sense that invariance determines dynamical interactions, and thus constitutes the unique natural way to detect a physical geometric spectrum as an outcome of a physical variation law.

0.2 Basic Working Notions

Ever since Leray and Cartan in the 1950's formulated the concept of a sheaf, the various examples and applications of sheaves have come to play a major role in such diverse fields of mathematics as several complex variables, algebraic geometry and algebraic topology. In particular, the work of Grothendieck in algebraic geometry and the generalization of the notion of sheaves beyond the realm of topological spaces, making it applicable over general category-theoretic sites (categories equipped with covering systems called Grothendieck topologies), has created a wealth of new models enriched with the power of the methods of homological algebra.

The concept of a *sheaf* expresses essentially *gluing* conditions, namely the way by which local structural algebraic data (like groups or modules) can be collated compatibly into global ones over a covering system, like the one defined by the open sets of a topological space. In general, a sheaf may be thought of as a *continuously variable* algebraic structure, whose continuous variation is considered over specified local covering domains interlocking together non-trivially. More recently it was discovered that sheaves also have important applications in logic and model theory. For example, the category of all sheaves on a fixed topological space can be viewed as a model of constructive set theory. One of the most interesting aspects of the sheaf logic formalism is that it provides a framework in which a sheaf can be viewed as a model of a space which does *not* have a local structure defined by points but a local structure which is intrinsically continuous with respect to a family of covers, for example the open sets covering a topological space. More concretely, from this perspective the main difference from the usual notion of a set (or an algebraic structure defined set-theoretically) is that in a sheaf we have an extra intrinsic *internal* structure of *granulation* of the elements (called sections) by the partially ordered set of the covers. Heuristically, we think of the covers as measures of definability and the elements of a sheaf restricted to a cover are exactly the members defined to the *extent* provided by this cover.

From a physical viewpoint, sheaf-theoretic ideas pertain to the problem of *extendibility* of *local* observable properties into *global* descriptions, to the problem of *gauge-invariance*, as well as to the problem of *structural order*. Initially, the foundational significance of sheaf theory for physics is related with the possibility of re-examining the problem of locality, together with the inter-relation between local and global physical descriptions under a new rigorous theoretical perspective. More precisely, sheaf-theoretic methods allow the *disassociation* of localization processes from *locally Euclidean* metrical spatiotemporal contexts, which are based on the *point*-localization structure of the real line. This becomes possible via the replacement of the classical metrical gauge of point localization on a background smooth manifold by a sheaf gauge of algebraic-topological localization on an appropriate categorical site. A site, for example a covering system of open sets on a topological space, should be solely used to *support* the relational local-to-global information pertaining to the corresponding scale of observable behavior. This strategy permits the effectuation of generalized localization schemes being capable of capturing even the microscopic non-smooth observable scale of quantum phenomena, where a locally Euclidean model is totally inapplicable. Of course, the proper implementation of these localization schemes should be supplemented by an appropriate sheaf-theoretic differential mechanism generalizing the usual analytic differential geometric framework of smooth manifolds. This requirement has been met by the successful integration of the methods of homological algebra into sheaf theory.

More concretely, since it is possible to interpret standard algebraic notions as homomorphism, kernel, image, subobject and quotient object for sheaves, in such a way that these concepts have essentially the same meaning as in algebra, one can consider them from a categorical standpoint and apply all the constructions of homological algebra to sheaf theory. The resulting category of sheaves has the same classical properties as the category of Abelian groups or the category of modules; in particular, one can define for sheaves direct sums, direct products, tensor products, inductive limits, and other concepts. In this way, the apparatus of sheaf theory has penetrated into various fields of mathematics providing an effective algebraic tool especially in those areas which ask for global solutions to problems whose hypotheses are local. This is due to the fact that there is a *natural* definition of the cohomology of a site (for example, a topological space) with *coefficients in a sheaf*. In particular, the development of algebraic geometry by Grothendieck led to the crystallization of the idea that the natural argument of a cohomology theory is a *pair* consisting of a space (or more generally a category-theoretic site) *together* with a sheaf defined over it, rather than just a space. This realization has been transferred to the field of complex analysis and more recently to the field of differential geometry by the development of the geometric theory of vector sheaves equipped with a connection *à la Mallios*, called Abstract Differential Geometry (*ADG*). Now, the sheaf gauge of algebraic-topological localization and extendibility sought-after is provided by *sheaf cohomology*.

Cohomology can be viewed as a method of assigning *global invariants* to a topological space (or more generally a categorical site) in a *homotopy-invariant* way. The cohomology groups measure the *global obstructions* for extending sections from the local to the global level (for example extending local solutions of a differential equation to a global solution). For instance, de Rham cohomology measures the extent that closed differential forms fail to be exact, and thus the obstruction to integrability (according to the lemma of Poincaré every closed differential form is locally exact). The *de Rham theorem* asserts that the homomorphism from the de Rham cohomology ring to the differentiable singular cohomology ring given by integration of closed forms over differentiable singular cycles is a ring isomorphism. The sheaf-theoretic understanding of this deep result came after the realization that both the de Rham cohomology and the differentiable singular cohomology are actually *special* isomorphic cases of sheaf cohomology. In particular, it has been also clarified that the de Rham cohomology of a differential manifold depends *only* on the property of paracompactness of the underlying topological space. The generalization of the usual de Rham theory on manifolds in sheaf-theoretic terms is most welcome also from a physical point of view, since it is a fact that the structure underlying an *intrinsic* approach to differential calculus, and in particular, differential equations in physics in terms of differential forms is actually *de Rham-cohomology*.

From an algebraic standpoint, the conceptual categorical skeleton of the *generative* mechanism of differential calculus and, *in extenso*, differential geometry is a consequence of the existence of a pair of adjoint functors, or equivalently an *adjunction* expressing the *inverse* algebraic processes of extending and restricting the scalars. Classically, the differential mechanism is based on the fundamental notion of a *connection* on a *projective module* over a commutative unital ring of scalars (or algebra over a field) considered together with the associated de Rham complex. A connection on a module induces a process of *infinitesimal extension* of the scalars of the underlying ring, which is interpreted geometrically as a process of *first-order parallel transport* along infinitesimally variable paths in the spectrum-space of this ring. The next stage of development of the differential mechanism involves the satisfaction of appropriate global requirements referring to the transition from the infinitesimal to the global level. These requirements are of a *homological* nature and characterize the integrability property of the variation process induced by a connection. Moreover, they are properly addressed by the construction of the *de Rham complex* associated to an integrable connection. The non-integrability of a connection is characterized by the notion of *curvature* bearing the semantics of observable disturbances to the process of cohomologically unobstructed variation induced by the corresponding connection.

It is instructive to emphasize that the conceptualization of the classical differential mechanism along these lines *does not* presuppose the existence of an underlying differential manifold. This observation is particularly important because it provides the possibility of abstracting in functorial terms both the definition of a connection and the associated de Rham complex as well. The functorial recasting of the differential mechanism becomes possible by the appropriate qualification of the information encoded in the *extension/restriction of scalars adjunction*. This information permits the definition of the universal object of differential 1-forms, and subsequently, the definition of a connection, together with the associated de Rham complex. The benefits we obtain from the functorial reformulation of the differential mechanism are multiple. First, the differential mechanism becomes explicitly *non-dependent* on the existence of a smooth underlying manifold. Second, it can be applied effectively in a variety of *topological* and *topos-theoretic* localization environments, conceived as categories of sheaves, under the satisfaction of the appropriate cohomological requirements. Finally, the functorial recasting of the generative mechanism of differential calculus and differential geometry is of paramount importance for theoretical physics, because it can be properly used for the functorial formulation of dynamics taking into account the particular *localization properties* of the *observables* in the regime of each physical theory, which are modeled by sheaf-theoretic means. It is precisely such a functorial formulation of differential geometry which has been accomplished by ADG.

The development of the differential geometry of vector sheaves over an abstract topological space has given rise to a fully-fledged theory, which has been conceived and formulated in detail by Mallios [5-7], see also Vassiliou [1]. The significance of ADG for physics has been also shown by an explicit reconstruction and generalization of the framework of gauge theories (Maxwell and Yang-Mills) in sheaf cohomological terms by Mallios [15, 21], see also Mallios [14, 16, 18]. The main conceptual lesson obtained by this generalization is that any physical problem, which is based on the principle of *local gauge invariance*, should be formulated directly in terms of the relevant physical-geometric objects (sections of sheaves) without the intervention of any locally Euclidean space. ADG generalizes the classical differential geometric mechanism of smooth manifolds, by proving explicitly that most of the usual differential geometric constructions on manifolds can be carried out by purely sheaf-theoretic means without any use of any sort of smoothness or any of the conventional analytic methods of spatial differential calculus that is associated with it. The technique used for the generalization of the analytic methods of Classical Differential Geometry (CDG) consists in the following: Initially, a concept of CDG is *suitable* for extension to a broader differential context (beyond the smooth context of manifolds) if it is liable to a process of sheaf-theoretic *localization* explicated by Mallios [13]. Then, this concept is expressed by means of a *natural transformation* of sheaves. The prominent role is played by the *algebra sheaf* of coefficients playing the role of a *"functional arithmetic"*, see e.g. Mallios [17-18], meaning that all geometric objects involved in the formalism are locally expressed in terms of its sections. In physical terms, we interpret the sheaf of coefficients as a *sheaf of observable algebras*, see Zafiris [3, 7]. In this way, an algebra sheaf of coefficients is *not* constrained *ab initio* to be a smooth one (which is used to model an underlying smooth manifold as its spectrum), but it may be for instance a *singular sheaf* of generalized functions including distributions, defined by Rosinger, see e.g. Mallios-Rosinger [1, 2]. In the most general case, we may consider as algebra sheaves of coefficients those of *soft topological algebra sheaves*, which are "closed" with respect to complete projective topological tensor products, see Mallios [4, 6]. We mention parenthetically that the interested reader may consult the monographs by Mallios [1] and Fragoulopoulou [1] for pure mathematical expositions of the theory of topological algebras, the details of which nevertheless will be mostly not needed in the present work. In particular, topological algebra sheaves fitting in the above characterization, which are also *self-adjoint*, are suitable for the development of an extended differential geometric mechanism generalizing the one based on the non-normed topological algebra sheaf of smooth coefficients with interesting physical applications, see e.g. Mallios-Zafiris [1-3], Mallios-Raptis [1-3] and Mallios-Rosinger [1, 2]. This generalization has been motivated to a large extent from previous work carried out in the context of Banach-Šilov algebras having the Banach approximation property by Selesnick [5, 6]. Thus, concerning physical issues, we may employ for example a soft self-adjoint Fréchet-Šilov locally m-convex topological algebra as an algebra of observables, whose *Gelfand spectrum* is then a (Hausdorff) paracompact space. Then, any vector sheaf of rank 1, called a line sheaf, on the topological spectrum of this algebra of observables can be equipped with a connection, whose curvature defines an *integral cohomology class*.

We stress again the prominent physical role of an appropriate sheaf of observable algebras by the fact that all vector sheaves (of states), group sheaves of automorphisms (symmetries) and sheaves of differential forms (gauge potentials and field strengths) partaking in ADG are typically required to be *locally free* sheaves of modules and of *finite rank* over the sheaf of observable algebras. The concept of differentiability of the sections of vector sheaves (covariant derivative of sheaf sections) is formulated in terms of connections. A connection on a vector sheaf bears the dynamical semantics of a potential locally, which causes the differentiability of its sections. In this way, a vector sheaf equipped with a connection may be thought of as a *coordinate-free* generalization of the notion of a differential equation. The integrability properties of a connection under extension of the differential mechanism from the local to the global level are considered from the prism of sheaf cohomology and not by the typical analytic methods. The curvature of a connection is measured by the strength of the associated potential covariantly with respect to the corresponding sheaf of observable algebras.

One of the most important conclusions crystallized by ADG is a new conceptual perspective on the notion of a *base geometric space* for the formulation of differential geometry. A base geometric space provides merely a *"localization basis"* for the formulation of the differential mechanism, whereas the mechanism itself is *not* subordinate to this basis, meaning that it is not dependent on any particular localization basis, although it should respect it. Thus, the object of primary significance in ADG is *not* the base space itself but the sheaf of observable algebras localized over it. Put differently, the differential geometric mechanism is completely expressed locally in terms of the sections of the observable algebra sheaf, whereas the base space itself is only the *carrier* of the sections defined over it, or else it provides only the *means of localization* of these sections. Then, it is precisely the functoriality of ADG with respect to cohomologically appropriate observable algebra sheaves, which permits the detection of dynamical interactions, identified for instance in the case of line sheaves in terms of the integral cohomology class of the curvature. This is a crucial insight and provides a *justification* of the gauge principle and its consequences used successfully in theoretical physics. A particular manifestation of this principle in quantum mechanics is related to the local abelian *phase invariance* in the specification of the state vector of a quantum system. The locality demand in the formulation of the gauge principle is crucial because although it is true that the phase of a quantum system can be always gauged away locally, this is not the case globally.

The use of the differential geometry of vector sheaves and in particular sheaf cohomology, according to the functorial framework of ADG, is novel among the techniques of theoretical physics and arises from a sheaf cohomological local-to-global representation of physical properties in contrast to the conventional description in terms of local differential equations on a smooth spacetime manifold. This departure is necessitated to a certain extent by problems related with the modeling of the spacetime continuum. The usual spacetime manifold model based locally on the properties of Euclidean spaces and ultimately on the properties of the *real line* presupposes an underlying structure of point-events which can be resolved with infinite precision. This model was associated operationally with idealized measurement procedures based on the assumption that it is the local Euclidean geometry that decides which objects do and which do not qualify as *infinitesimal measuring rods* at each point of a 4-d smooth spacetime manifold. Developments in quantum mechanics, quantum field theories, and attempts at quantizing gravity encounter serious difficulties in reconciling with the local Euclidean manifold structure, which seem to be related conceptually to this classically idealized small-scale structure. Thus, it is of great interest to consider theories which represent the physical *event continuum* in an essentially different way without losing at the same time the differential geometric framework of dynamics provided by smooth manifolds.

For instance, the general theory of relativity is formulated in terms of differential equations involving *smooth functions* (local components of the metric tensor participating in Einstein's field equations), interpreted as providing a description of the local *metrical* properties of the spacetime manifold around any specific point, depending on the distribution of the energy-momentum tensor. Notwithstanding this fact, all of the observable cosmological predictions of the theory derive not from local solutions of the field equations, per se, but from the inferred global metric spacetime structure generated by the method of *analytic continuation* of some local solution over an extended region. It is precisely from these global solutions that we become able to make predictions. Now, the technique of analytic continuation is mathematically of a sheaf-theoretic nature and *not* dependent at all on both the hypothesis of the local Euclidean structure of spacetime, and the use of smooth functions. This fact points to the conjecture that other non-smooth or distribution-like sheaves of coefficients should be relevant for the continuation of some local solution over extended regions in case that the smooth one is obstructed or becomes singular, *preventing* at the same time the field equations from breaking down at *singularities*, given that these equations can be formulated appropriately in terms of these generalized sheaves of coefficients.

Regarding the above demand, it is instructive to stress again that the development of ADG, viz. the geometric theory of vector sheaves equipped with a connective structure (covariant differentiation), in the form of a functorial generalization of the differential geometric framework of smooth manifolds, has led to the realization that the mechanism underlying an intrinsic approach to modeling and addressing dynamical problems *independently* of the local Euclidean structure is essentially de Rham cohomology, and in particular *sheaf cohomology*. More precisely, the validity of the de Rham complex in its sheaf-theoretic guise is *not* restricted to the consideration of smooth coefficients (at it is the case when the spectrum of the coefficients is a smooth manifold). Thus, we may consider more general distribution-like sheaves of coefficients fitting in the de Rham complex and formulate the field equations in terms of these sheaves instead of the smooth ones. In this way, there exists the possibility of *re-assessing* the global or large-scale problems of general relativity related with the intrinsic smooth spacetime manifold singularities and anomalies from the perspective of appropriate generalized sheaves of coefficients. From a physical viewpoint, this approach would allow to consider distribution-like sheaves corresponding to *non-punctual* localization properties, which would nevertheless still satisfy the field equations (in a sheaf-theoretic form). Beside the technical flexibility provided by the geometry of differential vector sheaves in tackling efficiently these physical problems, an important reason to consider as a clear indication that such an approach is worth pursuing further in relation to physical geometry, stems from its harmonic resonance with the already mentioned critical remark of the father of the *"gauge principle"* H. Weyl, which has actually played a major role in motivating this work: *"While topology has succeeded fairly well in mastering continuity, we do not yet understand the inner meaning of the restriction to differential manifolds. Perhaps one day physics will be able to discard it".*

0.3   Observables and States

0.3.1   *Sheaf-Theoretic Observable Localization*

Procedures of *physical measurement* presuppose the existence of *localization processes* for extracting information related with the local behavior of physical systems. A localization process is implemented by the preparation of *local reference frames* for the measurement of *observables*. It is important to realize that the effectuation of a localization process is not always equivalent to conferring a numerical identity to observables, expressed in terms of some value corresponding to a physical attribute. On the contrary, the latter is only a limited case under the assumption that all local reference frames can be contracted to points. In this limited case, points constitute unique measures of localization in the *physical continuum*. This is the crucial assumption underlying the employment of the set-theoretic structure of the real line as a model of the physical event continuum, understood as a set of independent points possessing the property of infinite distinguishability with absolute precision. The success of this localizing philosophy in classical physics is due to the association of the notion of physical continuum with the attribute of position and the fact that all classical observables can be potentially determined precisely and simultaneously at the unique measure of localization of this attribute, namely at a spatial point parameterized by the field of real numbers. Nevertheless, the major foundational difference of quantum systems from classical ones is a consequence of the incompatibility of precise determination of all quantum observables within a single local measurement frame, meaning that the simultaneous sharp specification of all observables is impossible within a single local measurement frame.

Thus, a foundational strategy implied by quantum theory should fulfill the following objectives: First, it should disassociate the physical meaning of localization from its *restricted spatial* connotation context. Secondly, it should allow the functional dependence of observables on generalized localization measures, not necessarily based on the existence of an underlying structure of points on the real line. For implementing this strategy, it should be essential to interpret any local observable as a relational algebraic *partial information carrier* with respect to a corresponding local frame of measurement. Next, it should be necessary to establish appropriate compatibility conditions for *gluing* the information content of local observables globally. Mathematically, the implementation of this localization strategy is precisely captured by the concept of a *sheaf-theoretic localization* structure referring to an *algebra of observables*. The nature of sheaf-theoretic localization induces i) a shift from point-set to topological localization models of quantum algebraic observable structures, and ii) a replacement of the classical metrical gauge of point localization on a background manifold by a sheaf gauge of algebraic topological localization on a categorical site capturing the relational information of observables.

Initially, it is important to indicate that both of our fundamental physical theories, namely general relativity and quantum mechanics, can be characterized in general terms as special instances in the process of replacement of the *constant* by the *variable*. More concretely, in the case of general relativity this process takes place through the negation of the fixed kinematical structure of spacetime, by making the *metric* into a *dynamical* object determined solely by the solution of *Einstein's field equations*. This essentially means that the geometrical relations defined on a four dimensional manifold, making it into a spacetime, become variable. The dynamically induced infinitesimal variability of the metric constitutes the means of making geometrical relations variable in the descriptive terms of general relativity. The intelligibility of the framework is enriched by the imposition of the principle of *general covariance* of the field equations under arbitrary smooth coordinate transformations of the points of the underlying spacetime manifold. As an immediate consequence, the points of the manifold are *not dynamically localizable* entities in the theory. Thus, it is *illegitimate* to use the set-theoretic point structure of the manifold for individuating directly *spacetime events*, independently of the localization properties of the metric determined dynamically. In this way, it can be asserted that manifold points bear only an indirect reference as indicators of spacetime events, and most importantly, only after the dynamical specification of variable geometrical relations among them, obtained as particular *solutions* of the generally covariant field equations in terms of the metric. We conclude that the conceptualization of the physical event continuum in general relativity requires explicit reference to the *relational localization* properties of the metric, which is determined dynamically in the context of this theory.

In the case of quantum theory, the process of replacement of the constant by the variable is signified by the imposition of *Heisenberg's uncertainty relations*, which determine the limits for *simultaneous* measurements of certain pairs of *conjugate* physical observables, like position and momentum. In a well-defined sense, the uncertainty relations may be interpreted as measures of the *valuation vagueness* associated with the simultaneous determination of conjugate pairs of observables within the same experimental context. This is in *contradistinction* to all classical theories, where the valuation algebra is fixed once and for all to be the field of real numbers, reflecting the fact that values admissible as measured results must be real numbers, *irrespectively* of the measurement context and simultaneously for all physical observables. The above problem of observable valuation vagueness in quantum theory can be resolved *topos-theoretically* via a relational process of topological localization of the *global non-commutative* algebra of quantum observables with respect to a *categorical diagram* of interlocking families of *local commutative* subalgebras of observables. Each maximal local commutative subalgebra contains all observables that can be simultaneously measured within the same experimental context.

From a mathematical point of view, the general process of semantic transition from the constant to the variable along the previous lines takes place by passing from rigid set-theoretic algebraic structures to *continuous sheaf-theoretic* algebraic structures. In general, a *sheaf* may be thought of as a *continuously variable set*, whose continuous variation is considered over specified local frames interlocking together non-trivially. The local frames are required to obey certain topological closure conditions, which generalize the definition of a topology formulated in terms of open sets in classical topological spaces. In this way, we obtain a *local relativization of physical representability* by shifting the semantics of observables from the topos of sets to a topos of sheaves for an appropriate topology. Intuitively speaking, this amounts to a relativization with respect to the local behavior of physical observables as opposed to their point behavior. We remind that in the guiding instance of a topological space, a *topology* is defined by a collection of *open loci* (to be thought of as *probes* or *covers* of a space) which is closed by the formation of arbitrary unions and finite intersections of open loci. The definition of a topology on a space provides the means to talk about the notion of a *continuous function* between topological spaces. A function is said to be continuous if and only if the inverse image of every open locus of the range is an open locus of the domain topological space. This is an attempt to capture the intuition that there are no breaks or separations in a continuous function. It is important to realize the following: i) The definition of a topology on a space is *solely* used for the formalization of what a continuous function is on that space, ii) the continuity of a function is a property which is determined *locally*, that is only by reference to open loci. This means that due to the variability of open loci in the topology the property of continuity of a function should respect the inverse algebraic operations of *restriction* and *extension* (uniquely) with respect to open loci. Thus, a continuous function can be restricted consistently to open subloci of any open locus in the topology and inversely extended by gluing uniquely together all its local restrictions. This is the *crucial conceptual insight* which is incorporated in the technical definition of a sheaf.

This insight referring to the precise formulation of the property of continuity may be generalized in two directions: Firstly, instead of open loci or open regions covering a topological space we may consider generalized local covering frames under the constraint that they collectively obey topological *closure* conditions *analogous* to the ones used for open loci. Secondly, instead of functions varying continuously over local frames we may consider *generalized functional relations*, called technically *sections* of a sheaf. Conceptually speaking, local sections of a sheaf describe relations relatively to a local frame. A sheaf is essentially the totality of its sections (both local and global if the latter exist) and their consistent interrelations with respect to the local-global distinctions subsumed by the *interlocking properties* of the underlying frames. In this sense, the operation of restriction of sections acquires a meaning with respect to *nesting* local frames, whereas the operation of extension or gluing of sections with respect to compatibly pairwise *overlapping* local frames. We note that if only the operation of restriction for nested local frames is satisfied, then we obtain a structure called a *presheaf.* In case that the operation of extension is also satisfied, then we obtain a *separated presheaf.* Only in case that the operation of extension can be *uniquely* carried out, that is local sections can be uniquely glued together, a separated presheaf becomes a sheaf. We note that, in general, there will be more locally defined or partial sections than globally defined ones, since not all partial sections need be extendible to global ones, but a compatible family of partial sections uniquely extends to a global one, or in other words, any presheaf uniquely defines a sheaf. More precisely, for the leading example of localization with respect to the open covers of a topological space we have the following basic notions:

A *presheaf* $\mathbb{F}$ of sets on a topological space X, consists of the following data:

I. For every open set U of X, *a set* denoted by $\mathbb{F}(U)$, and

II. For every inclusion $V \hookrightarrow U$ of open sets of X, a *restriction morphism* of sets in the opposite direction:

$$(0.3.1) \qquad r(U|V) : \mathbb{F}(U) \to \mathbb{F}(V)$$

such that:

a. $r(U|U) = $ identity at $\mathbb{F}(U)$ for all open sets U of X.

b. $r(V|W) \circ r(U|V) = r(U|W)$ for all open sets $W \hookrightarrow V \hookrightarrow U$.

Usually, the following simplifying notation is used: $r(U|V)(s) := s|_V$.

A presheaf $\mathbb{F}$ of sets on a topological space X, is defined to be a *sheaf* if it satisfies the following two conditions, for every family V_a, $a \in I$, of local open covers of V, where V open set in X, such that $V = \cup_a V_a$:

I. *Local identity* axiom of sheaf:

Given $s, t \in \mathbb{F}(V)$ with $s|_{V_a} = t|_{V_a}$ for all $a \in I$, then $s = t$.

II. *Gluing* axiom of sheaf:

Given $s_a \in \mathbb{F}(V_a)$, $s_b \in \mathbb{F}(V_b)$, $a, b \in I$, such that:

$$(0.3.2) \qquad s_a|_{V_a \cap V_b} = s_b|_{V_a \cap V_b},$$

for all $a, b \in I$, then there exists a *unique* $s \in \mathbb{F}(V)$, such that: $s|_{V_a} = s_a \in F(V_a)$ and $s|_{V_b} = s_b \in F(V_b)$.

If we consider the *partially ordered set* of open subsets of a topological space X as a category denoted by $\mathcal{O}(X)$, then $\mathbb{F}$ denotes the *contravariant* presheaf/sheaf functor that assigns to each open set $U \subset X$ a set in the category **Sets**. We note that the above definitions hold if instead of presheaves/sheaves of sets we consider presheaves/sheaves of algebraic structures, for example groups, algebras over a field, vector spaces over a field or modules over an algebra. For detailed pure mathematical expositions of sheaf theory the reader may consult Mallios [6], Mac Lane - Moerdijk [1], Tennison [1] and Vassiliou [1].

As a basic example, if $\mathbb{F}$ denotes the contravariant presheaf functor that assigns to each open set $U \subset X$, the *commutative unital* algebra of all real-valued continuous functions on U, then $\mathbb{F}$ is actually a sheaf. This is intuitively clear since the specification of a topology on X (and hence, of a topological localization system on $\mathcal{O}(X)$) is solely used for the definition of the continuous functions on X, thought of as observables on X. Thus, the continuity of each function can be determined *locally*. This means that continuity respects the operation of restriction to open sets, and moreover that continuous functions can be collated in a unique manner, as it is required for the satisfaction of the sheaf condition. More precisely, the sheaf condition in this case means that the following *sequence* of commutative algebras is *left exact*;

$$(0.3.3) \qquad 0 \to \mathbb{F}(U) \to \prod_a \mathbb{F}(U_a) \to \prod_{a,b} \mathbb{F}(U_a \cap U_b).$$

Let us further consider that x is a point of a topological space of control variables X. Moreover, let T be a set consisting of open subsets of X, containing x, such that the following condition holds: For any two open sets U, V, containing x, there exists an open set $W \in T$, contained in the intersection $U \cap V$. We may say that T constitutes *a basis* for the system of open sets around x. We form the *disjoint union* of all $\mathbb{F}(U)$, denoted by;

$$(0.3.4) \qquad \mathbb{D}(x) := \bigcup_{U \in T} \mathbf{F}(U).$$

Then we can define an *equivalence relation* in $\mathbb{D}(x)$, by requiring that $f \sim g$ for $f \in \mathbb{F}(U)$, $g \in \mathbb{F}(V)$, provided that they have the *same restriction* to a smaller open set contained in T. Then we define;

$$(0.3.5) \qquad colim_T[\mathbb{F}(U)] := \mathbb{D}(x)/\sim_T.$$

It is clear that the above definition is independent of the chosen basis of open covers T, and thus corresponds to an *inductive limit* construction. The inductive limit obtained is denoted by $\mathbb{F}_x$, and referred to as the *stalk* of $\mathbb{F}$ at the point $x \in X$. Physically, we identify an element of $\mathbb{F}$ of sort U with a *local section* of $\mathbb{F}$ over the open set U. Then the equivalence relation, used in the definition of the stalk $\mathbb{F}_x$ at the point $x \in X$ is interpreted as follows: Two local sections $f \in \mathbf{F}(U)$, $g \in \mathbf{F}(V)$, induce the *same contextual information* at x in X, provided that they have the same restriction to a smaller open set contained in the basis T. Then, the stalk $\mathbb{F}_x$ is the set containing all contextual information at x, that is the set of all equivalence classes. Moreover, the image of a local section $f \in \mathbf{F}(U)$ at the stalk $\mathbb{F}_x$, denoted by f_x, that is the equivalence class of this local section f, is precisely the *germ* of f at the point x. We deduce that the *fibration* corresponding to a sheaf of sets $\mathbb{F}$ is a *topological bundle*, called an *ètale bundle*, defined by the continuous mapping $\varphi : F \to X$, where;

$$(0.3.6) \qquad F = \bigcup_{x \in X} \mathbb{F}_x,$$

$$(0.3.7) \qquad \varphi^{-1}(x) = \mathbb{F}_x.$$

The mapping φ is locally a *homeomorphism* of topological spaces. The topology in F is defined as follows: for each local section $f \in \mathbb{F}(U)$, the set $\{f_x, x \in U\}$ is open, and moreover, an arbitrary open set is a union of sets of this form. Obviously, the same arguments hold in the case of a sheaf of sets $\mathbb{F}$ endowed with some algebraic structure, for example $\mathcal{F}$-algebras (where $\mathcal{F}$ is a field). Finally, the sheaf $\mathbb{F}$ can be canonically identified as the *sheaf of cross-sections* of the corresponding ètale bundle F.

We stress that the notion of a sheaf depends on the fact that we require the gluing condition with respect to all local covering frames, that is to *all covers* of any local frame. In principle, one could select some covers of a local frame and require the gluing condition only with respect to the selected covers. In this way, the notion of sheaf would be meant with respect to the selected family of covers. On the other hand, there is no restriction in considering only *hereditary covers*, that is covers containing all their subcovers. More precisely, any cover can be made hereditary (by adding to each cover all its considered subcovers) and compatible (uniquely glued) families of sections on the original cover are in *bijective* correspondence with compatible families of sections on the new one. This fact provides an important insight on the nature of the *topological localization process* of observables implicated by the concept of sheaf. First, let us think of local frames as partial information carriers related to measurement of observables. Now, the idea of considering hereditary covers can be implemented physically via a procedure of localization by *sieving*. More concretely, a *sieve* on a local frame can be thought of as consisting of *spectrum holes* distributed in different nested layers, such that every partial information carrier which is in a sense "smaller" than one of its holes *passes through* it but not otherwise. In this way, the notion of a sieve incorporates an *asymmetric logical order*. Moreover, since the partial information carriers should bear the semantics of local covers, certain conditions should hold distinguishing some sieves as *covering* ones within the class of all possible sieves. These conditions are the following: α) The covering sieves should be *stable under intersection*, meaning that the intersection of covering sieves should also be a covering sieve, for each local frame; β) the covering sieves should be *transitive*, such that covering sieves of subframes of a frame in covering sieves of this frame, should also be covering sieves of this frame themselves and, γ) the covering sieves should be *stable under change* of a local frame.

From a physical perspective, the consideration of covering sieves as hereditary covers used for localizing observable information with respect to local frames, gives rise to *localization systems* of global observable structures which induce a semantic transition of events from a set-theoretic to a sheaf-theoretic level. The operation which assigns to each local frame a collection of covering sieves satisfying the stability and transitivity conditions described qualitatively above, defines a topology, which is technically called a *Grothendieck topology*. The notion of a Grothendieck topology formulated in terms of covering sieves is significant for the following reasons: First, it elucidates precisely the topological concept of *locality* in relational *logical order*-theoretic terms, such that this concept becomes distinguished from its usual metrical spatiotemporal connotation. Second, it permits the *collation* of local observable information into global ones by utilization of the notion of sheaf with respect to a defined topology. In more detail, the extension of observable information from the local to the global takes place via a compatibly glued family of sections over a covering sieve (localization system) constituted of covering local frames (local covers). A sheaf assigns a set of sections to each local frame of a localization system, representing a set of observables which can be measured with respect to this local frame. A selection of sections from these sets, one for each local frame, forms a *compatibly glued family* with respect to a localization system, if the selection respects the operation of restriction, and additionally, if the sections selected agree whenever two local frames of the localization system overlap. If such a locally compatible selection of sections *extends uniquely* to a global one, then the sheaf conditions are satisfied. We note that in general, there will be more local sections than global ones (if they exist), since not all local observables need be extendible to global ones, but a compatible family of local observables uniquely extends to a global one with respect to a localization system. We conclude from the previous analysis that sheaf-theoretic gluing provides the means of *extensive connection*, that is extending a local section over a local frame to a higher level local frame and so on towards approximating the global with respect to some covering sieve.

0.3.2   *Vector Sheaves of States and Local Gauge Invariance*

From a physical viewpoint, the construction of a sheaf constitutes the natural outcome of a complete localization process. For example, in quantum mechanics we may consider the localization process of the set of the *Hamiltonian eigenstates* of an electron (energy eigenfunctions) constituting a *Hilbert space basis of vectors* with respect to some base space of control parameters.

We consider a pair $(X, \mathcal{A})$ consisting of a *paracompact* (*Hausdorff*) topological space X and a *soft sheaf* of commutative rings $\mathcal{A}$ localized over X. According to our general philosophy, we consider the above pair as the *Gelfand spectrum* of an appropriate algebra of observables A. We remind that if $\mathbb{C}$ is the field of complex numbers, then an $\mathbb{C}$-algebra A is a ring A together with a morphism of rings $\mathbb{C} \to A$ (making A into a vector space over $\mathbb{C}$) such that, the morphism $A \to \mathbb{C}$ is a linear morphism of vector spaces. Notice that the same holds if we substitute the field $\mathbb{C}$ with any other field, for instance, the field of real numbers $\mathbb{R}$. We also assume that the stalk $\mathcal{A}_x$ of germs is a *local commutative* $\mathbb{C}$-algebra for any point $x \in X$. A typical example is the case, where X is a smooth manifold of and $\mathcal{A}$ is the $\mathbb{C}$-algebra sheaf of germs of smooth functions localized over X. Together with a $\mathbb{C}$-algebra sheaf $\mathcal{A}$ we also consider the *abelian group* sheaf of invertible elements of $\mathcal{A}$, denoted by $\mathcal{A}^\bullet := \tilde{\mathcal{A}}$.

An $\mathcal{A}$-module $\mathcal{E}$ is called a *locally free $\mathcal{A}$-module of states of finite rank m*, or simply a *vector sheaf of states*, if for any point $x \in X$ there exists an open set U of X such that:

$$(0.3.8) \qquad \mathcal{E}\mid_U \cong \overset{m}{\bigoplus}(\mathcal{A}\mid_U) := (\mathcal{A}\mid_U)^m,$$

where $(\mathcal{A}\mid_U)^m$ denotes the m-terms *direct sum* of the sheaf of $\mathbb{C}$-observable algebras $\mathcal{A}$ restricted to U, for some $m \in \mathbb{N}$. We call $(\mathcal{A}\mid_U)^m$ the *local sectional frame* or equivalently *local gauge* of states of $\mathcal{E}$ associated via the open covering $\mathcal{U} = \{U\}$ of X.

In case that the rank is 1, the corresponding vector sheaf is called a *line sheaf of states*, that is locally for any point $x \in X$ there exists an open set U of X such that:

$$(0.3.9) \qquad \mathcal{E}\mid_U \cong \mathcal{A}\mid_U.$$

Furthermore, if for any point $x \in X$ there exists an open set U of X such that:

$$(0.3.10) \qquad \mathbb{S}\mid_U \cong \overset{m}{\bigoplus}(\mathbb{C}\mid_U) := (\mathbb{C}\mid_U)^m,$$

then we call any locally free $\mathbb{C}$-module $\mathbb{S}$ of finite rank m, for some $m \in \mathbb{N}$, a *complex linear local system* of rank m.

The notion of a vector sheaf of states generalizes the notion of a vector space of states when there exists for instance a *parametric dependence* of a system from a topological space of control variables. The crucial point is that locally every section of a finite rank vector sheaf can be written as a *finite linear combination of a basis of sections with coefficients from the local observable algebra*. For example, if X is a smooth manifold and $\mathcal{A}$ is the $\mathbb{C}$-algebra or $\mathbb{R}$-algebra sheaf of germs of smooth functions on X, then every section can be locally written as a *finite* linear combination or *superposition* of a basis of sections with coefficient being real-valued smooth functions. We note that the physical notion of *time dependence* may be implicitly introduced via *time-parameterized paths* on X. It is also the case that the set of sections of every *vector bundle* on a topological space (not necessarily a smooth manifold) forms a vector sheaf of sections localized over this space. Thus, the notion of a vector sheaf of states provides the most *general* conceptual and technical framework to consider in a unifying way all cases, where there exists a parametric dependence of some observable by a topological space of control variables X.

Given a vector sheaf of states $\mathcal{E}$, there is specified a *Čech 1-cocycle with respect to a covering* $\mathcal{U} = \{U\}$ of X, called a *coordinate 1-cocycle* in $Z^1(\mathcal{U}, GL(m, \mathcal{A}))$ (with values in the sheaf of germs of sections into the group $GL(m, \mathbb{C})$), as follows:

$$(0.3.11) \qquad \eta_\alpha : \mathcal{E}\mid_{U_\alpha} \cong (\mathcal{A}\mid_{U_\alpha})^m,$$

$$(0.3.12) \qquad \eta_\beta : \mathcal{E}\mid_{U_\beta} \cong (\mathcal{A}\mid_{U_\beta})^m.$$

Thus, for every $x \in U_\alpha$ we have a *stalk isomorphism*:

(0.3.13)
$$\eta_\alpha(x) : \mathcal{E}_x \cong \mathcal{A}_x{}^m,$$

and similarly for every $x \in U_\beta$. If we consider that $x \in U_\alpha \cap U_\beta$, then we obtain the isomorphism:

(0.3.14)
$$g_{\alpha\beta}(x) = \eta_\alpha(x)^{-1} \circ \eta_\beta(x) : \mathcal{A}_x{}^m \cong \mathcal{A}_x{}^m.$$

The $g_{\alpha\beta}(x)$ is thought of as an *invertible matrix of germs* at x. Consequently, $g_{\alpha\beta}$ is an invertible matrix section in the sheaf of germs $GL(m, \mathcal{A})$ (taking values in the *general linear group* $GL(m, \mathbb{C})$). Moreover, $g_{\alpha\beta}$ satisfy the *cocycle conditions* $g_{\alpha\beta} \circ g_{\beta\gamma} = g_{\alpha\gamma}$ on triple intersections whenever they are defined.

It is clear in this way that we obtain a vector bundle with typical fiber $\mathbb{C}^m$, structure group $GL(m, \mathbb{C})$, whose sections form the vector sheaf of states we started with. In particular, for $m = 1$, we obtain a *line bundle* L with fiber $\mathbb{C}$, structure group $GL(1, \mathbb{C}) \cong \dot{\mathbb{C}}$ (the non-zero complex numbers), whose sections form a *line sheaf* of states $\mathcal{L}$. Clearly, by imposing a *unitarity* condition the structure group is reduced to $U(1)$. Thus, particularly in the case of a line sheaf of states we have a bijective correspondence:

(0.3.15)
$$\mathcal{L} \leftrightarrow (g_{\alpha\beta}) \in Z^1(\mathcal{U}, \tilde{\mathcal{A}})$$

where $GL(1, \mathcal{A}) \cong \mathcal{A}^\bullet := \tilde{\mathcal{A}}$ is the group sheaf of invertible elements of $\mathcal{A}$ (taking values in $\dot{\mathbb{C}}$), and $Z^1(\mathcal{U}, \tilde{\mathcal{A}})$ is the set of coordinate 1-cocycles. In physical terminology, a coordinate 1-cocycle effects a *local frame transformation*, or equivalently a *local gauge transformation* of a vector sheaf of states $\mathcal{E}$.

Every 1-cocycle can be conjugated with a 0-*cochain* (t_α) in the set $C^0(\mathcal{U}, \tilde{\mathcal{A}})$ to obtain another equivalent 1-cocycle:

(0.3.16)
$$g'_{\alpha\beta} = t_\alpha \cdot g_{\alpha\beta} \cdot t_\beta^{-1}.$$

If we consider the *coboundary operator* :

(0.3.17)
$$\Delta^0 : C^0(\mathcal{U}, \tilde{\mathcal{A}}) \to C^1(\mathcal{U}, \tilde{\mathcal{A}}),$$

then the image of Δ^0 gives the set of *1-coboundaries* of the form $\Delta^0(t_\alpha^{-1})$ in $B^1(\mathcal{U}, \tilde{\mathcal{A}})$:

(0.3.18)
$$\Delta^0(t_\alpha^{-1}) := t_\alpha \cdot t_\beta^{-1}.$$

Thus, we obtain that the 1-cocycle $g'_{\alpha\beta}$ is equivalent to the 1-cocycle $g_{\alpha\beta}$ in $Z^1(\mathcal{U}, \tilde{\mathcal{A}})$ if and only if there exists a 0-cochain (t_α) in the set $C^0(\mathcal{U}, \tilde{\mathcal{A}})$, such that:

(0.3.19)
$$g'_{\alpha\beta} \cdot g_{\alpha\beta}^{-1} = \Delta^0(t_\alpha^{-1})$$

where, the multiplication above is meaningful in the *abelian group* of 1-cocycles $Z^1(\mathcal{U}, \tilde{\mathcal{A}})$.

Due to the bijective correspondence of line sheaves with coordinate 1-cocycles with respect to an open covering $\mathcal{U}$, we deduce immediately the following:

Theorem 0.1. *The set of isomorphism classes of line sheaves of states over the same topological space X, denoted by $Iso(\mathcal{L})(X)$, is in bijective correspondence with the set of cohomology classes $H^1(X, \tilde{\mathcal{A}})$:*

(0.3.20)
$$Iso(\mathcal{L})(X) \cong H^1(X, \tilde{\mathcal{A}}).$$

The above proposition interpreted in physical terms constitutes the *cohomological formulation of the principle of local gauge (phase) invariance* in the description of the state space of a system described by a line sheaf. Furthermore, each equivalence class $[\mathcal{L}] \equiv \mathcal{L}$ in $Iso(\mathcal{L})(X)$ has an inverse, defined by:

$$(0.3.21) \qquad \mathcal{L}^{-1} := Hom_{\mathcal{A}}(\mathcal{L}, \mathcal{A})$$

where, $Hom_{\mathcal{A}}(\mathcal{L}, \mathcal{A}) := \mathcal{L}^*$ denotes the *dual line sheaf* of $\mathcal{L}$. This is actually deduced from the fact that we can define the *tensor product* of two equivalence classes of line sheaves over $\mathcal{A}$ so that:

$$(0.3.22) \qquad \mathcal{L} \otimes_{\mathcal{A}} \mathcal{L}^* \cong Hom_{\mathcal{A}}(\mathcal{L}, \mathcal{L}) \equiv \mathcal{E}nd_{\mathcal{A}}\mathcal{L} \cong \mathcal{A}.$$

Hence, we obtain the following refinement in the cohomological formulation of local gauge invariance:

Theorem 0.2. *The set of isomorphism classes of line sheaves of states over the same topological space X, $Iso(\mathcal{L})(X)$, has an abelian group structure with respect to the tensor product over the observable algebra sheaf $\mathcal{A}$, and analogously the set of cohomology classes $H^1(X, \tilde{\mathcal{A}})$ is also an abelian group, where the tensor product of two line sheaves of states corresponds to the product of their respective coordinate 1-cocycles.*

0.3.3 *Exponential Short Exact Sequence*

In all cases of physical interest, the topological space X of control variables is considered to be paracompact, while the observable algebra sheaf $\mathcal{A}$ is a soft sheaf meaning that every section over some closed subset in X can be *extended* to a section over X. More importantly, the following *short sequence of abelian group sheaves is exact*, which models sheaf-theoretically the process of *exponentiation*, for details see Mallios [7, p.330]:

$$(0.3.23) \qquad 0 \to \mathbb{Z} \overset{\iota}{\hookrightarrow} \mathcal{A} \overset{e}{\to} \tilde{\mathcal{A}} \to 1,$$

where $\mathbb{Z}$ is the constant abelian group sheaf of integers (sheaf of locally constant sections valued in the group of integers), such that:

$$(0.3.24) \qquad Ker(e) = Im(\iota) \cong \mathbb{Z}.$$

Clearly, due to the *canonical imbedding* of the constant abelian group sheaf $\mathbb{C}$ into $\mathcal{A}$ as well as of $\tilde{\mathbb{C}}$ into $\tilde{\mathcal{A}}$, we have the validity of the *short exact sequence of constant abelian group sheaves*:

$$(0.3.25) \qquad 0 \to \mathbb{Z} \overset{\iota}{\hookrightarrow} \mathbb{C} \overset{exp}{\longrightarrow} \tilde{\mathbb{C}} \to 1,$$

$$(0.3.26) \qquad Ker(exp) = Im(\iota) \cong \mathbb{Z}.$$

The above exponential short exact sequence can be specialized further to the following short exact sequence of constant abelian group sheaves:

$$(0.3.27) \qquad 0 \to \mathbb{Z} \xrightarrow{\iota} \mathbb{R} \xrightarrow{exp(2\pi i)} \mathbb{U}(1) \to 1$$

where $\mathbb{R}$ is the constant abelian group sheaf of reals and $\mathbb{U}(1)$ is the abelian group sheaf of unit modulus complexes (phases).

The fundamental significance of these three exponential short exact sequences of abelian group sheaves cannot be overestimated. In particular, we obtain a further *refinement* of the cohomological formulation of local phase (gauge) invariance pertaining to the state space, according to the following:

Theorem 0.3. *Each equivalence classes of line sheaves of states in $Iso(\mathcal{L})(X)$ is in bijective correspondence with a cohomology class in the integral 2-dimensional cohomology group of X.*

If we consider the exponential short exact sequence of abelian group sheaves, we immediately obtain a *long exact sequence in sheaf cohomology*, which due to paracompactness of X is reduced to Čech cohomology. Thus, we have:

$$(0.3.28)$$
$$\ldots \to H^1(X, \mathbb{Z}) \to H^1(X, \mathcal{A}) \to H^1(X, \tilde{\mathcal{A}}) \to H^2(X, \mathbb{Z}) \to H^2(X, \mathcal{A}) \to H^2(X, \tilde{\mathcal{A}}) \to \ldots$$

Because of the fact that $\mathcal{A}$ is a soft sheaf, we have:

$$(0.3.29) \qquad H^1(X, \mathcal{A}) = H^2(X, \mathcal{A}) = 0$$

and consequently we obtain:

$$(0.3.30) \qquad 0 \to H^1(X, \tilde{\mathcal{A}}) \xrightarrow{\delta_c} H^2(X, \mathbb{Z}) \to 0.$$

Thus, equivalently we obtain the following isomorphism of abelian groups (*Chern isomorphism*) :

$$(0.3.31) \qquad \delta_c : H^1(X, \tilde{\mathcal{A}}) \xrightarrow{\cong} H^2(X, \mathbb{Z}).$$

Since we have shown previously that the set of isomorphism classes of line sheaves of states over X, viz. $Iso(\mathcal{L})(X)$, is in bijective correspondence with the abelian group of cohomology classes $H^1(X, \tilde{\mathcal{A}})$, we deduce that:

$$(0.3.32) \qquad Iso(\mathcal{L})(X) \cong H^2(X, \mathbb{Z}).$$

Thus, each equivalence classes of line sheaves of states is in bijective correspondence with a cohomology class in the integral 2-dimensional cohomology group of X.

0.4   Connections and Differential Analysis

0.4.1   *Kähler-de Rham Paradigm*

Differential geometry is traditionally associated with the notion of *smooth manifolds*. Smooth manifolds are geometric spaces which are *locally homeomorphic to*

Euclidean spaces of an appropriate dimension. The topological (actually metrical) linear vector space structure of a Euclidean space in the neighborhood of a manifold point is used for transferring locally the methods of analytic calculus to smooth manifolds, and thus generate locally the whole differential geometric framework. Conclusively, this *local linear structure* provides the seed of generation of *spatially determined differential analysis* on a smooth manifold, which is interpreted physically as a process of dynamical extensive connection caused by an interaction field. The localization scheme provided by the notion of a smooth manifold is very *restrictive* not only for the incorporation of quantum phenomena (where the local Euclidean linear space structure is not tenable) but even for an efficient treatment of spacetime *singularities* obstructing the validity of the field equations formulated in terms of the corresponding algebra of smooth functions.

In this manner, the pertinent question is the following: Is the generation of local differential analysis, and thus differential geometry, *solely dependent*, and thus spatially determined, by the local linear space structure of Euclidean spaces or there exists a *universal algebraic mechanism* of generation of differentiability not dependent on this particular metrical Euclidean localization scheme? If such an algebraic mechanism of differential analysis exists, then the whole differential geometric framework of smooth manifolds would constitute a particular manifestation shaped by the modeling Euclidean linear space localization scheme.

A crucial observation comes from the fact that Kähler's *purely algebraic* theory of generation of differential forms does not involve *any* assumptions involving Euclidean spaces and is solely based on the properties of the *inverse processes* of extension and restriction of commutative algebras. Concomitantly, De Rham's *algebraic-topological* theory of cohomology is based on the existence of modules of *exterior differential n-forms* (admitting an appropriate characterization as closed or exact by intrinsic means), and the formation of *complexes* by means of differential operators acting on these modules. The *Kähler-de Rham homological algebraic mechanism of differential analysis* is *not dependent* on any local Euclidean vector space structure for its function, and thus provides an excellent candidate for the generalization of the differential geometric framework of smooth manifolds to more general localization schemes specified by *dynamical* means (in analogy to the case of General Relativity). In order to be applicable for modeling the process of extensive connection dynamically (that is via the action of a physical field) in the observability scale of a physical theory (defined by the corresponding reference frames) it should be susceptible to *sheaf-theoretic localization*.

The fact that the homological Kähler-de Rham differential mechanism actually admits a sheaf-theoretic localization with respect to appropriate localization schemes is not a trivial issue, for details see Mallios[7, p.326], Vassiliou [1, p.57], Abel-Ntumba[1], Fragoulopoulou-Papatriantafillou [1], and for a physical interpretation and applications Mallios-Zafiris [1, 2]. What it shows is that the concepts of differential geometry which remain logically invariant under abstraction from the smooth manifold scaffolding are the ones which can be *localized by sheaf-theoretic means*. Accordingly, the sheaf gauge of localization of observables shows that the intrinsic mechanism of differential observability induced by the action of a physical field is of a *sheaf cohomological* nature. As we have already mentioned, cohomology can be conceived as an effective method of assigning *global invariants* to a localization scheme (for example to a topological space, or more generally to a categorical site) in a *homotopy-invariant* way. The cohomology groups measure the *global obstructions* for extending sections from the local to the global level. From a physical perspective, these obstructions have global observable effects and may be associated with *field sources* preventing the uniform and maximally undisturbed dynamical process of extensive connection. It is precisely these obstructions detected by cohomology groups, which make sheaf cohomology the proper *measurement tool* of global observability.

0.4.2   *Kähler's Algebraic Extension Method*

It is instructive to describe briefly the purely algebraic form of Kähler's construction from a physical viewpoint. In general algebraic terms, the process of extending the form of observation with respect to an algebra of scalars, for instance an $\mathbb{R}$-algebra $\mathcal{A}$, due to field interactions, is described by means of a *fibering*, defined as an injective morphism of $\mathbb{R}$-algebras $\iota : A \hookrightarrow B$. Thus, the $\mathbb{R}$-algebra B is considered as a module over the algebra A. A section of the fibering $\iota : A \hookrightarrow B$, is represented by a morphism of $\mathbb{R}$-algebras $s : B \to A$ such that $\iota \circ s = id_B$. The fundamental extension of scalars of the $\mathbb{R}$-algebra A is obtained by *tensoring* A with itself over the subalgebra of the base field (for instance the real numbers), that is $\iota : A \hookrightarrow A \otimes_{\mathbb{R}} A$. Trivial cases of scalars extensions, in fact isomorphic to A, induced by the fundamental one, are obtained by tensoring A with $\mathbb{R}$ from both sides, that is $\iota_1 : A \hookrightarrow A \otimes_{\mathbb{R}} \mathbb{R}$, $\iota_2 : A \hookrightarrow \mathbb{R} \otimes_{\mathbb{R}} A$.

The basic idea of Riemann that has been incorporated for instance in the context of Einstein's general theory of relativity is that *geometry* should be built from the *infinitesimal to the global*. Geometry in this context is understood in terms of metric structures that can be defined on a differential spacetime manifold. If we adopt Kähler's algebraic viewpoint, then the effectuation of geometry as a result of interactions, requires first of all the *extension of scalars* of the algebra A by *infinitesimal quantities*, defined as a fibration:

$$(0.4.1) \qquad\qquad d_* : A \hookrightarrow A \oplus M \cdot \epsilon,$$

$$(0.4.2) \qquad\qquad f \mapsto f + d(f) \cdot \epsilon$$

where $d(f) =: df$ is considered as the infinitesimal part of the extended scalar, and ϵ denotes the infinitesimal unit obeying $\epsilon^2 = 0$. The algebra of infinitesimally extended scalars $A \oplus M \cdot \epsilon$ is called the algebra of *dual numbers* over A with coefficients in the A-module M. It is immediate to see that the algebra $A \oplus M \cdot \epsilon$, as an abelian group is just the direct sum $A \oplus M$, whereas the multiplication is defined by:

$$(0.4.3) \qquad (f + df \cdot \epsilon) \bullet (f' + df' \cdot \epsilon) = f \cdot f' + (f \cdot df' + f' \cdot df) \cdot \epsilon.$$

Note that, we further require that the composition of the augmentation $A \oplus M \cdot \epsilon \to A$, with d_* is the identity. Equivalently, the above fibration, viz., the homomorphism of algebras $d_* : A \hookrightarrow A \oplus M \cdot \epsilon$, can be formulated as a *derivation*, that is in terms of an additive $\mathbb{R}$-linear morphism:

$$(0.4.4) \qquad\qquad\qquad d : A \to M,$$

$$(0.4.5) \qquad\qquad\qquad f \mapsto df$$

that, moreover, satisfies the Leibniz rule :

$$(0.4.6) \qquad\qquad d(f \cdot g) = f \cdot dg + g \cdot df.$$

Since the formal symbols of differentials $\{df, f \in A\}$, are reserved for the *universal derivation*, the A-module M is identified as the free A-module $\Omega := \Omega^1(A)$ of 1-*forms* generated by these formal symbols, modulo the Leibniz constraint, where the scalars of the distinguished subalgebra $\mathbb{R}$, that is the real numbers, are treated as constants.

Kähler's fundamental insight consists in the realization that the free A-module Ω can be constructed explicitly from the fundamental form of scalars extension of A, that is $\iota : A \hookrightarrow A \otimes_{\mathbb{R}} A$ by considering the morphism:

$$(0.4.7) \qquad\qquad\qquad \mu : A \otimes_{\mathbb{R}} A \to A,$$

$$(0.4.8) \qquad\qquad\qquad \sum_i f_i \otimes g_i \mapsto \sum_i f_i \cdot g_i.$$

Then, by taking the *kernel* of this morphism of algebras, that is, the ideal :

$$(0.4.9) \qquad I = Ker\mu = \{\theta \in A \otimes_{\mathbb{R}} A : \mu(\theta) = 0\} \subset A \otimes_{\mathbb{R}} A$$

we obtain the following:

Theorem 0.4. *The morphism of A-modules:*

$$(0.4.10) \qquad\qquad\qquad \Sigma : \Omega \to \frac{I}{I^2},$$

$$(0.4.11) \qquad\qquad\qquad df \mapsto 1 \otimes f - f \otimes 1$$

is an isomorphism.

In order to show this, we notice that the fractional object $\frac{I}{I^2}$ has an A-module structure defined by:

$$(0.4.12) \qquad\quad f \cdot (f_1 \otimes f_2) = (f \cdot f_1) \otimes f_2 = f_1 \otimes (f \cdot f_2)$$

for $f_1 \otimes f_2 \in I$, $f \in A$. We can check that the second equality is true by proving that the difference of $(f \cdot f_1) \otimes f_2$ and $f_1 \otimes (f \cdot f_2)$ belonging to I, is actually an element of I^2, viz., the equality is true modulo I^2. So we have:

$$(0.4.13) \qquad (f \cdot f_1) \otimes f_2 - f_1 \otimes (f \cdot f_2) = (f_1 \otimes f_2) \cdot (f \otimes 1 - 1 \otimes f).$$

The first factor of the above product of elements belongs to I by assumption, whereas the second factor also belongs to I, since we have that:

$$(0.4.14) \qquad \mu(f \otimes 1 - 1 \otimes f) = 0.$$

Hence, the product of elements above belongs to $I \cdot I = I^2$. Consequently, we can define a morphism of A-modules:

$$(0.4.15) \qquad \Sigma : \Omega \to \frac{I}{I^2},$$

$$(0.4.16) \qquad df \mapsto 1 \otimes f - f \otimes 1.$$

Now, we construct the inverse of that morphism as follows: The A-module Ω can be made an *ideal* in the algebra of dual numbers over A, viz., $A \oplus \Omega \cdot \epsilon$. Moreover, we can define the morphism of algebras:

$$(0.4.17) \qquad A \times A \to A \oplus \Omega \cdot \epsilon,$$

$$(0.4.18) \qquad (f_1, f_2) \mapsto f_1 \cdot f_2 + f_1 \cdot df_2 \epsilon.$$

This is an $\mathbb{R}$-bilinear morphism of algebras, and thus, it gives rise to a morphism of algebras:

$$(0.4.19) \qquad \Theta : A \otimes_{\mathbb{R}} A \to A \oplus \Omega \cdot \epsilon.$$

Then, by definition we have that $\Theta(I) \subset \Omega$, and also, $\Theta(I^2) = 0$. Hence, there is obviously induced a morphism of A-modules:

$$(0.4.20) \qquad \Omega \leftarrow \frac{I}{I^2},$$

which is the *inverse* of Σ. Consequently, we conclude that:

$$(0.4.21) \qquad \Omega \cong \frac{I}{I^2}.$$

Thus, the free A-module Ω of 1-forms is isomorphic with the free A-module $\frac{I}{I^2}$ of *Kähler differentials* of the algebra of scalars A over $\mathbb{R}$, conceived as distinguished ideals within the algebra of infinitesimally extended scalars $[A \oplus \Omega \cdot \epsilon]$ due to interaction, according to the following split short exact sequence :

$$(0.4.22) \qquad \Omega \hookrightarrow A \oplus \Omega \cdot \epsilon \twoheadrightarrow A,$$

or equivalently formulated as:

$$(0.4.23) \qquad 0 \to \Omega \to A \otimes_{\mathbb{R}} A \to A.$$

By dualizing, we obtain the *dual* A-module of Ω, that is, $\Xi := Hom(\Omega, A)$. Consequently, we have at our disposal, expressed in terms of infinitesimal scalars extension of the algebra $\mathcal{A}$, semantically intertwined with the generation of geometry as a result of interaction, new types of observables related with the incorporation of differentials and their duals, called vectors.

Let us now explain the functional role of geometry in relation to the infinitesimally extended rings of scalars. This is of particular significance for *dynamical physical theories*, where the ring of scalars stands for an algebra of *observable attributes* taking values in the field of real numbers. This *absolute valuation requirement* necessitates that our form of observation is tautological with real numbers representability. This means that all types of physical quantities should possess *uniquely* defined dual types, such that their representability can be made possible by means of real numbers. This is exactly the role of a *geometry induced by a metric*. Concretely, a metric structure assigns a unique dual to each physical quantity, by effectuating an isomorphism $\tilde{g}$ between the A-module Ω and its dual A-module $\Xi = Hom(\Omega, A)$, that is:

$$(0.4.24) \qquad\qquad \tilde{g} : \Omega \longrightarrow \Xi,$$

such that:

$$(0.4.25) \qquad\qquad \tilde{g} : \Omega \cong \Xi,$$

$$(0.4.26) \qquad\qquad df \mapsto v_f := \tilde{g}(df).$$

Equivalently, a metric g stands for an $\mathbb{R}$-valued *symmetric bilinear form* on Ω, that is $g : \Omega \times \Omega \to \mathbb{R}$, yielding an invertible $\mathbb{R}$-linear morphism $\tilde{g} : \Omega \to \Xi$. Notice that for df, $dh \in \Omega$, a symmetric bilinear form g acts, via $\tilde{g}$, on df to give an element of the dual, $\tilde{g}(df) \in \Xi$, which then acts on dh to give $(\tilde{g}(df))(dh) = (\tilde{g}(dh))(df)$, or equivalently, $v_f(dh) = v_h(df) \in \mathbb{R}$. Also note that the invertibility of $\tilde{g}$ amounts to the property of *non-degeneracy* of g, meaning that for each $df \in \Omega$, there exists $dh \in \Omega$, such that $(\tilde{g}(df))(dh) = v_f(dh) \neq 0$.

Thus the functional role of a metric geometry induced by a bilinear form forces the observation of extended scalars by means of representability in the field of real numbers and is reciprocally conceived as the result of interactions causing infinitesimal variations in the scalars of an $\mathbb{R}$-algebra A of observables.

In order that Kähler's algebraic extension method becomes suitable as a generator of a universal mechanism of differential analysis it should be susceptible to the process of sheaf-theoretic localization of an observable algebra. This would allow the applicability of this mechanism in conjunction with the principle of local gauge invariance in a sheaf-theoretic context suited to the localization requirements of a corresponding dynamical physical theory. Most important, it would allow the complete disassociation of the differential mechanism of any underlying spatial substratum opening up the way for a *functorial* and *topos-theoretic* formulation of differential geometry.

0.4.3   Connections and the Sheaf-Theoretic de Rham Complex

The localization of Kähler's algebraic extension method in sheaf theoretic terms requires first of all the notion of a *universal derivation* of the observable algebra sheaf $\mathcal{A}$ considered for generality as an algebra sheaf over the constant sheaf of complexes $\mathbb{C}$.

The *universal $\mathbb{C}$-derivation* of the observable algebra sheaf $\mathcal{A}$ to the universal $\mathcal{A}$-module sheaf $\Omega^1(\mathcal{A}) := \Omega^1$, called the *$\mathcal{A}$-module sheaf of differential 1-forms*, is the universal $\mathbb{C}$-linear sheaf morphism (natural transformation) $\partial := d^0$:

$$(0.4.27) \qquad\qquad d^0 : \mathcal{A} \to \Omega^1$$

such that the *Leibniz condition* is satisfied:

$$(0.4.28) \qquad\qquad d^0(s \cdot t) = s \cdot d^0(t) + t \cdot d^0(s),$$

for any continuous local sections s, t belonging to $\mathcal{A}(U)$, with U open set in X. We may also use the following notational convention: $\mathcal{A} := \Omega^0$.

Associated with the universal $\mathbb{C}$-derivation d^0 on $\mathcal{A}$, there also exists the *logarithmic universal derivation* defined as follows:

The *logarithmic universal derivation* of the observable abelian group sheaf $\tilde{\mathcal{A}}$ to the abelian group sheaf of 1-forms Ω^1 is the universal morphism of abelian group sheaves $\tilde{\partial} := \tilde{d}^0$:

$$(0.4.29) \qquad\qquad \tilde{d}^0 : \tilde{\mathcal{A}} \to \Omega^1,$$

defined by the relation:

$$(0.4.30) \qquad\qquad \tilde{d}^0(u) := u^{-1} \cdot d^0(u)$$

for any continuous *invertible* local section $u \in \tilde{\mathcal{A}}(U)$, with U open set in X.

Given the validity of the *Poincaré Lemma*, we also have:

$$(0.4.31) \qquad\qquad Ker(\tilde{d}^0) = \tilde{\mathbb{C}}.$$

If we consider a complex vector sheaf of states $\mathcal{E}$, then we define an $\mathcal{A}$-connection on $\mathcal{E}$ as follows:

A connection $\mathcal{D}_{\mathcal{E}} := \nabla_{\mathcal{E}}$ on the vector sheaf $\mathcal{E}$ is a $\mathbb{C}$-*linear* sheaf morphism:

$$(0.4.32) \qquad\qquad \nabla_{\mathcal{E}} : \mathcal{E} \to \Omega^1(\mathcal{A}) \otimes_{\mathcal{A}} \mathcal{E},$$

referring to $\mathbb{C}$-vector space sheaves, such that the corresponding Leibniz condition is satisfied:

$$(0.4.33) \qquad\qquad \nabla_{\mathcal{E}}(a \cdot s) = a \cdot \nabla_{\mathcal{E}}(s) + s \otimes d^0(a).$$

Moreover, for each $n \in N$, $n \geq 2$, the n-fold *exterior product* is defined as follows: $\Omega^n(\mathcal{A}) = \wedge^n \Omega^1(\mathcal{A})$ where $\Omega(\mathcal{A}) := \Omega^1(\mathcal{A})$. We notice that there exists a $\mathbb{C}$-linear sheaf morphism:

$$(0.4.34) \qquad\qquad d^n : \Omega^n(\mathcal{A}) \to \Omega^{n+1}(\mathcal{A})$$

for all $n \geq 0$. Let $\omega \in \Omega^n(\mathcal{A})$, then ω has the form:

$$(0.4.35) \qquad \omega = \sum f_i(dl_{i1} \wedge \ldots \wedge dl_{in})$$

with f_i, l_{ij}, $\in \mathcal{A}$ for all integers i, j. Further, we define:

$$(0.4.36) \qquad d^n(\omega) = \sum df_i \wedge dl_{i1} \wedge \ldots \wedge dl_{in}.$$

From the above, we immediately obtain that the composition of two consecutive $\mathbb{C}$-linear sheaf morphisms vanishes, that is $d^{n+1} \circ d^n = 0$.

The sequence of $\mathbb{C}$-linear sheaf morphisms:

$$(0.4.37) \qquad \mathcal{A} \to \Omega^1(\mathcal{A}) \to \ldots \to \Omega^n(\mathcal{A}) \to \ldots$$

is a *complex* of $\mathbb{C}$-vector space sheaves, called the *sheaf-theoretic de Rham complex* of $\mathcal{A}$.

Moreover, given the validity of the Poincaré Lemma, $Ker(d^0) = \mathbb{C}$, and the fact that $\mathcal{A}$ is a soft observable algebra sheaf, we obtain:

The sequence of $\mathbb{C}$-vector space sheaves is exact:

$$(0.4.38) \qquad \mathbf{0} \to \mathbb{C} \to \mathcal{A} \to \Omega^1(\mathcal{A}) \to \ldots \to \Omega^n(\mathcal{A}) \to \ldots$$

Thus, the sheaf-theoretic de Rham complex of the observable algebra sheaf $\mathcal{A}$ constitutes a *resolution of the constant sheaf* $\mathbb{C}$.

If we assume that the pair $(\mathcal{E}, \nabla_{\mathcal{E}})$ denotes a complex vector sheaf of states equipped with a connective structure, defined by a connection $\nabla_{\mathcal{E}}$ on $\mathcal{E}$, then $\nabla_{\mathcal{E}}$ induces a sequence of $\mathbb{C}$-linear morphisms:

$$(0.4.39) \qquad \mathcal{E} \to \Omega^1(\mathcal{A}) \otimes_{\mathcal{A}} \mathcal{E} \to \ldots \to \Omega^n(\mathcal{A}) \otimes_{\mathcal{A}} \mathcal{E} \to \ldots$$

or equivalently:

$$(0.4.40) \qquad \mathcal{E} \to \Omega^1(\mathcal{E}) \to \ldots \to \Omega^n(\mathcal{E}) \to \ldots$$

where the morphism:

$$(0.4.41) \qquad \nabla^n : \Omega^n(\mathcal{A}) \otimes_{\mathcal{A}} \mathcal{E} \to \Omega^{n+1}(\mathcal{A}) \otimes_{\mathbf{A}} \mathcal{E},$$

is given by the formula:

$$(0.4.42) \qquad \nabla^n(\omega \otimes v) = d^n(\omega) \otimes v + (-1)^n \omega \wedge \nabla(v),$$

for all $\omega \in \Omega^n(\mathcal{A})$, $v \in \mathcal{E}$. It is immediate to see that $\nabla^0 = \nabla_{\mathcal{E}}$.

The composition of $\mathbb{C}$-linear morphisms $\nabla^1 \circ \nabla^0$ is called the *curvature* of the connection $\nabla_{\mathcal{E}}$:

$$(0.4.43) \qquad \nabla^1 \circ \nabla^0 := R_{\nabla} : \mathcal{E} \to \Omega^2(\mathcal{A}) \otimes_{\mathcal{A}} \mathcal{E} = \Omega^2(\mathcal{E}).$$

As a straightforward consequence of the above definition we obtain:

The curvature R_{∇} of a connection $\nabla_{\mathcal{E}}$ on the vector sheaf of states $\mathcal{E}$ is an $\mathcal{A}$-linear sheaf morphism, that is a $\mathcal{A}$-covariant, or equivalently an $\mathcal{A}$-*tensor*.

The $\mathcal{A}$-covariant nature of the curvature R_{∇} is to be contrasted with the connection $\nabla_{\mathcal{E}}$ which is only $\mathbb{C}$-covariant.

The sequence of $\mathbb{C}$-linear sheaf morphisms

(0.4.44) $$\mathcal{E} \to \Omega^1(\mathcal{A}) \otimes_\mathcal{A} \mathcal{E} \to \ldots \to \Omega^n(\mathcal{A}) \otimes_\mathcal{A} \mathcal{E} \to \ldots$$

is a *complex* of $\mathbb{C}$-vector space sheaves if and only if:

(0.4.45) $$R_\nabla = 0.$$

Thus, the curvature $\mathcal{A}$-linear sheaf morphism R_∇ expresses the *obstruction* for the above sequence to qualify as a complex. We say that the connection $\nabla_\mathcal{E}$ is *integrable* if $R_\nabla = 0$, and we refer to the obtained complex as the sheaf-theoretic de Rham complex of the integrable connection $\nabla_\mathcal{E}$ on the vector sheaf $\mathcal{E}$ in that case. It is also usual to call a connection $\nabla_\mathcal{E}$ *flat* if $R_\nabla = 0$. We note that the universal $\mathbb{C}$-derivation d^0 on $\mathcal{A}$ defines an integrable or flat connection.

A flat connection defines a *maximally undisturbed process of dynamical variation* caused by the corresponding physical field. In this sense, a non-vanishing curvature signifies the existence of disturbances from the maximally symmetric state of that variation. These disturbances can be cohomologically identified as obstructions to deformation caused by physical sources. In this case, the sheaf-theoretic de Rham complex is *not acyclic*, viz. it has non-trivial cohomology groups. These groups measure the obstructions caused by sources and are responsible for a non-vanishing covariantly observable curvature of the connection.

In the context of a dynamical physical theory, where the representability principle over the field of real numbers is demanded, the existence of uniquely defined duals is necessary. Thus, the action of a field is identified with a *linear connection* on the $\mathcal{A}$-vector sheaf of states $\Xi = Hom(\Omega^1, \mathcal{A})$, made isomorphic with Ω^1, by means of a bilinear form playing the role of a *metric*:

(0.4.46) $$g : \Omega^1 \simeq \Xi = \Omega^{1^*}.$$

We stress the fact that a physical observable geometry in this case is considered with respect to a metric. Consequently, the *physical field* may be represented by the pair (Ξ, ∇_Ξ). The required *metric compatibility* of the connection is expressed as follows:

(0.4.47) $$\nabla_{Hom_\mathcal{A}(\Xi, \Xi^*)}(g) = 0.$$

Taking into account the requirement of representability over the real numbers, and thus considering the relevant *evaluation trace operator* by means of the metric, we arrive at the analogue of *Einstein's field equations*, which in the absence of matter sources with respect to $\mathcal{A}$ read:

(0.4.48) $$\mathcal{R}(\nabla_\Xi)(\Xi) = 0.$$

where $\mathcal{R}(\nabla_\Xi)$ denotes the relevant *Ricci scalar curvature*. In more detail, we first define the curvature endomorphism $\mathfrak{R}_\nabla \in End(\Xi)$, called *Ricci curvature operator*. Since the Ricci curvature $\mathfrak{R}_\nabla$ is locally matrix-valued, by taking its trace using the metric, that is by considering its evaluation or contraction by means of the metric, we arrive at the definition of the Ricci scalar curvature $\mathcal{R}(\nabla_\Xi)(\Xi)$ obeying the above equation.

Thus, the metric describing the physical geometry, as a result of field interactions, is *dynamically* determined as a solution of the above equation in relation to the metric compatible connection on the vector sheaf $Hom_\mathcal{A}(\Xi, \Xi^*)$.

0.4.4 Local Forms of Connection and Curvature on Vector Sheaves of States

We proceed by a direct calculation to show that every connection $\nabla_\mathcal{E}$, where $\mathcal{E}$ is a finite rank-n vector sheaf of states on X, can be *decomposed locally* as follows:

$$(0.4.49) \qquad \nabla_\mathcal{E} = d^0 + \omega,$$

where $\omega = \omega_{\alpha\beta}$ denotes an $n \times n$ matrix of sections of local 1-forms, called the *matrix potential* of $\nabla_\mathcal{E}$. Moreover, under a change of local frame matrix $g = g_{\alpha\beta}$ the matrix potentials transform as follows:

$$(0.4.50) \qquad \acute{\omega} = g^{-1}\omega g + g^{-1}d^0 g.$$

If we consider a coordinatizing basis of sections defined over an open cover U of X, denoted by:

$$(0.4.51) \qquad e^U \equiv \{U; \ (e_\alpha)_{1 \leq \alpha \leq n}\}$$

of the vector sheaf $\mathcal{E}$ of rank-n, called a local sectional frame or a local gauge of $\mathcal{E}$, then every continuous local section $s \in \mathcal{E}(U)$, where, $U \in \mathcal{U}$, can be expressed uniquely with respect to this local frame as a *superposition*:

$$(0.4.52) \qquad s = \sum_{\alpha=1}^n s_\alpha e_\alpha,$$

with coefficients s_α in $\mathcal{A}(U)$. The action of $\nabla_\mathcal{E}$ on these sections of $\mathcal{E}$ is expressed as follows:

$$(0.4.53) \qquad \nabla_\mathcal{E}(s) = \sum_{\alpha=1}^n (s_\alpha \nabla_\mathcal{E}(e_\alpha) + e_\alpha \otimes d^0(s_\alpha)),$$

$$(0.4.54) \qquad \nabla_\mathcal{E}(e_\alpha) = \sum_{\alpha=1}^n e_\alpha \otimes \omega_{\alpha\beta}, \ 1 \leq \alpha, \beta \leq n,$$

where $\omega = \omega_{\alpha\beta}$ denotes an $n \times n$ matrix of sections of local 1-forms. Consequently we have;

$$(0.4.55) \qquad \nabla_\mathcal{E}(s) = \sum_{\alpha=1}^n e_\alpha \otimes (d^0(s_\alpha) + \sum_{\beta=1}^n s_\beta \omega_{\alpha\beta}) \equiv (d^0 + \omega)(s).$$

Thus, every connection $\nabla_\mathcal{E}$, where $\mathcal{E}$ is a finite rank-n vector sheaf on X, can be decomposed locally as follows:

$$(0.4.56) \qquad\qquad \nabla_\mathcal{E} = d^0 + \omega.$$

In this context, $\nabla_\mathcal{E}$ is identified as a *covariant derivative* operator acting on sections of the vector sheaf of states $\mathcal{E}$, and being decomposed locally as a sum consisting of a flat or integrable part identical with d^0, and a generally non-integrable part ω, called the local frame (gauge) matrix potential of the connection.

The behavior of the local potential ω of $\nabla_\mathcal{E}$ under local frame transformations constitutes the 'transformation law of local potentials' and is obtained as follows: Let $e^U \equiv \{U;\ e_{\alpha=1\cdots n}\}$ and $h^V \equiv \{V;\ h_{\beta=1\cdots n}\}$ be two local frames of $\mathcal{E}$ over the open sets U and V of X, such that $U \cap V \neq \emptyset$. Let us denote by $g = g_{\alpha\beta}$ the following change of local frame matrix:

$$(0.4.57) \qquad\qquad h_\beta = \sum_{\alpha=1}^{n} g_{\alpha\beta} e_\alpha.$$

Under such a local frame transformation $g_{\alpha\beta}$, it is straightforward to obtain that the local potential ω of $\nabla_\mathcal{E}$ transforms as follows in matrix form:

$$(0.4.58) \qquad\qquad \acute\omega = g^{-1}\omega g + g^{-1}d^0 g.$$

Furthermore, it is instructive to find the local form of the curvature R_∇ of a connection $\nabla_\mathcal{E}$, where $\mathcal{E}$ is a locally free finite rank-n sheaf of modules (vector sheaf) $\mathcal{E}$ on X, defined by the following $\mathcal{A}$-linear morphism of sheaves:

$$(0.4.59) \qquad R_\nabla := \nabla^1 \circ \nabla^0 : \mathcal{E} \to \Omega^2(\mathcal{A}) \otimes_\mathcal{A} \mathcal{E} := \Omega^2(\mathcal{E}).$$

Due to its property of $\mathcal{A}$-*covariance*, a non-vanishing curvature represents in this context, the $\mathcal{A}$-covariant, and thus *geometric observable* deviation from the *monodromic* form of variation corresponding to an integrable connection. Moreover, since the curvature R_∇ is an $\mathcal{A}$-linear morphism of sheaves of $\mathcal{A}$-modules, that is an $\mathcal{A}$-tensor, R_∇ may be thought of as an element of $End(\mathcal{E}) \otimes_\mathcal{A} \Omega^2(\mathcal{A}) := \Omega^2(End(\mathcal{E}))$, that is:

$$(0.4.60) \qquad\qquad R_\nabla \in \Omega^2(End(\mathcal{E})).$$

Hence, the local form of the curvature R_∇ of a connection $\nabla_\mathcal{E}$, consists of local $n \times n$ matrices having for entries local 2-forms. In particular, the local form of the curvature $R_\nabla|_U$, where U open in X, in terms of the local potentials ω is expressed by:

$$(0.4.61) \qquad\qquad R_\nabla|_U = d^1\omega + \omega \wedge \omega,$$

at it can be easily shown by substitution of the local potentials in the composition $\nabla^1 \circ \nabla^0$. Furthermore, by application of the differential operator d^2 on the above we obtain:

$$(0.4.62) \qquad\qquad d^2 R_\nabla|_U = R_\nabla|_U \wedge \omega - \omega \wedge R_\nabla|_U.$$

The behavior of the curvature R_∇ of a connection $\nabla_\mathcal{E}$ under local frame transformations constitutes the '*transformation law of potentials' strength*'. If we agree that $g = g_{\alpha\beta}$ denotes the change of local frame matrix, we have previously considered in the discussion of the transformation law of local connection potentials, we deduce the following local transformation law:

$$(0.4.63) \qquad R_\nabla \overset{g}{\mapsto} \acute{R}_\nabla = g^{-1}(R_\nabla)g,$$

that is they transform covariantly by *conjugation* with respect to a local frame transformation. It is useful to collect the above findings in the form of the following proposition:

Theorem 0.5.

I. The local form of the curvature $R_\nabla|_U$, where U open in X, in terms of the local matrix potentials ω is given by:

$$(0.4.64) \qquad R_\nabla|_U = d^1\omega + \omega \wedge \omega.$$

II. Under a change of local frame matrix $g = g_{\alpha\beta}$ the local form of the curvature transforms by conjugation:

$$(0.4.65) \qquad R_\nabla \overset{g}{\mapsto} \acute{R}_\nabla = g^{-1}(R_\nabla)g.$$

We note that the above holds for any complex vector sheaf $\mathcal{E}$. We may now specialize in the case of a *line sheaf* of states $\mathcal{L}$ equipped with a connection, denoted by the pair $(\mathcal{L}, \nabla)$. In this case, due to the isomorphism:

$$(0.4.66) \qquad \mathcal{L} \otimes_A \mathcal{L}^* \cong Hom_A(\mathcal{L}, \mathcal{L}) \equiv End_A\mathcal{L} \cong \mathcal{A},$$

we obtain the following simplifications: the local form of a connection over an open set is just a local 1-form or *local potential* (that is a local continuous section of Ω^1), whence the local form of the curvature of the connection over an open set is a local 2-form. The significant result obtained by the local transformation law in this case is that the curvature is *local frame invariant*, that is it does not change under any local frame transformation.

$$(0.4.67) \qquad R_\nabla \overset{g}{\mapsto} R_\nabla' = R_\nabla = d^1\omega.$$

Thus, we obtain a *global* 2-*form* R_∇ defined over X, which is also a *closed* 2-form because of the fact that:

$$(0.4.68) \qquad d^2 R_\nabla = 0.$$

Therefore, we conclude that for a line sheaf of states $\mathcal{L}$ equipped with a connection ∇, denoted by the pair $(\mathcal{L}, \nabla)$, the curvature R_∇ of the connection is a global closed 2-form.

0.4.5   *Gauge Equivalence Classes of Differential Line Sheaves*

We consider two line sheaves which are equivalent via an isomorphism $h : \mathcal{L} \xrightarrow{\cong} \mathcal{L}'$. We would like to extend the notion of equivalence for two line sheaves equipped with a connective structure $(\mathcal{L}, \nabla)$ and $(\mathcal{L}', \nabla')$. We proceed as follows:

Given an isomorphism $h : \mathcal{L} \xrightarrow{\cong} \mathcal{L}'$ of line sheaves of states, we say that ∇ is *frame or gauge equivalent* to ∇' if they are *conjugate connections* under the action of h:

$$(0.4.69) \qquad\qquad \nabla' = h \cdot \nabla \cdot h^{-1}.$$

Thus, we may consider the set of equivalence classes on pairs of the form $(\mathcal{L}, \nabla)$ under an isomorphism h as previously, denoted by $Iso(\mathcal{L}, \nabla)$. It is now important to find the relation between $Iso(\mathcal{L}, \nabla)$ and the *abelian group* $Iso(\mathcal{L})$. For this purpose, we need to explore the local form of the pair $(\mathcal{L}, \nabla)$.

We call a line sheaf $\mathcal{L}$ equipped with a connection ∇ a *differential line sheaf* and we denote it by the pair $(\mathcal{L}, \nabla)$. Then, we have the following:

Theorem 0.6. *I. The local form of a differential line sheaf is given by:*

$$(0.4.70) \qquad\qquad (\mathcal{L}, \nabla) \leftrightarrow (g_{\alpha\beta}, \omega_\alpha) \in Z^1(\mathcal{U}, \tilde{A}) \times C^0(\mathcal{U}, \Omega^1).$$

II. An arbitrary pair $(g_{\alpha\beta}, \omega_\alpha) \in Z^1(\mathcal{U}, \tilde{A}) \times C^0(\mathcal{U}, \Omega^1)$ *determines a differential line sheaf if*

$$(0.4.71) \qquad\qquad \delta^0(\omega_\alpha) = \tilde{d}^0(g_{\alpha\beta}).$$

In order to elucidate the above, we note that a line sheaf is expressed in local coordinates bijectively in terms of a Čech coordinate 1-cocycle $g_{\alpha\beta}$ in $Z^1(\mathcal{U}, \tilde{A})$ associated with the covering $\mathcal{U}$. A connection ∇ is expressed bijectively in terms of a 0-cochain of 1-forms, called the local potentials of the connection, and denoted by ω_α with respect to the covering $\mathcal{U}$ of X, that is $\omega_\alpha \in C^0(\mathcal{U}, \Omega^1)$. We deduce that the local form of a differential line sheaf is the following:

$$(0.4.72) \qquad\qquad (\mathcal{L}, \nabla) \leftrightarrow (g_{\alpha\beta}, \omega_\alpha) \in Z^1(\mathcal{U}, \tilde{A}) \times C^0(\mathcal{U}, \Omega^1).$$

Conversely, an arbitrary pair $(g_{\alpha\beta}, \omega_\alpha) \in Z^1(\mathcal{U}, \tilde{A}) \times C^0(\mathcal{U}, \Omega^1)$ determines a differential line sheaf if the 'transformation law of local potentials' is satisfied by this pair, that is:

$$(0.4.73) \qquad\qquad \omega_\beta = g_{\alpha\beta}^{-1} \omega_\alpha g_{\alpha\beta} + g_{\alpha\beta}^{-1} d^0 g_{\alpha\beta},$$

$$(0.4.74) \qquad\qquad \omega_\beta = \omega_\alpha + g_{\alpha\beta}^{-1} d^0 g_{\alpha\beta}.$$

Thus, given an open covering $\mathcal{U} = \{U_\alpha\}$, a 0-cochain (ω_α) valued in the sheaf Ω^1 determines the local form of a connection ∇ on the line sheaf $\mathcal{L}$, where the latter is expressed in local coordinates bijectively in terms of a Čech coordinate 1-cocycle $(g_{\alpha\beta})$ valued in $\tilde{A}$ with respect to $\mathcal{U}$, if and only if the corresponding local 1-forms

ω_α of the 0-cochain with respect to $\mathcal{U}$ are pairwise inter-transformable (local frame-equivalent) on overlaps $U_{\alpha\beta}$ via the local frame transition functions (isomorphisms) $g_{\alpha\beta} \in \tilde{\mathcal{A}}(U_{\alpha\beta})$ according to the 'transformation law of local potentials':

$$(0.4.75) \qquad \delta^0(\omega_\alpha) = \tilde{d}^0(g_{\alpha\beta}),$$

where,

$$(0.4.76) \qquad \tilde{d}^0(g_{\alpha\beta}) = g_{\alpha\beta}^{-1} \cdot d^0 g_{\alpha\beta},$$

and δ^0 denotes the 0-th coboundary operator $\delta^0 : C^0(\mathcal{U}, \Omega^1) \to C^1(\mathcal{U}, \Omega^1)$, such that:

$$(0.4.77) \qquad \delta^0(\omega_\alpha) = \omega_\beta - \omega_\alpha.$$

Next, we consider two line sheaves which are equivalent via an isomorphism $h : \mathcal{L} \xrightarrow{\cong} \mathcal{L}'$, such that their corresponding connections are conjugate under the action of h:

$$(0.4.78) \qquad \nabla' = h\nabla h^{-1}.$$

Under these conditions the differential line sheaves $(\mathcal{L}, \nabla)$ and $(\mathcal{L}', \nabla')$ are called *gauge or frame equivalent*. Thus, we may consider the set of *gauge equivalence classes* $[(\mathcal{L}, \nabla)]$ of differential line sheaves as above, denoted by $Iso(\mathcal{L}, \nabla)$. Remarkably, the following holds:

Theorem 0.7. *The set of gauge equivalence classes of differential line sheaves* $Iso(\mathcal{L}, \nabla)$, *is an abelian subgroup of the abelian group* $Iso(\mathcal{L})$.

If we consider the local form of the tensor product of two gauge equivalent differential line sheaves we have:

$$(0.4.79) \qquad (\mathcal{L}, \nabla) \otimes_\mathcal{A} (\mathcal{L}', \nabla') \leftrightarrow (g_{\alpha\beta} \cdot g'_{\alpha\beta}, \omega_\alpha + \omega'_\alpha),$$

which satisfies the 'transformation law of local potentials':

$$(0.4.80) \quad \tilde{d}^0(g_{\alpha\beta} \cdot g'_{\alpha\beta}) = \tilde{d}^0(g_{\alpha\beta}) + \tilde{d}^0(g'_{\alpha\beta}) = \delta^0(\omega_\alpha) + \delta^0(\omega'_\alpha) = \delta^0(\omega_\alpha + \omega'_\alpha).$$

Moreover, the inverse of a pair $(g_{\alpha\beta}, \omega_\alpha)$ is given by $(g_{\alpha\beta}^{-1}, -\omega_\alpha)$, whereas the neutral element in the group $Iso(\mathcal{L}, \nabla)$ is given by $(id_{\alpha\beta}, 0)$, which corresponds to the trivial differential line sheaf $(\mathcal{A}, d^0)$.

Now, if we consider a line sheaf $\mathcal{L}$ of states, it is classified up to isomorphism by a *cohomology class* $\delta_c(\mathcal{L})$ in the abelian group $H^2(X, \mathbb{Z})$. A natural question arising in this setting is if it is possible to express the *global invariant information* provided by the equivalent line sheaves' classifying integer cohomology class by means of a *differential cohomological invariant*. This differential invariant should be associated with a *differential equation*, or more concretely with the *parallel transport* rule imposed by a physically induced connectivity structure on the line sheaf. The possibility of expressing the classifying integer cohomology class of isomorphic line sheaves of states by means of a differential invariant would also provide a cohomological justification of the physical gauge principle, according to which local gauge invariance implies the existence of local gauge potentials, and thus of a connection on a line sheaf of states. Given that local gauge invariance is depicted by means of coordinate 1-cocycles with respect to a covering $\mathcal{U}$ of X and the fact that the localization of a connection is expressed in terms of local gauge potentials with respect to $\mathcal{U}$, it is natural to seek for a relation between the invariant of the former, that is an integer cohomology class, and the differential invariant of the latter, that is the class of the curvature of the connection.

0.4.6   Quantization Condition via Cohomology

An immediate important consequence of the above characterization of gauge equivalent differential line sheaves in local terms with respect to an open covering $\mathcal{U}$, is that all of them have the same curvature. Indeed, by applying the differential operator d^1 on the above relation and using the fact that:

$$(0.4.81) \qquad\qquad d^1 \circ d^0 = 0,$$

from which we have that:

$$(0.4.82) \qquad\qquad d^1 \circ \tilde{d}^0 = 0,$$

we obtain:

$$(0.4.83) \qquad\qquad d^1(\omega'_\alpha) = d^1(\omega_\alpha).$$

Hence, we formulate the following:

Theorem 0.8. *Gauge equivalent differential line sheaves always have the same curvature.*

We denote the *curvature of a gauge equivalence class* of differential line sheaves by R. We notice that R is a global 2-form on X since it is local frame-change invariant, which is also closed because of the fact that:

$$(0.4.84) \qquad\qquad d^2 \circ d^1\omega_\alpha = d^2R = 0.$$

Thus, the global 2-form R, which belongs to $Ker(d^2) : \Omega^2 \to \Omega^3$, called Ω^2_c, being a $\mathbb{C}$-vector sheaf subspace of Ω^2, determines a global differential invariant of gauge equivalent differential line sheaves. This the case because the global 2-form R determines a *2-dimensional de Rham cohomology class* $[R]$, identified as a *2-dimensional complex Čech cohomology class* in $H^2(X, \mathbb{C})$. In particular, if we consider a differential line sheaf the differential invariant de Rham cohomology class $[R]$ is *independent of the connection* used to represent R locally. In other words, a particular connection of a differential line sheaf provides the means to express this global differential invariant locally, whereas the latter is independent of the particular means used to represent it locally.

Furthermore, since any two gauge equivalent differential sheaves have the same curvature, we conclude that they are *physically indistinguishable*. This means that the abelian group $Iso(\mathcal{L}, \nabla)$ is partitioned into *orbits* over the image of $Iso(\mathcal{L}, \nabla)$ into Ω_c^2, where each orbit (fiber) is labeled by a closed 2-form R of Ω_c^2:

$$(0.4.85) \qquad\qquad Iso(\mathcal{L}, \nabla) = \sum_R Iso(\mathcal{L}, \nabla)_R.$$

Thus, the abelian group of equivalence classes of differential line sheaves fibers over those elements of Ω_c^2, viz. over those closed global 2-forms in Ω_c^2 which can be identified as curvatures. In this way, a pertinent problem is the concrete identification of the image of $Iso(\mathcal{L}, \nabla)$ into Ω_c^2. Put differently, we are looking for an *intrinsic* characterization of those global closed 2-forms in Ω_c^2, which are instantiated as curvatures of gauge equivalence classes of differential line sheaves. The resolution of this problem is provided by the sheaf-theoretic formulation of the *Chern-Weil integrality theorem*. In physical terms, the integrality theorem provides the general and intrinsic cohomological formulation of *Dirac's quantization condition*. According to this, the de Rham cohomology class $[R]$ of any differential line sheaf in $H^2(X, \mathbb{C})$ is in the image of a cohomology class in the integral 2-dimensional cohomology group $H^2(X, \mathbb{Z})$ into $H^2(X, \mathbb{C})$. More precisely, the Chern-Weil integrality theorem may be formulated as follows in our context:

Theorem 0.9. *A global closed 2-form is the curvature R of a differential line sheaf if and only if its 2-dimensional de Rham cohomology class is integral, viz. $[R] \in Im(H^2(X, \mathbb{Z}) \to H^2(X, \mathbb{C}))$.*

It is instructive to discuss briefly the basic concepts of this fundamental theorem providing the cohomological justification of the quantization condition. We know that each equivalence classes of line sheaves is in bijective correspondence with a cohomology class in the integral 2-dimensional cohomology group of X. Thus, we have to show that a 2-dimensional de Rham cohomology class is a curvature differential invariant class of gauge equivalent differential line sheaves if and only if it is an integral 2-dimensional cohomology class. First, by the natural injection $\mathbb{Z} \hookrightarrow \mathbb{C}$ we obtain:

$$(0.4.86) \qquad\qquad H^2(X, \mathbb{Z}) \to H^2(X, \mathbb{C}),$$

where any cohomology class belonging to the image of the above map is called an *integral cohomology class*. Next, we consider the following exact sequences:

$$(0.4.87) \qquad\qquad 0 \to \mathbb{Z} \xrightarrow{\iota} \mathcal{A} \xrightarrow{exp} \tilde{\mathcal{A}} \to 1,$$

which is the exponential short exact sequence of abelian group sheaves, such that:

$$(0.4.88) \qquad\qquad Ker(exp) = Im(\iota) \cong \mathbb{Z}$$

and the short exact sequence of $\mathbb{C}$-vector sheaves,

$$(0.4.89) \qquad\qquad 0 \to \mathbb{C} \xrightarrow{\varepsilon} \mathcal{A} \xrightarrow{d^0} d^0\mathcal{A} \to 0,$$

which is a fragment of the *de Rham resolution of the constant sheaf* $\mathbb{C}$, such that:

$$(0.4.90) \qquad\qquad Ker(d^0) = Im(\varepsilon) \cong \mathbb{C}.$$

Furthermore, we consider the following commutative diagram:

$$\begin{array}{ccc}
\mathcal{A} & \xrightarrow{\;exp\;} & \tilde{\mathcal{A}} \\
id \downarrow & & \downarrow \frac{1}{2\pi i}\tilde{d}^0 \\
\mathcal{A} & \xrightarrow{\;d^0\;} & d^0\mathcal{A}
\end{array}$$

Thus we obtain the relation:

(0.4.91) $$2\pi i \cdot d^0 = \tilde{d}^0 \circ exp.$$

If we take the relevant fragments of the corresponding long exact sequences in cohomology we obtain the following commutative diagram:

$$\begin{array}{ccc}
H^1(X, \tilde{A}) & \xrightarrow{\;\delta_c\;} & H^2(X, \mathbb{Z}) \\
\frac{1}{2\pi i}\tilde{d}^0 \downarrow & & \downarrow \iota^* \\
H^1(X, d^0\mathcal{A}) & \xrightarrow{\;\cong\;} & H^2(X, \mathbb{C})
\end{array}$$

Thus, the image of a cohomology class of $H^2(X, \mathbb{Z})$ into an integral cohomology class of $H^2(X, \mathbb{C})$ corresponds to the cohomology class specified by the image of the 1-cocycle $g_{\alpha\beta}$ into $H^1(X, d^0\mathcal{A})$, viz. by the 1-cocycle $\frac{1}{2\pi i}\tilde{d}^0(g_{\alpha\beta})$.

In more detail, we first have that due to the exactness of the exponential sheaf sequence of abelian group sheaves:

(0.4.92) $$0 \to \mathbb{Z} \xrightarrow{\iota} \mathcal{A} \xrightarrow{\;exp\;} \tilde{\mathcal{A}} \to 1,$$

(0.4.93) $$g_{\alpha\beta} = exp(w_{\alpha\beta}),$$

where $(w_{\alpha\beta}) \in C^1(\mathcal{U}, \mathcal{A})$ as well as

(0.4.94) $$\delta_c(w_{\alpha\beta}) = (z_{\alpha\beta\gamma}) \in Z^2(\mathcal{U}, \mathbb{Z}).$$

Explicitly, we may consider $w_{\alpha\beta} = ln(g_{\alpha\beta})$, so that:

(0.4.95) $$\delta_c(w_{\alpha\beta}) = (z_{\alpha\beta\gamma}) := \frac{1}{2\pi i}(ln(g_{\alpha\beta}) + ln(g_{\beta\gamma}) - ln(g_{\alpha\gamma})) \in Z^2(\mathcal{U}, \mathbb{Z}).$$

Now, we need to locate the curvature 2-dimensional complex de Rham cohomology class differential invariant, viz. the complex 2-dimensional cohomology class $[R]$ in $H^2(X, \mathbb{C})$. For this purpose, regarding the curvature we have that:

(0.4.96) $$R = (d\omega_\alpha) \in Z^0(\mathcal{U}, d\Omega^1).$$

By the transformation law of local potentials, it holds:

(0.4.97)
$$\delta^0(\omega_\alpha) = \tilde{d}^0(g_{\alpha\beta}),$$

where,

(0.4.98)
$$\tilde{d}^0(g_{\alpha\beta}) = g_{\alpha\beta}^{-1} \cdot d^0 g_{\alpha\beta}$$

and $g_{\alpha\beta} \in Z^1(\mathcal{U}, \tilde{\mathcal{A}})$, whence δ^0 denotes the 0-th coboundary operator δ^0 : $C^0(\mathcal{U}, \Omega^1) \to C^1(\mathcal{U}, \Omega^1)$, such that:

(0.4.99)
$$\delta^0(\omega_\alpha) = \omega_\beta - \omega_\alpha.$$

Thus, we obtain:

(0.4.100)
$$\delta^0(\omega_\alpha) = 2\pi i \cdot d^0(w_{\alpha\beta}),$$

where $d^0(w_{\alpha\beta}) \in Z^1(\mathcal{U}, d^0\mathcal{A})$. Therefore, we deduce that:

(0.4.101)
$$\frac{1}{2\pi i}(\omega_\beta - \omega_\alpha) = d^0(ln(g_{\alpha\beta})).$$

If we inspect the short exact sequence of $\mathbb{C}$-vector sheaves:

(0.4.102)
$$0 \to \mathbb{C} \xrightarrow{\varepsilon} \mathcal{A} \xrightarrow{d^0} d^0\mathcal{A} \to 0,$$

we immediately deduce that:

(0.4.103)
$$\delta(ln(g_{\alpha\beta})) = (ln(g_{\alpha\beta}) + ln(g_{\beta\gamma}) - ln(g_{\alpha\gamma})) \in Z^2(\mathcal{U}, \mathbb{C}).$$

Thus, finally we obtain:

(0.4.104)
$$[R] = 2\pi i \cdot [(z_{\alpha\beta\gamma})] \in H^2(X, \mathbb{C}),$$

(0.4.105)
$$[(g_{\alpha\beta})] = \frac{1}{2\pi i}[R] = [(z_{\alpha\beta\gamma})] \in H^2(X, \mathbb{Z}),$$

(0.4.106)
$$[R] \in Im(H^2(X, \mathbb{Z}) \to H^2(X, \mathbb{C})).$$

Thus, we have obtained an intrinsic characterization of the subset of those global closed 2-forms in Ω_c^2, which are instantiated as curvatures of gauge equivalence classes of differential line sheaves, denoted by $\Omega_{c,\mathbb{Z}}^2$. Therefore, a global closed 2-form is the curvature R of a differential line sheaf if and only if its 2-dimensional de Rham cohomology class is integral, viz. $[R] \in Im(H^2(X, \mathbb{Z}) \to H^2(X, \mathbb{C}))$. This constitutes the generalized cohomological formulation of Dirac's quantization condition.

0.4.7   *Integrable Differential Line Sheaves*

A very important observation is that a *constant* 1-*cocycle* $(\xi_{\alpha\beta}) \in Z^1(\mathcal{U}, \tilde{\mathbb{C}})$ can be interpreted as the coordinate 1-cocycle of a line sheaf of states with respect to an open covering $\mathcal{U}$. Since the coordinate 1-cocycle $(\xi_{\alpha\beta})$ is constant, the bijectively corresponding to it line sheaf is a complex linear local system of rank 1, abbreviated as a *line local system*. The significance of this notion in relation to the integrability properties of a differential line sheaf lies on the following theorem:

Theorem 0.10. *There exists a bijective correspondence between differential line sheaves equipped with an integrable connection and line local systems.*

In order to clarify the above assertion, let us consider any differential line sheaf $(\mathcal{L}, \nabla)$ whose connection ∇ is integrable, viz. its curvature is zero. Then, the set of sections of $\mathcal{L}$ being in the kernel of the connection morphism ∇, that is:

$$(0.4.107) \qquad Ker(\nabla) := \{ s \in \mathcal{L} : \nabla(s) = 0 \}$$

forms a line local system. We call the sections of $Ker(\nabla)$ *covariantly constant sections* of $\mathcal{L}$ with respect to ∇. Inversely, given a line local system, denoted by Λ, we may define a differential line sheaf $(\mathcal{L}, \nabla)$ as follows:

$$(0.4.108) \qquad\qquad \mathcal{L} := \mathcal{A} \otimes_{\mathbb{C}} \Lambda,$$

and for every pair of local sections $a \in \mathcal{A}(U)$, $s \in \Lambda(U)$:

$$(0.4.109) \qquad\qquad \check{\nabla}(a \otimes s) := da \otimes s.$$

Clearly, the above defined connection is integrable and leads to the conclusion that there exists a *bijective* correspondence between differential line sheaves with an integrable connection, denoted by $(\mathcal{L}, \check{\nabla})$ and line local systems. Moreover, every line local system is identified with the sheaf of covariantly constant sections of an integrable differential line sheaf, meaning of a line sheaf $\mathcal{L}$ with respect to an integrable connection $\check{\nabla}$ on $\mathcal{L}$, viz. with $Ker(\check{\nabla})_{\mathcal{L}}$.

Therefore, we immediately obtain the following equivalences:

$$(0.4.110) \qquad Z^1(\mathcal{U}, \tilde{\mathbb{C}}) \ni (\xi_{\alpha\beta}) \cong (\mathcal{L}, \check{\nabla}) \cong Ker(\check{\nabla})_{\mathcal{L}} \cong \Lambda,$$

$$(0.4.111) \qquad H^1(X, \tilde{\mathbb{C}}) \cong [(\mathcal{L}, \check{\nabla})] \cong [Ker(\check{\nabla})_{\mathcal{L}}] \cong [\Lambda].$$

If we also assume that the underlying topological space X of control variables is locally path-connected, then the above series of equivalences is refined as follows:

$$(0.4.112) \qquad Hom(\pi_1(X), \tilde{\mathbb{C}}) \cong H^1(X, \tilde{\mathbb{C}}) \cong [(\mathcal{L}, \check{\nabla})] \cong [Ker(\check{\nabla})_{\mathcal{L}}] \cong [\Lambda],$$

where the first term denotes the set of representations of the *fundamental group* of the topological space X of control variables to $\tilde{\mathbb{C}}$.

0.4.8   Quantum Unitary Rays

For reasons implicated by the probabilistic interpretation of states in quantum mechanics, which utilizes the *inner product* structure of the state space, we need to focus on the case of isomorphism classes of Hermitian line sheaves, defined as follows:

Given a line sheaf $\mathcal{L}$ on X, an $\mathcal{A}$-*valued Hermitian inner product* on $\mathcal{L}$ is a skew-$\mathcal{A}$-bilinear sheaf morphism :

$$(0.4.113) \qquad \varrho : \mathcal{L} \oplus \mathcal{L} \to \mathcal{A},$$

$$(0.4.114) \qquad \varrho(\alpha s, \beta t) = \alpha \cdot \bar{\beta} \cdot \varrho(s, t),$$

for any $s, t \in \mathcal{L}(U)$, $\alpha, \beta \in \mathcal{A}(U)$, U open in X. Moreover, $\varrho(s, t)$ is skew-symmetric, viz. $\varrho(s, t) = \overline{\varrho(s, t)}$.

A line sheaf $\mathcal{L}$ on X, together with an $\mathcal{A}$-valued Hermitian inner product on $\mathcal{L}$ constitute a *Hermitian line sheaf.*

A line sheaf is expressed in local coordinates bijectively in terms of a Čech coordinate 1-cocycle $g_{\alpha\beta}$ in $Z^1(\mathcal{U}, \tilde{\mathcal{A}})$ associated with the open covering $\mathcal{U}$. A Čech coordinate 1-cocycle $g_{\alpha\beta}$ corresponding to a Hermitian line sheaf consists of local sections of $\mathcal{SU}(1, \mathcal{A})$, viz. the *special unitary* group sheaf of $\mathcal{A}$ of order 1. This is simply a coordinate 1-cocycle $g_{\alpha\beta}$ in $Z^1(\mathcal{U}, \tilde{\mathcal{A}})$, such that the unitarity condition $|g_{\alpha\beta}| = 1$ is satisfied. Clearly, in the case that a coordinate 1-cocycle $g_{\alpha\beta}$ is constant, we have $g_{\alpha\beta}$ in $Z^1(\mathcal{U}, U(1))$, or equivalently $g_{\alpha\beta}$ in $Z^1(\mathcal{U}, \mathbb{S}^1)$.

Next, we need to define the notion of a connection which is compatible with the $\mathcal{A}$-valued Hermitian inner product on $\mathcal{L}$. A connection ∇ on $\mathcal{L}$ is called *Hermitian* if it is compatible with ϱ:

$$(0.4.115) \qquad d^0 \varrho(s, t) = \varrho(\nabla s, t) + \varrho(s, \nabla t)$$

for any $s, t \in \mathcal{L}(U)$, U open in X.

A Hermitian differential line sheaf of states of a quantum system, or equivalently a *quantum unitary ray*, denoted by $(\mathcal{L}, \nabla, \varrho)$, is a Hermitian line sheaf equipped with a Hermitian connection.

First, we note that although we use the same symbol ϱ in the second part of the above, it refers to an extension of the $\mathcal{A}$-valued Hermitian inner product on $\mathcal{L}$, defined as follows:

$$(0.4.116) \qquad \varrho : \Omega(\mathcal{L}) \oplus \mathcal{L} \to \Omega(\mathcal{L}),$$

$$(0.4.117) \qquad \varrho(s \otimes t, s') := \varrho(s, s') \cdot t$$

for any $s, s' \in \mathcal{L}(U)$, $t \in \Omega(U)$, U open in X. Furthermore, Ω is considered to be a vector sheaf on X.

Second, if we have a Hermitian line sheaf $(\mathcal{L}, \varrho)$, which is also equipped with a connection ∇, described in terms of local potentials $\omega = (\omega_\alpha)$, it is straightforward to see that the compatibility condition of the connection with the Hermitian inner product ϱ is satisfied, such that $(\mathcal{L}, \nabla, \varrho)$ is a unitary ray, if and only if:

$$(0.4.118) \qquad \omega + \overline{\omega} = \tilde{d}^0(\varrho),$$

where $\tilde{d}^0$ denotes the logarithmic universal derivation of the observable abelian group sheaf $\tilde{\mathcal{A}}$ to the abelian group sheaf of 1-forms Ω^1, that is the universal morphism of abelian group sheaves:

$$(0.4.119) \qquad \tilde{d}^0 : \tilde{\mathcal{A}} \to \Omega^1$$

defined by the relation:

$$(0.4.120) \qquad \tilde{d}^0(u) := u^{-1} \cdot d^0(u)$$

for any continuous invertible local section $u \in \tilde{\mathcal{A}}(U)$, with U open set in X. Given the validity of the Poincaré Lemma, we also have:

$$(0.4.121) \qquad Ker(\tilde{d}^0) = \tilde{\mathbb{C}}.$$

Third, by considering the Hermitian line sheaf $(\mathcal{L}, \varrho)$ and a local frame of $\mathcal{L}$, we may apply locally the Gram-Schmidt *orthonormalization* procedure in our context, such that $\mathcal{L}$ has locally an *orthonormal frame*. Thus, with respect to this orthonormal frame of $\mathcal{L}$, we obtain:

$$(0.4.122) \qquad \tilde{d}^0(\varrho) = 0,$$

and finally we deduce that $\omega = -\overline{\omega}$.

Consequently, given that every Hermitian line sheaf $(\mathcal{L}, \varrho)$ of states admits a Hermitian connection, we have specified completely the notion of a quantum unitary ray, denoted by $(\mathcal{L}, \nabla, \varrho)$, according to the above. In order to simplify further the notation, we denote a quantum unitary ray, viz. a Hermitian differential line sheaf of states, by $(\mathcal{L}, \nabla_\varrho)$. The next important task is to establish the local form of a quantum unitary ray with respect to an open covering of X, and inversely determine the conditions that local components have to satisfy so that they constitute a quantum unitary ray. This is a corollary of the theorem formulated in 0.4.5.

Theorem 0.11. *i. The local form of a unitary ray is given by:*

$$(0.4.123) \qquad (\mathcal{L}, \nabla) \leftrightarrow (g_{\alpha\beta}, \omega_\alpha) \in Z^1(\mathcal{U}, \tilde{\mathcal{A}}) \times C^0(\mathcal{U}, \Omega^1),$$

where $|g_{\alpha\beta}| = 1$.

ii. An arbitrary pair $(g_{\alpha\beta}, \omega_\alpha) \in Z^1(\mathcal{U}, \tilde{\mathcal{A}}) \times C^0(\mathcal{U}, \Omega^1)$ *determines a unitary ray if*

$$(0.4.124) \qquad \delta^0(\omega_\alpha) = \tilde{d}^0(g_{\alpha\beta}),$$

$$(0.4.125) \qquad \omega + \overline{\omega} = \tilde{d}^0(\varrho).$$

A Čech coordinate 1-cocycle $g_{\alpha\beta}$ of a unitary ray consists of local sections of $\mathcal{SU}(1, \mathcal{A})$, viz. the special unitary group sheaf of $\mathcal{A}$ of order 1. This is simply a coordinate 1-cocycle $g_{\alpha\beta}$ in $Z^1(\mathcal{U}, \tilde{\mathcal{A}})$, such that the unitarity condition $|g_{\alpha\beta}| = 1$ is satisfied. Clearly, in the case that a coordinate 1-cocycle $g_{\alpha\beta}$ is constant, we have $g_{\alpha\beta}$ in $Z^1(\mathcal{U}, U(1))$, or equivalently $g_{\alpha\beta}$ in $Z^1(\mathcal{U}, \mathbb{S}^1)$. Moreover, given a Hermitian line sheaf $(\mathcal{L}, \varrho)$, which is also equipped with a connection ∇, described in terms of local potentials $\omega = (\omega_\alpha) \in C^0(\mathcal{U}, \Omega^1)$, the compatibility condition of the connection with the Hermitian inner product ϱ is satisfied if and only if $\omega + \overline{\omega} = \tilde{d}^0(\varrho)$.

0.4.9 *Gauge Equivalence of Quantum Unitary Rays*

We remind that if we consider two line sheaves which are equivalent via an isomorphism $h : \mathcal{L} \xrightarrow{\cong} \mathcal{L}'$, such that their corresponding connections are conjugate under the action of h:

$$(0.4.126) \qquad \nabla' = h\nabla h^{-1},$$

then the corresponding differential line sheaves $(\mathcal{L}, \nabla)$ and $(\mathcal{L}', \nabla')$ are *gauge or frame equivalent*, where the set of *gauge equivalence classes* $[(\mathcal{L}, \nabla)]$ of differential line sheaves is denoted by $Iso(\mathcal{L}, \nabla)$. Further, the following proposition holds:

Theorem 0.12. *The set of gauge equivalence classes of unitary rays $Iso(\mathcal{L}, \nabla_\varrho)$ is an abelian subgroup of the set of gauge equivalence classes of differential line sheaves $Iso(\mathcal{L}, \nabla)$, which in turn, is an abelian subgroup of the abelian group $Iso(\mathcal{L})$.*

This is an immediate corollary of Theorem 0.7. More precisely, since a unitary ray is a differential line sheaf such that $|g_{\alpha\beta}| = 1$ the set of gauge equivalence classes of quantum unitary rays $Iso(\mathcal{L}, \nabla_\varrho)$ is an abelian subgroup of the set of gauge equivalence classes of differential line sheaves $Iso(\mathcal{L}, \nabla)$.

Next, we need to specify the local conditions under which pairs of the form $(g_{\alpha\beta}, \omega_\alpha)$ and $(g'_{\alpha\beta}, \omega'_\alpha)$ determine gauge equivalent differential line sheaves, and concomitantly gauge equivalent quantum unitary rays.

Theorem 0.13. *For any two gauge equivalent differential line sheaves $(\mathcal{L}, \nabla)$ and $(\mathcal{L}', \nabla')$ in $Iso(\mathcal{L}, \nabla)$, the following holds: $(\mathcal{L}, \nabla)$ is equivalent to $(\mathcal{L}', \nabla')$, meaning that $h : \mathcal{L} \xrightarrow{\cong} \mathcal{L}'$ and $\nabla' = h\nabla h^{-1}$, if and only if there exists a 0-cochain $(t_a) \in C^0(\mathcal{U}, \tilde{\mathcal{A}}) = \prod_\alpha \tilde{\mathcal{A}}(U_\alpha)$, such that:*

$$(0.4.127) \qquad g'_{\alpha\beta} \cdot g_{\alpha\beta}^{-1} = \Delta^0(t_a^{-1}),$$

$$(0.4.128) \qquad \omega'_\alpha - \omega_\alpha = \tilde{d}^0(t_a^{-1}),$$

where by definition the coordinate 1-cocycles of the line sheaves $\mathcal{L}$ and $\mathcal{L}'$ associated with the common open covering $\mathcal{U}$ are given by $g_{\alpha\beta} = \phi_\alpha \circ \phi_\beta^{-1} \in Z^1(\mathcal{U}, \tilde{\mathcal{A}})$ and $g'_{\alpha\beta} = \psi_\alpha \circ \psi_\beta^{-1} \in Z^1(\mathcal{U}, \tilde{\mathcal{A}})$, according to the following diagram:

$$
\begin{array}{ccc}
\mathcal{L}|_{U_\alpha} & \xrightarrow{\ h_\alpha\ } & \mathcal{L}'|_{U_\alpha} \\[2pt]
\phi_\alpha \downarrow & & \downarrow \psi_\alpha \\[2pt]
\mathcal{A}|_{U_\alpha} & \xrightarrow{\ t_\alpha\ } & \mathcal{A}|_{U_\alpha}
\end{array}
$$

It is clear that an 1-cocycle $g'_{\alpha\beta}$ is equivalent to another 1-cocycle $g_{\alpha\beta}$ in $Z^1(\mathcal{U}, \tilde{A})$ if and only if there exists a 0-cochain (t_α) in the set $C^0(\mathcal{U}, \tilde{A})$, such that:

$$(0.4.129) \qquad g'_{\alpha\beta} = t_\alpha \cdot g_{\alpha\beta} \cdot t_\beta^{-1},$$

that is $g'_{\alpha\beta}$ is similar to $g_{\alpha\beta}$ under conjugation with (t_α). This is equivalently written as an action of the image of the coboundary operator, that is of the abelian group of 1-coboundaries in $B^1(\mathcal{U}, \tilde{A})$ on the abelian group of 1-cocycles in $Z^1(\mathcal{U}, \tilde{A})$:

$$(0.4.130) \qquad \Delta^0 : C^0(\mathcal{U}, \tilde{A}) \to C^1(\mathcal{U}, \tilde{A}),$$

$$(0.4.131) \qquad g'_{\alpha\beta} = \Delta^0(t_\alpha^{-1}) \cdot g_{\alpha\beta},$$

$$(0.4.132) \qquad g'_{\alpha\beta} \cdot g_{\alpha\beta}^{-1} = \Delta^0(t_\alpha^{-1}).$$

Thus, the hypothesis of an isomorphism $h : \mathcal{L} \xrightarrow{\cong} \mathcal{L}'$ is equivalent to the existence of a 0-cochain $(t_a) \in C^0(\mathcal{U}, \tilde{A}) = \prod_\alpha \tilde{A}(U_\alpha)$ such that the above condition is satisfied. At a further stage, the gauge equivalence of the differential line sheaves $(\mathcal{L}, \nabla)$ and $(\mathcal{L}', \nabla')$ induced by $h \leftrightarrow (t_\alpha)$ implies that $\nabla' = h\nabla h^{-1}$. Hence, by using the local expressions of the connections ∇' and ∇ by means of the local 1-forms ω'_α and ω_α correspondingly, we immediately obtain:

$$(0.4.133) \qquad \omega'_\alpha = t_\alpha \omega_\alpha t_\alpha^{-1} + t_\alpha d^0(t_\alpha^{-1}),$$

$$(0.4.134) \qquad \omega'_\alpha = \omega_\alpha + \tilde{d}^0(t_\alpha^{-1}).$$

Thus, we have shown that:

$$(0.4.135) \qquad \omega'_\alpha - \omega_\alpha = \tilde{d}^0(t_a^{-1}).$$

0.4.10   *Spectral Beams and Polarization Symmetry*

The global 2-form R, which belongs to $Ker(d^2) : \Omega^2 \to \Omega^3$, called Ω_c^2, being a $\mathbb{C}$-vector sheaf subspace of Ω^2, determines a global differential invariant of gauge equivalent quantum unitary rays. This the case because the global 2-form R determines a 2-dimensional de Rham cohomology class $[R]$, identified as a 2-dimensional complex Čech cohomology class in $H^2(X, \mathbb{C})$. The Hermitian connection of a quantum unitary ray provides the means to express this global differential invariant locally, whereas the latter is actually independent of the connection utilized to represent it locally.

From the *curvature recognition integrality theorem*, we have concluded that the abelian group $Iso(\mathcal{L}, \nabla)$ is partitioned into orbits over $\Omega_{c,\mathbb{Z}}^2$, where each orbit (fiber) is labeled by an integral global closed 2-form R of $\Omega_{c,\mathbb{Z}}^2$, providing the differential invariant $[R]$ of this orbit in de Rham cohomology:

$$(0.4.136) \qquad Iso(\mathcal{L}, \nabla) = \sum_{R \in \Omega_{c,\mathbb{Z}}^2} Iso(\mathcal{L}, \nabla)_R.$$

If we restrict the abelian group $Iso(\mathcal{L}, \nabla)$ to gauge equivalent quantum unitary rays we obtain an abelian subgroup of the former, denoted by $Iso(\mathcal{L}, \nabla_\varrho)$. Clearly, the latter abelian group is also partitioned into orbits over $\Omega^2_{c,\mathbb{Z}}$, where each *orbit* is labeled by an *integral global closed* 2-form R of $\Omega^2_{c,\mathbb{Z}}$, where R is the curvature of the corresponding gauge equivalence class of quantum unitary rays.

We define each Hermitian differential line sheaf of states $(\mathcal{L}, \nabla_\varrho)$ belonging to an orbit $Iso(\mathcal{L}, \nabla_\varrho)_R$ a *quantum unitary R-ray*, whence the orbit itself is called a *spectral [R]-beam* and is characterized by the integral differential invariant $[R]$.

Each spectral $[R]$-beam consists of gauge equivalent quantum unitary R-rays, which are *indistinguishable* from the perspective of their common curvature integral differential invariant $[R]$. A natural question arising in the context of gauge equivalent quantum unitary R-rays is how they are related to each other. In other words, although all gauge equivalent unitary R-rays cannot be distinguished from the perspective of their curvature differential invariant, is there any other intrinsic way that we can distinguish among them? From a quantum physical perspective, if such an intrinsic and invariant way of distinguishing among gauge equivalent unitary R-rays exists, it means that there exists a *global kinematical symmetry* of spectral $[R]$-beams.

Theorem 0.14. *There exists a free group action of the abelian group $H^1(X, \mathbb{S}^1)$ on the abelian group of unitary rays $Iso(\mathcal{L}, \nabla_\varrho)$, where $\mathbb{S}^1$ denotes the abelian group sheaf $\mathbb{S}^1 \equiv \mathcal{U}(1) \equiv \mathcal{SU}(1, \mathbb{C})$, which is restricted to a free group action on each spectral $[R]$-beam.*

There exists a free group action of $\mathbb{S}^1 \hookrightarrow \mathbb{C} \hookrightarrow \mathcal{A}$ on the group sheaf $\tilde{\mathcal{A}}$ of invertible elements of $\mathcal{A}$:

$$(0.4.137) \qquad \mathbb{S}^1 \times \tilde{\mathcal{A}}(U) \to \tilde{\mathcal{A}}(U),$$

$$(0.4.138) \qquad (\xi, f) \mapsto \xi \cdot f.$$

with $\xi \in \mathbb{S}^1$ and $f \in \tilde{\mathcal{A}}(U)$ for any open U in X. This action is transferred naturally as a free action to the corresponding groups of coordinate 1-cocycles of the respective abelian group sheaves:

$$(0.4.139) \qquad Z^1(\mathcal{U}, \mathbb{S}^1) \times Z^1(\mathcal{U}, \tilde{\mathcal{A}}) \to Z^1(\mathcal{U}, \tilde{\mathcal{A}}),$$

$$(0.4.140) \qquad (\xi_{\alpha\beta}) \cdot (g_{\alpha\beta}) = (\xi_{\alpha\beta} \cdot g_{\alpha\beta}),$$

where $(\xi_{\alpha\beta}) \in Z^1(\mathcal{U}, \mathbb{S}^1)$, $(g_{\alpha\beta}) \in Z^1(\mathcal{U}, \tilde{\mathcal{A}})$. This free action can be also extended to the corresponding cohomology groups being still a free action:

$$(0.4.141) \qquad H^1(X, \mathbb{S}^1) \otimes H^1(X, \tilde{\mathcal{A}}) \to H^1(X, \tilde{\mathcal{A}}),$$

$$(0.4.142) \qquad [(\xi_{\alpha\beta})] \cdot [(g_{\alpha\beta})] = [(\xi_{\alpha\beta} \cdot g_{\alpha\beta})],$$

where $[(\xi_{\alpha\beta})] \in H^1(X, \mathbb{S}^1)$, $[(g_{\alpha\beta})] \in H^1(X, \tilde{\mathcal{A}}) \cong Iso(\mathcal{L})$. Now, we can finally define a group action of $H^1(X, \mathbb{S}^1)$ on the abelian group $Iso(\mathcal{L}, \nabla)$ as follows: We consider $\xi \equiv [(\xi_{\alpha\beta})] \in H^1(X, \mathbb{S}^1)$, $[(\mathcal{L}, \nabla)] \in Iso(\mathcal{L}, \nabla)$, and we define the sought group action as follows:

(0.4.143)
$$\xi \cdot [(\mathcal{L}, \nabla)] := [(\xi \cdot \mathcal{L}, \nabla)] \equiv [(\mathcal{L}', \nabla)],$$

(0.4.144)
$$\mathcal{L}' = \xi \cdot \mathcal{L} \leftrightarrow (\xi_{\alpha\beta}) \cdot (g_{\alpha\beta}) = (\xi_{\alpha\beta} \cdot g_{\alpha\beta}).$$

It is immediate to show that the pair $(\xi_{\alpha\beta} \cdot g_{\alpha\beta}, \omega_\alpha)$ actually satisfies the 'transformation law of local potentials', that is:

(0.4.145)
$$\delta(\omega_\alpha) = \tilde{d}^0(\xi_{\alpha\beta} \cdot g_{\alpha\beta}).$$

Now, given that:

(0.4.146)
$$Ker(\tilde{d}^0) = \tilde{\mathbb{C}},$$

as a consequence of the Poincaré Lemma, we can easily show that the above defined group action of $H^1(X, \mathbb{S}^1)$ on the abelian group $Iso(\mathcal{L}, \nabla)$ is actually free.

For this purpose, we consider the equivalent differential line sheaves $(\mathcal{L}, \nabla)$ and $(\mathcal{L}', \nabla) = \xi \cdot [(\mathcal{L}, \nabla)] = [(\xi \cdot \mathcal{L}, \nabla)]$ in the abelian group $Iso(\mathcal{L}, \nabla)$, where $\xi := [(\xi_{\alpha\beta})] \in H^1(X, \mathbb{S}^1)$, and thus $(\xi_{\alpha\beta}) \in Z^1(\mathcal{U}, \mathbb{S}^1)$, according to the above. Due to the above equivalence, we conclude that there exists a 0-cochain $(t_a) \in C^0(\mathcal{U}, \tilde{\mathcal{A}}) = \prod_\alpha \tilde{\mathcal{A}}(U_\alpha)$, such that:

(0.4.147)
$$\xi_{\alpha\beta} = \Delta^0(t_\alpha^{-1}),$$

and hence, equivalently:

(0.4.148)
$$[(\xi_{\alpha\beta})] = 1 \in H^1(X, \tilde{\mathcal{A}}),$$

where $(\xi_{\alpha\beta}) \in Z^1(\mathcal{U}, \mathbb{S}^1) \hookrightarrow (\xi_{\alpha\beta}) \in Z^1(\mathcal{U}, \tilde{\mathcal{A}})$. Furthermore, by hypothesis we have:

(0.4.149)
$$\omega_\alpha' = \omega_\alpha,$$

and thus:

(0.4.150)
$$\tilde{d}^0(t_a^{-1}) = -\tilde{d}^0(t_a) = 0,$$

such that $t_a \in Ker(\tilde{d}^0) = \tilde{\mathbb{C}}$. Thus, we obtain that an automorphism h of a line sheaf $\mathcal{L}$ is also an automorphism of the differential line sheaf $(\mathcal{L}, \nabla)$ if and only if $t_a \in Ker(\tilde{d}^0) = \tilde{\mathbb{C}}$. Hence, in the unitary case considered, we deduce that the 0-cochain $(t_a) \in C^0(\mathcal{U}, \tilde{\mathcal{A}})$ actually belongs to $C^0(\mathcal{U}, \mathbb{S}^1)$. Therefore, we conclude that:

(0.4.151)
$$[(\xi_{\alpha\beta})] = 1 \in H^1(X, \mathbb{S}^1),$$

which proves the freeness of the above action.

We note that, if we do not assume the unitarity condition, the same argument shows that the group action of $H^1(X, \tilde{\mathbb{C}})$, induced by the natural injection $\tilde{\mathbb{C}} \hookrightarrow \tilde{\mathcal{A}}$ on the abelian group $Iso(\mathcal{L}, \nabla)$ is actually free, where in this more general case we have that $[(\xi_{\alpha\beta})] = 1 \in H^1(X, \tilde{\mathbb{C}})$, where $(\xi_{\alpha\beta}) \in Z^1(\mathcal{U}, \tilde{\mathbb{C}})$.

Consequently, the free group action of $H^1(X, \mathbb{S}^1)$ on $Iso(\mathcal{L}, \nabla)$ is *restricted* to a free group action on its abelian subgroup of quantum unitary rays $Iso(\mathcal{L}, \nabla_\varrho)$. Since the latter abelian group is partitioned into spectral $[R]$-beams over $\Omega^2_{c,\mathbb{Z}}$, we conclude that the above free group action is finally transferred as a *free group action* of $H^1(X, \mathbb{S}^1)$ on each spectral $[R]$-beam.

A cohomology class in the abelian group $H^1(X, \mathbb{S}^1)$ is called a *polarization phase germ* of a spectral $[R]$-beam.

The intuition behind the above definition is that a cohomology class in the abelian group $H^1(X, \mathbb{S}^1) \cong H^1(X, U(1))$ can be evaluated at a homology cycle $\gamma \in H_1(X)$ by means of the pairing:

$$(0.4.152) \qquad H_1(X) \times H^1(X, U(1)) \to U(1)$$

to obtain a global observable gauge-invariant phase factor in $U(1)$. Thus, gauge equivalent unitary R-rays may be intrinsically distinguished by means of a polarization phase germ, identified as a cohomology class in $H^1(X, \mathbb{S}^1)$.

If we assume that the underlying space X of control variables is a *locally path-connected* space, then we obtain the following:

Theorem 0.15. *A polarization phase germ of a spectral $[R]$-beam is realized by a representation of the fundamental group of the topological space X of control variables to $\mathbb{S}^1$.*

As an immediate consequence of the Hurewicz isomorphism we obtain:

$$(0.4.153) \qquad Hom(\pi_1(X), \tilde{\mathbb{C}}) \cong H^1(X, \tilde{\mathbb{C}}).$$

By restriction to the unitary case we obtain:

$$(0.4.154) \qquad Hom(\pi_1(X), \mathbb{S}^1) \cong H^1(X, \mathbb{S}^1).$$

Thus, a polarization phase germ expressed in terms of a cohomology class in $H^1(X, \mathbb{S}^1)$ is realized by a *representation of the fundamental group* of the topological space X to $\mathbb{S}^1$.

0.4.11 *Affine Structure of Spectral Beams*

We have shown that there exists a free group action of $H^1(X, \mathbb{S}^1) \cong Hom(\pi_1(X), \mathbb{S}^1)$ on each spectral $[R]$-beam. A natural problem in this context is the specification of the appropriate conditions, which would qualify this free action as a *transitive* one as well.

For this purpose, we consider the following sequence of abelian group sheaves:

$$(0.4.155) \qquad 1 \to \tilde{\mathbb{C}} \xrightarrow{\epsilon} \tilde{\mathcal{A}} \xrightarrow{\tilde{d}^0} \Omega^1 \xrightarrow{d^1} d^1\Omega^1 \to 0.$$

As a consequence of the Poincaré Lemma we have that:

$$(0.4.156) \qquad Ker(\tilde{d}^0) = \tilde{\mathbb{C}}.$$

Moreover, we have that $d^1 \circ \tilde{d}^0 = 0$, and thus $Im(\tilde{d}^0) \subseteq Ker(d^1)$. Now, we consider those closed 1-forms θ_α of Ω^1, for which the following condition is satisfied:

$$(0.4.157) \qquad Im(\tilde{d}^0) = Ker(d^1).$$

The closed 1-forms θ_α of Ω^1, satisfying $Im(\tilde{d}^0) = Ker(d^1)$ are called *logarithmically exact closed 1-forms*.

Theorem 0.16. *The following sequence of abelian group sheaves is an exact sequence if restricted to logarithmically exact closed 1-forms:.*

$$(0.4.158) \qquad 1 \to \tilde{\mathbb{C}} \xrightarrow{\epsilon} \tilde{A} \xrightarrow{\tilde{d}^0} \Omega^1 \xrightarrow{d^1} d^1\Omega^1 \to 0.$$

Theorem 0.17. *A spectral $[R]$-beam is an affine space with structure group the characters of the fundamental group with respect to logarithmically exact closed 1-forms.*

By the exact sequence of abelian group sheaves of the previous proposition, we obtain a 0-cochain (θ_α) of logarithmically exact closed 1-forms:

$$(0.4.159) \qquad (\theta_\alpha) \in C^0(\mathcal{U}, Ker(d^1)) = (\theta_\alpha) \in C^0(\mathcal{U}, Im(\tilde{d}^0)) = \tilde{d}^0(C^0(\mathcal{U}, \tilde{A})).$$

Hence, for a 0-cochain (θ_α) of logarithmically exact closed 1-forms θ_α, there exists a 0-cochain t_α in $\tilde{A}$, such that:

$$(0.4.160) \qquad \theta_\alpha = \tilde{d}^0(t_\alpha^{-1}).$$

This 0-cochain (θ_α) may be considered as representing an integrable connection $\check{\nabla}$ on a line sheaf **K**, whose coordinate 1-cocycle with respect to an open covering $\mathcal{U}$ is given by:

$$(0.4.161) \qquad \zeta_{\alpha\beta} = t_\beta^{-1} t_\alpha.$$

Clearly, in this case the transformation law of local potentials is satisfied as follows:

$$(0.4.162) \qquad \delta^0(\theta_\alpha) = \theta_\beta - \theta_\alpha = \tilde{d}^0(t_\beta^{-1}) - \tilde{d}^0(t_\alpha^{-1}) = \tilde{d}^0(t_\beta^{-1} t_\alpha) = \tilde{d}^0(\zeta_{\alpha\beta}).$$

Next, we consider a spectral $[R]$-beam, viz. an orbit $Iso(\mathcal{L}, \nabla_\varrho)_R$, consisting of gauge equivalent unitary R-rays, which are indistinguishable from the perspective of their common differential invariant $[R]$. Again, we also consider those closed 1-forms θ_α of Ω^1, for which the following condition is satisfied:

$$(0.4.163) \qquad Im(\tilde{d}^0) = Ker(d^1),$$

that is the logarithmically exact closed 1-forms. We take a pair of equivalent unitary R-rays, denoted by $(\mathcal{L}, \nabla_\varrho)$, $(\mathcal{L}', \nabla'_\varrho)$ correspondingly. Then, we have that:

$$(0.4.164) \qquad R = (d\omega_\alpha) = (d\omega'_\alpha),$$

$$(0.4.165) \qquad d(\omega_\alpha - \omega'_\alpha) = 0.$$

We conclude that $(\omega_\alpha - \omega'_\alpha)$ is of the form θ_α of Ω^1, viz. it is a logarithmically exact closed 1-form. This means that we obtain a 0-cochain $(\omega_\alpha - \omega'_\alpha)$ of logarithmically exact closed 1-forms:

(0.4.166)
$$(\omega_\alpha - \omega'_\alpha) \in C^0(\mathcal{U}, Ker(d^1)) = (\omega_\alpha - \omega'_\alpha) \in C^0(\mathcal{U}, Im(\tilde{d}^0)) = \tilde{d}^0(C^0(\mathcal{U}, \tilde{A})).$$

Hence, for a 0-cochain $(\omega_\alpha - \omega'_\alpha)$ of logarithmically exact closed 1-forms, there exists a 0-cochain t_α in $\tilde{A}$, such that:

(0.4.167)
$$\omega_\alpha - \omega'_\alpha = \tilde{d}^0(t_\alpha^{-1}),$$

(0.4.168)
$$\omega_\alpha = \omega'_\alpha + \tilde{d}^0(t_\alpha^{-1}).$$

Furthermore, we may consider another unitary R-ray in the same orbit, denoted by $(\mathcal{L}'', \nabla''_\varrho)$ characterized locally as follows:

(0.4.169)
$$g''_{\alpha\beta} := \zeta_{\alpha\beta} \cdot g'_{\alpha\beta} = t_\alpha \cdot g'_{\alpha\beta} \cdot t_\beta^{-1},$$

(0.4.170)
$$\omega''_\alpha = \omega_\alpha = \omega'_\alpha + \tilde{d}^0(t_\alpha^{-1}),$$

such that:

(0.4.171)
$$\tilde{d}^0(g''_{\alpha\beta}) = \delta^0(\omega''_\alpha).$$

Thus, if we consider any two unitary R-rays $(\mathcal{L}, \nabla_\varrho)$, $(\mathcal{L}', \nabla'_\varrho)$, we can always find another unitary R-ray $(\mathcal{L}'', \nabla''_\varrho = \nabla_\varrho)$ equivalent to both of them, viz. belonging to the *same orbit* over their common differential invariant $[R]$, or equivalently belonging to the same spectral $[R]$-beam. Therefore, for any two unitary R-rays $(\mathcal{L}, \nabla_\varrho)$, $(\mathcal{L}', \nabla'_\varrho)$, we can always *substitute*, for instance, the second one of them $(\mathcal{L}', \nabla'_\varrho)$ with an equivalent unitary R-ray characterized by the same 0-cochain of potentials like the first one, viz. by $(\mathcal{L}'', \nabla_\varrho)$ whose local form is given by the above defined pair $(g''_{\alpha\beta}, \omega_\alpha)$. Consequently, we obtain:

(0.4.172)
$$\tilde{d}^0(g''_{\alpha\beta}) = \delta^0(\omega''_\alpha) = \delta^0(\omega_\alpha) = \tilde{d}^0(g_{\alpha\beta}).$$

We conclude that:

(0.4.173)
$$\tilde{d}^0(g''_{\alpha\beta}) - \tilde{d}^0(g_{\alpha\beta}) = 0,$$

(0.4.174)
$$\tilde{d}^0(g''_{\alpha\beta} \cdot g_{\alpha\beta}^{-1}) = 0.$$

Thus, $(g''_{\alpha\beta} \cdot g_{\alpha\beta}^{-1}) \in Z^1(\mathcal{U}, Ker(\tilde{d}^0))$, or equivalently by the Poincaré Lemma:

(0.4.175)
$$(g''_{\alpha\beta} \cdot g_{\alpha\beta}^{-1}) \in Z^1(\mathcal{U}, \tilde{\mathbb{C}}),$$

which in the quantum unitary case considered reduces to:

(0.4.176)
$$(g''_{\alpha\beta} \cdot g_{\alpha\beta}^{-1}) \in Z^1(\mathcal{U}, \mathbb{S}^1).$$

The above, can be equivalently formulated as follows:

(0.4.177)
$$g''_{\alpha\beta} = \xi_{\alpha\beta} \cdot g_{\alpha\beta},$$

where $\xi_{\alpha\beta} \in Z^1(\mathcal{U}, \mathbb{S}^1)$. By considering the corresponding cohomology classes we obtain:

$$(0.4.178) \qquad [g''_{\alpha\beta}] = [(\xi_{\alpha\beta})] \cdot [(g_{\alpha\beta})] = [(\xi_{\alpha\beta} \cdot g_{\alpha\beta})],$$

where $[(\xi_{\alpha\beta})] \in H^1(X, \mathbb{S}^1)$, $[(g_{\alpha\beta})] \in H^1(X, \tilde{\mathcal{A}}) \cong Iso(\mathcal{L})$. Thus, we finally deduce that:

$$(0.4.179) \qquad [(\mathcal{L}', \nabla'_{\varrho})] = [(\mathcal{L}'', \nabla_{\varrho})] = [(\xi \cdot \mathcal{L}, \nabla)] = [(\xi)] \cdot [(\mathcal{L}, \nabla)],$$

where $[(\xi)] = [(\xi_{\alpha\beta})] \in H^1(X, \mathbb{S}^1)$. Therefore, we finally arrive at the following conclusion:

The free group action of $H^1(X, \mathbb{S}^1)$ on a spectral $[R]$-beam is also transitive with respect to logarithmically exact closed 1-forms, and therefore a spectral $[R]$-beam becomes a $H^1(X, \mathbb{S}^1) \cong Hom(\pi_1(X), \mathbb{S}^1)$-affine space, or equivalently an affine space with structure group the characters of the fundamental group.

Thus, each partition block or fiber $Iso(\mathcal{L}, \nabla_{\varrho})_R$ labeled by the curvature differential invariant $[R]$, viz. each spectral $[R]$-beam is an *affine space* with structure group $H^1(X, \mathbb{S}^1) \cong Hom(\pi_1(X), \mathbb{S}^1)$. In this manner, any two quantum unitary R-rays differ by an element of $H^1(X, \mathbb{S}^1)$, and conversely any two quantum unitary rays which differ by an element of $H^1(X, \mathbb{S}^1)$ are characterized by the same differential invariant $[R]$, or equivalently are R-rays of the same spectral $[R]$-beam. Hence, although all gauge equivalent quantum unitary R-rays cannot be distinguished from the perspective of their common curvature differential invariant, there exists a *free and transitive action* of the group $H^1(X, \mathbb{S}^1) \cong Hom(\pi_1(X), \mathbb{S}^1)$, characterized as the global kinematical symmetry group of a $[R]$-beam, which completely distinguishes among them by means of characters of the fundamental group of X. Inversely, from any one quantum unitary R-ray we can obtain intrinsically its whole equivalence class by means of the free and transitive action of the abelian group $H^1(X, \mathbb{S}^1)$ on the depicted one. We conclude that whenever two unitary rays are characterized by the same differential invariant $[R]$, viz. they belong to the same equivalence class (orbit) under the action of $H^1(X, \mathbb{S}^1)$ on $Iso(\mathcal{L}, \nabla_{\varrho})_R$ (which is actually the only class due to transitivity of this action, identified as a spectral $[R]$-beam), then they differ by a character of the fundamental group of X.

0.4.12   *Monodromy Group and Integrable Phase Factors*

We have shown in the previous Section that a polarization phase germ of a spectral $[R]$-beam is realized by a representation of the fundamental group of the topological space X of control variables to $\mathbb{S}^1$. As a consequence, we obtain:

Theorem 0.18. *The global polarization symmetry group of a spectral $[R]$-beam is realized in terms of unitary line local systems. Equivalently, there exists a bijective correspondence between the polarization phase germs of a spectral $[R]$-beam and isomorphism classes of unitary line local systems.*

This is an immediate corollary of the following equivalences:

$$(0.4.180) \qquad Z^1(\mathcal{U}, \tilde{\mathbb{C}}) \ni (\xi_{\alpha\beta}) \cong (\mathcal{L}, \check{\nabla}) \cong Ker(\check{\nabla})_\mathcal{L} \cong \Lambda,$$

$$(0.4.181) \qquad H^1(X, \tilde{\mathbb{C}}) \cong [(\mathcal{L}, \check{\nabla})] \cong [Ker(\check{\nabla})_\mathcal{L}] \cong [\Lambda].$$

The above equivalences restrict as follows in the case of Hermitian integrable differential line sheaves, or equivalently integrable unitary rays:

$$(0.4.182) \qquad Z^1(\mathcal{U}, \mathbb{S}^1) \ni (\xi_{\alpha\beta}) \cong (\mathcal{L}, \check{\nabla}_\varrho) \cong Ker(\check{\nabla}_\varrho)_\mathcal{L} \cong \Lambda_\varrho,$$

$$(0.4.183) \qquad H^1(X, \mathbb{S}^1) \cong [(\mathcal{L}, \check{\nabla}_\varrho)] \cong [Ker(\check{\nabla}_\varrho)_\mathcal{L}] \cong [\Lambda_\varrho].$$

A natural question in this setting refers to the *realization* of the polarization symmetry group of a spectral [R]-beam at a point of the base space of control variables X.

Theorem 0.19. *At each point of the topological space X, the polarization symmetry group of a spectral [R]-beam is realized via the monodromy group of a unitary line local system at the depicted point, or equivalently by the group of monodromies of the corresponding integrable unitary ray whose covariantly constant sections form this unitary line local system.*

We already know that a polarization phase germ of a spectral [R]-beam is realized by a representation of the fundamental group of X to $\mathbb{S}^1$. Moreover, $\tilde{\mathbb{C}}$ is identified locally with the group of automorphisms of a line local system, $GL(1, \mathbb{C}) \cong \tilde{\mathbb{C}}$. Thus, in the unitary case, $\mathbb{S}^1$ is identified locally with the group of automorphisms of a unitary line local system Λ_ϱ, which may be thought of as the locally constant sheaf of covariantly constant sections of a corresponding Hermitian integrable line sheaf (integrable unitary ray) $(\mathcal{L}, \check{\nabla}_\varrho)$.

For clarity, we may fix a base point x_0 of X, and consider a path $\gamma : [0, 1] \to X$, such that $\gamma(0) = x_0$, $\gamma(1) = x_1$. Thus, if Λ is a line local system on X, then it is pulled back to $[0, 1]$ as a constant sheaf, denoted by $\gamma^\star(\Lambda)$. Hence, we have that: $(\gamma^\star(\Lambda))_0 \cong (\gamma^\star(\Lambda))_1$. Therefore, due to the isomorphisms $(\gamma^\star(\Lambda))_0 \cong \Lambda_{\gamma(0)} := \Lambda_{x_0}$ and $(\gamma^\star(\Lambda))_1 \cong \Lambda_{\gamma(1)} := \Lambda_{x_1}$, we obtain a $\mathbb{C}$-vector space isomorphism $\Lambda_{x_0} \cong \Lambda_{x_1}$, which depends only on the homotopy class of γ. Furthermore, this isomorphism may be thought of as being induced by the parallel transport condition of the corresponding integrable connection of the Hermitian integrable line sheaf $(\mathcal{L}, \check{\nabla}_\varrho)$. Now, if we consider a loop based at the point x_0 of X, we obtain an abelian group homomorphism:

$$(0.4.184) \qquad \mu : \pi_1(X, x_0) \to GL(\Lambda_{x_0}) \cong (GL(1, \mathbb{C}))_{x_0} \cong \tilde{\mathbb{C}}_{x_0}.$$

In the case of a unitary line local system, we obtain the abelian group homomorphism:

$$(0.4.185) \qquad \mu : \pi_1(X, x_0) \to \mathbb{S}^1{}_{x_0}.$$

The image of μ in $\mathbb{S}^1{}_{x_0}$ is the *monodromy group of the unitary local system* Λ_ϱ. Thus, for each *homotopy class* of loops based at x_0, we obtain an *integrable phase factor*, identified with the *monodromy* of the unitary local system $\Lambda_{\varrho_{x_0}}$. Equivalently, this is the same as the monodromy of the corresponding integrable quantum unitary ray $(\mathcal{L}, \check{\nabla}_\varrho)$, acquired by *parallel transport* along a loop based at x_0 and belonging to a homotopy class in $\pi_1(X, x_0)$.

0.4.13   *Aharonov-Bohm Effect*

A concrete observable manifestation of global integrable phase factors in quantum theory is provided by the *Aharonov-Bohm effect*, see Aharonov-Bohm [1, 2]. This effect demonstrates the significance of the local electromagnetic potentials of a *zero curvature spectral beam*. More precisely, the global relative phase factor observed due to the Aharonov-Bohm effect is of a *topological* origin and expressed via an integrable connection. This is the case because the existence of a very long solenoid restricting the magnetic field flux within its borders and making the region it occupies inaccessible to a charged particle, makes the topology of the base localization space *multiple-connected* having the homotopical symmetry of a circle. The evolving states (wave functions), depicted as implicitly parameterized sections of a fibration by a time parameter, are transported by an integrable connection because the propagation takes place in a field-free region. The global phase factor measures the monodromy of the integrable connection due to the topological obstruction imposed to the motion by the inaccessible region enclosed by the boundary of the solenoid, where there exists the flux of the magnetic field.

The Aharonov-Bohm effect constitutes a perfect example of demonstrating the nature and significance of global observable topological phase factors in quantum theory. In particular, the two most important points clarified by the analysis of this effect, pertaining to the whole issue of global quantum topological phases are the following: [1] The *local gauge freedom* of the phase notion, which necessitates the consideration of sheaf-theoretic models of parametric dependence over non-trivial topological spaces; [2] The *mutually implicative* roles of the local and global levels depicted differentially, viz. by means of a connection as follows: (i) An extensive integration process of the contributions of all local gauge potentials to monodromies from the local to the global, and (ii) inversely as a differential localization process of some global topological invariant in terms of a multiplicity of local gauge potentials expressing the allowed contextual variability of the connection with respect to this invariant.

Before we examine in detail this effect from our perspective we formulate the following:

Theorem 0.20. *The abelian group of quantum unitary rays $Iso(\mathcal{L}, \nabla_\varrho)$ is a central extension of the abelian group of integral global closed 2-forms R of $\Omega^2_{c,\mathbb{Z}}$ by the abelian group of polarization phase germs $H^1(X, \mathbb{S}^1)$.*

The proof of the proposition is straightforward by the exactness of the following sequence of abelian groups:

$$(0.4.186) \qquad 1 \to H^1(X, \mathbb{S}^1) \xrightarrow{\sigma} Iso(\mathcal{L}, \nabla_\varrho) \xrightarrow{\kappa} \Omega^2_{c,\mathbb{Z}} \to 0.$$

We restrict our focus on spectral $[R]$-beams. Each spectral $[R]$-beam $Iso(\mathcal{L}, \nabla_\varrho)_R$ is an affine space with structure group $H^1(X, \mathbb{S}^1)$ and from Dirac's quantization condition we know that $(2\pi i)^{-1}[R]$ is a 2-dimensional integral cohomology class of X.

Theorem 0.21. *A zero curvature spectral $[R]$-beam $Iso(\mathcal{L}, \nabla_\varrho)_0$ is isomorphic to the abelian group of polarization phase germs $H^1(X, \mathbb{S}^1)$.*

We have that $Iso(\mathcal{L}, \nabla_\varrho)_R = \kappa^{-1}(R)$, where $R \in \Omega^2_{c,\mathbb{Z}}$. Hence, by the fact that $H^1(X, \mathbb{S}^1) \hookrightarrow Iso(\mathcal{L}, \nabla_\varrho)$ and the previous proposition we deduce that:

$$(0.4.187) \qquad \kappa^{-1}(0) = Iso(\mathcal{L}, \nabla_\varrho)_0 \cong H^1(X, \mathbb{S}^1).$$

Thus, we obtain immediately the following corollary:

Theorem 0.22. *A spectral $[R]$-beam is an $Iso(\mathcal{L}, \nabla_\varrho)_0$-torsor with respect to logarithmically exact closed 1-forms.*

In physical terminology, an *Aharonov-Bohm phase factor* refers to the realization of a zero curvature spectral $[R]$-beam $Iso(\mathcal{L}, \nabla_\varrho)_0$. By this we mean that an Aharonov-Bohm phase is the experimentally realized global gauge-invariant phase characteristic of a *gauge equivalence class* of integrable Hermitian differential line sheaves, or equivalently integrable quantum unitary rays. From the above, we have that:

$$(0.4.188) \qquad Iso(\mathcal{L}, \nabla_\varrho)_0 \cong H^1(X, \mathbb{S}^1) \cong Hom(\pi_1(X), \mathbb{S}^1).$$

Thus, for each point x_0 of the topological space X, we have:

$$(0.4.189) \qquad \mu : \pi_1(X, x_0) \to \mathbb{S}^1{}_{x_0} \cong U(1).$$

Therefore, an Aharonov-Bohm phase factor is realized as the global monodromy group element $\mu(\gamma) \in U(1)$ for each homotopy class of loops γ based at x_0. In the original formulation, the topological space X is homotopically contractible to the circle, and hence its second integer cohomology is trivial. So we have a gauge equivalence class of integrable Hermitian differential line sheaves, or equivalently a gauge equivalence class of line local systems on the circle, which form a zero curvature spectral beam. The global gauge-invariant phase factors via which this beam is realized is the monodromy group of the beam, which is identified with the image of the fundamental group of the circle, viz. the integers $\mathbb{Z}$ into $U(1)$. The monodromy depends only on the *winding number* and is observed as a shift in the *interference pattern* of the beam.

We may equivalently provide a simple cohomological argument, which explains the realization of Aharonov-Bohm type of phases for non-simply connected X, which for simplicity here assume to be a manifold. This argument stresses again the fact that the Aharonov-Bohm type of phases is the global gauge-invariant observable factor of a zero curvature beam, viz. of a gauge equivalence class of zero curvature quantum unitary rays. For this purpose, we use the fact that a zero curvature beam is isomorphic to the abelian group $H^1(X, U(1))$. Thus, in classical differential geometric terms may be expressed as follows:

$$(0.4.190) \qquad [\Psi(\theta_\alpha, -)] = exp\left(\frac{ie}{\hbar c} \oint_{[-]} \theta_\alpha\right),$$

where we have inserted the corresponding physical units, and we consider θ_α real-valued in the Lie algebra of $U(1)$. Notice that $[\Psi(\theta_\alpha, -)]$ is an element of $H^1(X, U(1))$, which is evaluated at a homology cycle $\gamma \in H_1(X)$ by means of the pairing:

$$(0.4.191) \qquad H_1(X) \times H^1(X, U(1)) \to U(1),$$

$$(0.4.192) \qquad (\gamma, [\Psi(\theta_\alpha, -)]) \mapsto [\Psi(\theta_\alpha, \gamma)] = exp\left(\frac{ie}{\hbar c} \oint_\gamma \theta_\alpha\right),$$

where $[\Psi(\theta_\alpha, \gamma)]$ is identified as a global Aharonov-Bohm observable gauge-invariant phase factor of the beam in $U(1)$. Notice that if we consider any unitary ray of this beam, then we only obtain a real-valued phase, defined by:

$$(0.4.193) \qquad \Psi(\theta_\alpha, \gamma) = \frac{e}{\hbar c} \oint_\gamma \theta_\alpha.$$

Due to the isomorphism of groups $\mathbb{R}/\mathbb{Z} \cong U(1)$, we have to take the quotient of the set of all $\Psi((\theta_\alpha)_i, \gamma)$ for all unitary rays by the equivalence relation: $\Psi((\theta_\alpha)_1, \gamma) \sim \Psi((\theta_\alpha)_2, \gamma)$ if $((\theta_\alpha)_1 - (\theta_\alpha)_2) \in \mathbb{Z}$. In physical terms, this means that the interference phase patterns of quantum unitary rays differing by an integer *cannot be distinguished* experimentally, and thus the physically meaningful global gauge-invariant information is only the Aharonov-Bohm phase factors of the form $[\Psi(\theta_\alpha, \gamma)] = exp(\frac{ie}{\hbar c} \oint_\gamma \theta_\alpha)$ referring to the *global realization* of the beam, viz. to the global realization of the whole gauge equivalence class of quantum unitary rays.

0.4.14   *Holonomy of Spectral Beams*

In the original *Berry-Simon* formulation of the notion of a global quantum *geometric phase factor* acquired by a quantum system after completion of an *adiabatic cyclic evolution*, the parametric localization of a quantum state from a base space is induced by the functional dependence of the Hamiltonian of the system by a set of *control variables*, see Berry [1], Simon [1], Wilczek-Shapere [1]. Thus, the dynamical evolution is driven via the *implicit temporal dependence* of the Hamiltonian through the control variables. Under the assumption that the set of these variables forms a smooth manifold the time dependence is depicted by means of differentiable paths on this space. The cyclic evolution signifies the *periodic property* of a quantum state with respect to the control variables, whence the adiabatic hypothesis is equivalent to the specification of a parallel transport condition on the evolution of normalized state vectors, which in general is considered to be *path-dependent*. The fibers of the induced *spectral line bundle* stand for the *eigenspaces* of the Hamiltonian operator. The adiabatic rule gives rise to a non-integrable connection, which in turn defines a covariant derivative operator on the sections of the corresponding Hermitian line sheaf, that is the eigenstates of the Hamiltonian. We note that an eigenstate of the Hamiltonian is required to remain in the eigenspace of the same instantaneous eigenvalue during the adiabatic evolution. The non-dynamical (non-Hamiltonian dependent) global phase assembled during a cyclic evolution along a closed path on the base space is thought of as *"memory"* of the evolution, since it encodes the global geometric features of the space of control variables in the algebraic (group-theoretic) form of the *holonomy* of the connection.

The observable global phase factor is of a geometric origin because it depends *solely* on the geometry of the base space pathway along which the state is transported. If the eigenvalues of the Hamiltonian are degenerate or close to each other, then the adiabatic transportation constraint is not realistic and is substituted by another appropriate connection depending on the context considered. Moreover, the gauge freedom of a state vector localized at a fiber over an eigenspace of the Hamiltonian is not an one-dimensional complex phase any more but an n-dimensional complex matrix of phases, called a non-abelian complex phase. Thus, even the adiabatic transportation condition is *not necessary* for the experimental detection of global phase factors. This has been also demonstrated by the line bundle formulation of the complex Hilbert space of states over the complex projective Hilbert space. In this way, the one-dimensional projection operators play the role of control variables. This line bundle is endowed with a natural connection obtained by differentiation of the inner product of normalized sections (quantum state vectors) of the bundle and the adiabatic hypothesis is not involved at all. Then, the global phase factor is obtained as the holonomy transformation of this connection with respect to a closed path on the complex projective space.

We note that most of the discussions about geometric phase factors in quantum mechanics are expressed in the language of *holonomies*. From the abstract perspective of this work, all observable geometric phase factors are consequences of the curvature of spectral $[R]$-beams and in particular of the fact that $(2\pi i)^{-1}[R]$ is a 2-dimensional integral cohomology class of X. As we have already explained, global observable topological phase factors can be completely understood in terms of monodromies of zero curvature beams. Again it needs to be emphasized that observable holonomies refer to spectral $[R]$-beams and *not* to individual unitary R-rays. For simplicity, we restrict our discussion to the case that X is a paracompact smooth manifold.

We notice that for any real valued form ϕ of degree k we may define:

$$(0.4.194) \qquad \hat{\phi}(\eta) := \Psi(\phi, \eta) = \left(\int_\eta \phi \right) + \mathbb{Z},$$

where η is a k-chain of X and $\hat{\phi}$ is a k-cochain of X with values in $\mathbb{R}/\mathbb{Z}$. Now, we consider the group morphism:

$$(0.4.195) \qquad \Xi : Z_k(X) \to \mathbb{R}/\mathbb{Z},$$

such that there exists a $k+1$-form τ, which satisfies $\Xi \circ \partial = \hat{\tau}$, or equivalently:

$$(0.4.196) \qquad \Xi(\partial\zeta) = \left(\int_\zeta \tau \right) + \mathbb{Z} = \hat{\tau}(\zeta) \in \mathbb{R}/\mathbb{Z},$$

for any smooth map $\zeta : \Delta_{k+1} \to X$. We note that Δ_n stands for the standard n-dimensional *simplex* and the space of n-chains is generated by Δ_n. We also notice that the following holds:

$$(0.4.197) \qquad d(\hat{\tau}) = \widehat{d\tau} = \Xi \circ \partial^2 = 0,$$

and therefore τ is a closed $k+1$-form.

Next, we consider a unitary R-ray $(\mathcal{L}, \nabla_\varrho)$ and use the fact that R is an integral global closed 2-form R of $\Omega^2_{c,\mathbb{Z}}$. Similarly to the definition of gauge potentials in the case of Hermitian differential line sheaves, we consider R as a *purely imaginary* closed 2-form, such that $R = i \cdot \Theta$. Thus, according to the above, and since Θ is an integral global real-valued form of degree 2, we may consider the following group morphism, which we call *holonomy morphism*:

$$(0.4.198) \qquad \mathbb{H} : Z_1(X) \to \mathbb{R}/\mathbb{Z},$$

$$(0.4.199) \qquad \mathbb{H}(\partial S) = \left(\int_S \Theta \right) + \mathbb{Z} = \hat{\Theta}(S) \in \mathbb{R}/\mathbb{Z},$$

for any smooth map $S : \Delta_2 \to X$. Equivalently, we have:

$$(0.4.200) \qquad \mathbb{H} \circ \partial = \hat{\Theta}.$$

We notice that for a fixed curvature R of $\Omega^2_{c,\mathbb{Z}}$, the same holonomy morphism $\mathbb{H}$ is defined for any other unitary R-ray. Thus, it provides a characterization of the whole gauge equivalence class of unitary R-rays classified by the differential invariant $[R]$. Equivalently stated, it provides a characterization of a spectral $[R]$-beam, and therefore we obtain a *holonomy cohomology class* in $H^1(X, U(1))$, defined by:

$$(0.4.201) \qquad Hol(\partial S) = exp\left(i \int_S \Theta \right).$$

From the above, it is straightforward to conclude that given the *Chern morphism*:

$$(0.4.202) \qquad \delta_c : H^1(X, U(1)) \to H^2(X, \mathbb{Z}),$$

and the natural morphism:

$$(0.4.203) \qquad \iota : H^2(X, \mathbb{Z}) \to H^2(X, \mathbb{R}),$$

the holonomy cohomology class in $H^1(X, U(1))$ of a spectral beam for fixed Θ, is in the inverse image of the Chern characteristic class:

$$(0.4.204) \qquad c_1 = -\frac{1}{2\pi i}[\Theta] \in H^2(X, \mathbb{Z}),$$

under δ_c, or equivalently in the inverse image of the cohomology class $[\Theta]$ in $H^2(X, \mathbb{R})$, such that:

$$(0.4.205) \qquad (\iota \circ \delta_c)(Hol) = [\Theta].$$

Thus, we conclude that an $\mathbb{R}/\mathbb{Z} \cong U(1)$-observable holonomy morphism value, which formalizes the notion of a *non-integrable geometric phase factor*, is a global observable gauge-invariant characteristic of a spectral beam, that is it characterizes the whole gauge equivalence class of quantum unitary rays having the same curvature, such that in cohomology the above relation is satisfied.

0.5 The Functorial Imperative

0.5.1 *Representable Functors and Natural Transformations*

The definition of a category of algebraic objects of some kind with arrows being homomorphisms between them, constitutes an abstraction of the behavior of functions closed under the associative operation of *composition*. More precisely, the notion of a function is generalized to the notion of a homomorphism, whereas the associative operation of composition becomes an operation on sets of homomorphisms between algebraic objects of the same kind satisfying the same properties that functions and compositions satisfy. Notice that the composition of two functions f, g, denoted as $f \circ g$ is defined only in case that the codomain of g is the domain of f. Moreover, the composition of a function f with the identity of either its domain or codomain gives f again.

In a nutshell, the notion of a *category* of algebraic objects of some kind, for instance observables or events, is a conception based on the behavior of functions *closed under the associative operation of composition*, abstracted in terms of homomorphisms, which in turn, have been idealized as algebraic "generalized elements" (or their —*duals*) determining completely the algebraic objects themselves. Because there exists an obvious *duality* between incoming and outgoing homomorphisms with respect to an algebraic object, this is built into the definition of a category so that the operation of *arrows reversal* leaves invariant the concept of a category, meaning that this operation gives again a category being in dual or opposite relation in comparison to the given one. A consequence of this fact is that all categorical constructions come in dual pairs corresponding to the dual viewpoints of considering incoming or outgoing arrows with respect to a constituted algebraic object, see Mac Lane [1, 3].

Now, the establishment of the notion of a category of algebraic objects naturally rises the problem of defining the notion of a function whose domain and codomain are categories. Obviously such a function should *preserve* the composition operation binding a category as an associatively closed universe of discourse. This is precisely the notion of *functor* between categories. A *covariant functor* is a functor which preserves the directionality of an arrow in the domain category, whereas a *contravariant functor* is a functor which reverses it.

Each object A of a category $\mathcal{A}$ determines a covariant functor of homomorphisms emanating from A, denoted by $\curvearrowright_A : \mathcal{A} \to \mathcal{S}ets$, called the covariant $Hom_{\mathcal{A}}$-functor *represented* by A, and defined as follows:

[1]. For all objects X in $\mathcal{A}$, $\curvearrowright_A(X) := Hom_{\mathcal{A}}(A, X)$.

[2]. For all homomorphisms $f : X \to Y$ in $\mathcal{A}$,

$$(0.5.1) \qquad \curvearrowright_A(f) : Hom_{\mathcal{A}}(A, X) \to Hom_{\mathcal{A}}(A, Y)$$

is defined as post-composition with f, viz., $\curvearrowright_A(f)(g) := f \circ g$.

The covariant representable functor $\curvearrowright_A : \mathcal{A} \to \mathcal{S}ets$, can be thought of as constructing an image of $\mathcal{A}$ in the category of $\mathcal{S}ets$ in a covariant way.

Now, let us consider the *opposite category* $\mathcal{A}^{op}$, and let A be an object in this category. The object A determines a contravariant functor of homomorphisms targeting A. Then, the contravariant $Hom_{\mathcal{A}}$-functor *represented* by A is the contravariant functor $\curvearrowright^A : \mathcal{A}^{op} \to \mathcal{S}ets$, defined as follows:

[1]. For all objects B in $\mathcal{A}^{op}$, $\curvearrowright^A(B) := Hom_{\mathcal{A}^{op}}(B, A)$.

[2]. For all homomorphisms $f : C \to B$ in $\mathcal{A}^{op}$,

$$(0.5.2) \qquad \curvearrowright^A(f) : Hom_{\mathcal{A}^{op}}(B, A) \to Hom_{\mathcal{A}^{op}}(C, A)$$

is defined as pre-composition with f, viz., $\curvearrowright^A(f)(g) := g \circ f$.

The contravariant $Hom_{\mathcal{A}}$-functor represented by A, viz. $\curvearrowright^A : \mathcal{A}^{op} \to \mathcal{S}ets$, is called the *functor of generalized elements* (incoming homomorphisms) of A. Moreover, the information contained in A is *classified* completely by its functor of generalized elements $\curvearrowright^A$. Dually the covariant $Hom_{\mathcal{A}}$-functor represented by A, viz. $\curvearrowright_A : \mathcal{A} \to \mathcal{S}ets$ is called the *functor of generalized co-elements* (outgoing homomorphisms) of A. Similarly, the information contained in A is classified completely by its functor of generalized co-elements $\curvearrowright_A$.

Now, given a small category $\mathcal{A}$, viz. a category such that for all objects B, A, the Hom-class (class of homomorphisms) $Hom_{\mathcal{A}}(B, A)$ is a set, we may consider the $Hom_{\mathcal{A}}$-*bifunctor*, see Mac Lane [1]:

$$(0.5.3) \qquad Hom_{\mathcal{A}} := \curvearrowright_B^A = \mathcal{A}^{op} \times \mathcal{A} \to \mathcal{S}ets$$

from the *product category* $\mathcal{A}^{op} \times \mathcal{A}$ to the category of sets, such that for objects B, A of $\mathcal{A}$, $\curvearrowright_B : \mathcal{A} \to \mathcal{S}ets$ is the covariant representable functor (represented by B), and $\curvearrowright^A : \mathcal{A}^{op} \to \mathcal{S}ets$ is the contravariant representable functor (represented by A).

Continuing in the same frame of thought, the next question arising is the following: Which is the proper notion of morphism which captures the notion of a *transformation from some functor to another functor* having both the same domain and the same codomain categories? Defining the proper notion of morphism between such functors is important because it would give us the possibility to consider legitimately the notion of a *functor category* $\mathcal{C}^{\mathcal{A}}$, where the algebraic objects would be functors $\mathbb{F} : \mathcal{A} \to \mathcal{C}$ and the morphisms would be the sought transformations between such functors. The leading idea has to do with the requirement that a transformation of the sought form should *compare two functorial processes* having the same domain and the same codomain in a way that is *not dependent* on the specific objects and arrows involved, that is it should relate the processes themselves without the intervention of ad hoc choices. This is precisely the notion required for the formalization of the concept of *naturality* referring to the relation or comparison of two functorial processes sharing the same source and the same target categories. Concomitantly the corresponding notion of morphism between functors of the above form is captured in the notion of a *natural transformation*.

More precisely, if $\mathbb{F}$, $\mathbb{G}$, are functors from the category $\mathcal{A}$ to the category $\mathcal{C}$, a *natural transformation* τ from $\mathbb{F}$ to $\mathbb{G}$ is a function assigning to each object A in $\mathcal{A}$ a morphism τ_A from $\mathbb{F}(A)$ to $\mathbb{G}(A)$ in $\mathcal{C}$, such that for every arrow $f : A \to B$ in $\mathcal{A}$ the following diagram in $\mathcal{C}$ commutes:

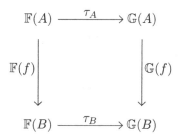

That is, for every arrow $f : A \to B$ in $\mathcal{A}$ we have:

$$(0.5.4) \qquad\qquad \mathbb{G}(f) \circ \tau_A = \tau_B \circ \mathbb{F}(f).$$

A natural transformation $\tau : \mathbb{F} \to \mathbb{G}$ is called a *natural isomorphism (or natural equivalence)* if every component τ_A is *invertible*. Obviously, a natural isomorphism is an invertible natural transformation in the functor category $\mathcal{C}^{\mathcal{A}}$.

This is a key concept and a posteriori justifies the whole categorical framework, since it captures precisely the *criterion of naturality* referring to the comparison of any two functorial processes of the same kind.

0.5.2 *Adjoint Functors: Universals and Equivalence*

The fundamental fact about representable functors is that an object A in $\mathcal{A}$ is *entirely retrievable up to equivalence* (canonical isomorphism) from knowledge of the representable functor $\curvearrowright^A : \mathcal{A}^{op} \to \mathcal{S}ets$ (or equivalently from $\curvearrowright_A : \mathcal{A} \to \mathcal{S}ets$).

This fact, a consequence of a categorical proposition known as the *Yoneda Lemma*, see Mac Lane-Moerdijk [1], can be expressed this way:

Theorem 0.23. *Let A, B be objects in a category $\mathcal{A}$. Suppose we are given an isomorphism of their associated functors: $\curvearrowright^A \cong \curvearrowright^B$. Then there exists a unique isomorphism of the objects themselves, that is $A \cong B$, which gives rise to this isomorphism of functors.*

The conceptual significance of this result is that it bears the following relational interpretation:

An object A of a category $\mathcal{A}$ is *determined uniquely up to equivalence* by the *network of all internal relations* that the object A has with all the other objects in $\mathcal{A}$.

A natural further important question arising in this categorical setting is the following:

Is it possible for an object of a category to be determined uniquely up to equivalence by networks of relations among *partially congruent objects* targeting it that belong to another category functorially correlated with the former one?

It turns out that from a physical viewpoint this is the most *significant function* of the *whole categorical imperative*. This kind of *heteronymous determination* becomes possible in a functorial way if the network of these relations in the determining category becomes appropriately *internalized* in the former one, such that the determination factors uniquely via *a universal*, which exemplifies this determination in a paradigmatic or archetypal form. In technical terms, there should exist a *pair of adjoint functors* between these categories, which should be interpreted as *conceptual inverses* to each other, associated with a bidirectional process of *encoding/decoding* information associated with their structural form. It is instructive to explain this process by starting from the notion of *representation* of an arbitrary $\mathcal{S}ets$-valued functor as follows:

A *representation of a contravariant $\mathcal{S}ets$-valued functor* of the form $\mathbb{X} : \mathcal{A}^{op} \to \mathcal{S}ets$, where $\mathcal{A}$ is a (locally) small category consists of an object K in $\mathcal{A}$ and a natural isomorphism:

$$(0.5.5) \qquad\qquad \mathbb{X} \cong Hom_{\mathcal{A}^{op}}(-, K),$$

or equivalently:

$$(0.5.6) \qquad\qquad \mathbb{X} \cong \curvearrowright^K,$$

where K is called the *representing object* of the functor $\mathbb{X} : \mathcal{A}^{op} \to \mathcal{S}ets$. Thus, $\mathbb{X}$ is representable, if and only if such a representing object exists. Notice that, representations of contravariant $\mathcal{S}ets$-valued functors are unique up to a unique isomorphism. Evidently, there exists a dual formulation referring to a representation of a covariant $\mathcal{S}ets$-valued functor of the form $\mathbb{Z} : \mathcal{A} \to \mathcal{S}ets$. Namely, the covariant functor $\mathbb{Z}$ is representable if and only if there exists an object N in $\mathcal{A}$ and a natural isomorphism:

$$(0.5.7) \qquad\qquad \mathbb{Z} \cong Hom_{\mathcal{A}}(N, -),$$

or equivalently:

(0.5.8) $$\mathbb{Z} \cong \curlywedge_N.$$

By analogy, we may consider the general case of a *bifunctor*:

(0.5.9) $$\bigwedge(-,-) : \mathcal{C}^{op} \times \mathcal{L} \to \mathcal{S}ets$$

from the product category $\mathcal{C}^{op} \times \mathcal{L}$ to the category of sets. Using the conceptual approach of birepresentability, see Ellerman [1], we formulate the following proposition:

Theorem 0.24. *The bifunctor* $\bigwedge(-,-) : \mathcal{C}^{op} \times \mathcal{L} \to \mathcal{S}ets$ *is birepresentable, viz. it is representable within both the categories* $\mathcal{C}^{op}$, $\mathcal{L}$, *if and only if there exist two functors pointing into opposite directions, that is,* $\overrightarrow{\mathbb{A}} : \mathcal{C}^{op} \to \mathcal{L}$ *and* $\overleftarrow{\mathbb{A}} : \mathcal{L} \to \mathcal{C}^{op}$ *and a series of isomorphisms:*

(0.5.10) $$Hom_{\mathcal{L}}(\overrightarrow{\mathbb{A}}(C), L) \cong \bigwedge(C, L) \cong Hom_{\mathcal{C}^{op}}(C, \overleftarrow{\mathbb{A}}(L)),$$

which is natural in both the arguments C *in* $\mathcal{C}^{op}$ *and* L *in* $\mathcal{L}$.

Equivalently, we state that the bifunctor: $\bigwedge(-,-) : \mathcal{C}^{op} \times \mathcal{L} \to \mathcal{S}ets$ is birepresentable, if and only if there exist two functors $\overrightarrow{\mathbb{A}} : \mathcal{C}^{op} \to \mathcal{L}$, $\overleftarrow{\mathbb{A}} : \mathcal{L} \to \mathcal{C}^{op}$ together with *natural isomorphisms of bifunctors*:

(0.5.11) $$Hom_{\mathcal{L}}(\overrightarrow{\mathbb{A}}(-), -) \cong \bigwedge(-,-) \cong Hom_{\mathcal{C}^{op}}(-, \overleftarrow{\mathbb{A}}(-)).$$

In this case, we say that the two oppositely pointing functors $\overrightarrow{\mathbb{A}}$, $\overleftarrow{\mathbb{A}}$ form a *categorical adjunction*, induced by the requirement of birepresentability of the bifunctor: $\bigwedge(-,-) : \mathcal{C}^{op} \times \mathcal{L} \to \mathcal{S}ets$, where $\overrightarrow{\mathbb{A}} : \mathcal{C}^{op} \to \mathcal{L}$ is the *left adjoint functor* of the adjunction, and symmetrically, $\overleftarrow{\mathbb{A}} : \mathcal{L} \to \mathcal{C}^{op}$ is the *right adjoint functor* of the adjunction, denoted usually by $\mathbb{L}$ and $\mathbb{R}$ respectively. By ignoring the middle term we obtain the natural isomorphism of the *Hom*-bifunctors:

(0.5.12) $$\Psi : Hom_{\mathcal{L}}(\overrightarrow{\mathbb{A}}(-), -) \cong Hom_{\mathcal{C}^{op}}(-, \overleftarrow{\mathbb{A}}(-)).$$

Equivalently, we obtain the isomorphism:

(0.5.13) $$\Psi_{C,L} : Hom_{\mathcal{L}}(\overrightarrow{\mathbb{A}}(C), L) \cong Hom_{\mathcal{C}^{op}}(C, \overleftarrow{\mathbb{A}}(L)),$$

which is natural in both the arguments C in $\mathcal{C}^{op}$ and L in $\mathcal{L}$.

The requirement of birepresentability of a bifunctor of the form: $\bigwedge(-,-) : \mathcal{C}^{op} \times \mathcal{L} \to \mathcal{S}ets$, satisfied by the existence of a categorical adjunction, consisting of the functors $\overrightarrow{\mathbb{A}} : \mathcal{C}^{op} \to \mathcal{L}$ (left adjoint functor of the adjunction), and $\overleftarrow{\mathbb{A}} : \mathcal{L} \to \mathcal{C}^{op}$ (right adjoint functor of the adjunction), is tantamount to the establishment of a bidirectional functorial process of *information encoding/decoding* between the categories $\mathcal{L}$, $\mathcal{C}^{op}$. It is important to notice the directionality build in the process of information circulation by means of functorial encoding/decoding, viz. it is understood as an information-encoding of an object of $\mathcal{L}$ by an object of $\mathcal{C}^{op}$ via the functor $\overleftarrow{\mathbb{A}} : \mathcal{L} \to \mathcal{C}^{op}$, and inversely, as an information-decoding from an object of $\mathcal{C}^{op}$ back to an object of $\mathcal{L}$ via the functor $\overrightarrow{\mathbb{A}} : \mathcal{C}^{op} \to \mathcal{L}$, such that the structural relations of the respective categories are being preserved.

It is also important to emphasize that the bidirectional process of information encoding/decoding between the categories $\mathcal{L}, \mathcal{C}^{op}$ should be *functorial*. This is due to the fact that the appropriate information-encoders/decoders between objects are only those that preserve their structures; in this case, the objects, arrows, domains, codomains, compositions, and identities of the respective categories. Consequently, an information-encoding of an object of $\mathcal{L}$ by an object of $\mathcal{C}^{op}$ takes place via the functor $\overleftarrow{\mathbb{A}} : \mathcal{L} \to \mathcal{C}^{op}$, and inversely, an information-decoding from an object of $\mathcal{C}^{op}$ back to an object of $\mathcal{L}$ takes place via the functor $\overrightarrow{\mathbb{A}} : \mathcal{C}^{op} \to \mathcal{L}$. The functorial character of this bidirectional process of information encoding/decoding means precisely that the structural relations of the respective categories are being preserved.

The morphisms in $\mathcal{L}, \mathcal{C}^{op}$ related to each other by the above isomorphism are called *adjoint transposes*. Hence, if we consider a morphism:

$$(0.5.14) \qquad h : \overrightarrow{\mathbb{A}}(C) \to L$$

in $\mathcal{L}$, then by virtue of the adjunction isomorphism, it has an adjoint transpose morphism in $\mathcal{C}^{op}$, namely:

$$(0.5.15) \qquad h^* : C \to \overleftarrow{\mathbb{A}}(L).$$

Dually, if we consider a morphism:

$$(0.5.16) \qquad g : C \to \overleftarrow{\mathbb{A}}(L)$$

in $\mathcal{C}^{op}$, then by virtue of the adjunction isomorphism, it has an adjoint transpose morphism in $\mathcal{L}$, namely:

$$(0.5.17) \qquad g^* : \overrightarrow{\mathbb{A}}(C) \to L.$$

Now, we may consider the *identity* morphism at $\overrightarrow{\mathbb{A}}(C)$ in $\mathcal{L}$, that is $id_{\overrightarrow{\mathbb{A}}(C)}$ belonging to the set $Hom_{\mathcal{L}}(\overrightarrow{\mathbb{A}}(C), \overrightarrow{\mathbb{A}}(C))$. The adjoint transpose of $id_{\overrightarrow{\mathbb{A}}(C)}$ is called the *unit* morphism at C, that is:

$$(0.5.18) \qquad (id_{\overrightarrow{\mathbb{A}}(C)})^* = \eta_C : C \to \overleftarrow{\mathbb{A}}(\overrightarrow{\mathbb{A}}(C))$$

belonging to the set $Hom_{\mathcal{C}^{op}}(C, \overleftarrow{\mathbb{A}}(\overrightarrow{\mathbb{A}}(C)))$. Since the above is natural on the argument C in $\mathcal{C}^{op}$, we obtain a *natural transformation of the identity functor* on $\mathcal{C}^{op}$, via *self-referential information circulation* realized by functorially decoding and then encoding information, called the *unit natural transformation* of the adjunction:

$$(0.5.19) \qquad \eta : id_{\mathcal{C}^{op}} \to \overleftarrow{\mathbb{A}}\,\overrightarrow{\mathbb{A}}.$$

Dually, we may consider the identity morphism at $\overleftarrow{\mathbb{A}}(L)$ in $\mathcal{C}^{op}$, that is $id_{\overleftarrow{\mathbb{A}}(L)}$ belonging to the set $Hom_{\mathcal{C}^{op}}(\overleftarrow{\mathbb{A}}(L), \overleftarrow{\mathbb{A}}(L))$. The adjoint transpose of $id_{\overleftarrow{\mathbb{A}}(L)}$ is called the *counit* morphism at L, that is:

$$(0.5.20) \qquad (id_{\overleftarrow{\mathbb{A}}(L)})^* = \varepsilon_L : \overrightarrow{\mathbb{A}}(\overleftarrow{\mathbb{A}}(L)) \to L$$

belonging to the set $Hom_\mathcal{L}(\overrightarrow{\mathbb{A}}(\overleftarrow{\mathbb{A}}(L)), L)$. Since the above is natural on the argument L in $\mathcal{L}$, we obtain a natural transformation of the identity functor on $\mathcal{L}$, via self-referential information circulation realized by functorially encoding and then decoding information, called the *counit natural transformation* of the adjunction:

$$(0.5.21) \qquad\qquad \varepsilon : \overrightarrow{\mathbb{A}}\,\overleftarrow{\mathbb{A}} \to id_\mathcal{L}.$$

The categorical adjunction being formed by $\overrightarrow{\mathbb{A}}$, $\overleftarrow{\mathbb{A}}$ can be equivalently represented in terms of the unit and counit natural transformations.

Furthermore, we notice that if we consider a morphism $h : \overrightarrow{\mathbb{A}}(C) \to L$ in $\mathcal{L}$, then by virtue of the unit natural transformation, its adjoint transpose morphism in $\mathcal{C}^{op}$ *factors uniquely* via the unit morphism at C as follows:

$$(0.5.22) \qquad h^* = \overleftarrow{\mathbb{A}}(h) \circ \eta_C : C \to \overleftarrow{\mathbb{A}}(\overrightarrow{\mathbb{A}}(C)) \to \overleftarrow{\mathbb{A}}(L).$$

Since the above holds for every object L in $\mathcal{L}$, the pair $(\overrightarrow{\mathbb{A}}(C), \eta_C)$ constitutes a *universal* for the functor:

$$(0.5.23) \qquad Hom_{\mathcal{C}^{op}}(C, \overleftarrow{\mathbb{A}}(-)) = \curvearrowright_C \circ \overleftarrow{\mathbb{A}} : \mathcal{L} \to \mathcal{S}ets.$$

Thus, for every object C in $\mathcal{C}^{op}$, the above composite encoding $\mathcal{S}ets$-valued functor becomes representable by the decoding of C in $\mathcal{L}$, viz. by the object $\overrightarrow{\mathbb{A}}(C)$ in $\mathcal{L}$:

$$(0.5.24) \qquad\qquad (\curvearrowright_C \circ \overleftarrow{\mathbb{A}}) \cong Hom_\mathcal{L}(\overrightarrow{\mathbb{A}}(C), -).$$

Dually, if we consider a morphism $g : C \to \overleftarrow{\mathbb{A}}(L)$ in $\mathcal{C}^{op}$, then by virtue of the counit natural transformation, its adjoint transpose morphism in $\mathcal{L}$ factors uniquely via the counit morphism at L as follows:

$$(0.5.25) \qquad \varepsilon_L \circ \overrightarrow{\mathbb{A}}(g) : \overrightarrow{\mathbb{A}}(C) \to \overrightarrow{\mathbb{A}}(\overleftarrow{\mathbb{A}}(L)) \to L.$$

Since the above holds for every object C in $\mathcal{C}^{op}$, the pair $(\overleftarrow{\mathbb{A}}(L), \varepsilon_L)$ constitutes a universal for the functor:

$$(0.5.26) \qquad Hom_\mathcal{L}(\overrightarrow{\mathbb{A}}(-), L) = \curvearrowright^L \circ \overrightarrow{\mathbb{A}} : \mathcal{C}^{op} \to \mathcal{S}ets.$$

Thus, for every object L in $\mathcal{L}$, the above composite decoding $\mathcal{S}ets$-valued functor becomes representable by the encoding of L in $\mathcal{C}^{op}$, viz. by the object $\overleftarrow{\mathbb{A}}(L)$ in $\mathcal{C}^{op}$:

$$(0.5.27) \qquad\qquad (\curvearrowright^L \circ \overrightarrow{\mathbb{A}}) \cong Hom_{\mathcal{C}^{op}}(-, \overleftarrow{\mathbb{A}}(L)).$$

Now, if both the unit and the counit natural transformations of the adjunction formed by the functors $\overrightarrow{\mathbb{A}}$, $\overleftarrow{\mathbb{A}}$, are natural isomorphisms then we obtain an *equivalence* of the categories $\mathcal{C}^{op}$, $\mathcal{L}$, that is we obtain a *duality*, $\mathcal{C}^{op} \simeq \mathcal{L}$. Thus, the notion of *functorial duality* is a consequence of the existence of a pair of adjoint functors between two categories when both the unit and counit natural transformations of the formed adjunction are natural isomorphisms.

We conclude that the three most fundamental concepts *epitomizing* the function of the whole categorical framework are the concepts of *naturality, universality and equivalence*. More concretely, category theory provides the precise formal means of identifying a universal in mathematical terms, that is the *archetypic form of instantiation* of some property which exemplify the property in such a paradigmatic way that all other instances have this property by virtue of *factorizing uniquely* through the corresponding universal. In this sense, objects are specified functorially via some universality property and only up to equivalence (canonical isomorphism) in a natural manner. All categorical universals are defined up to canonical isomorphism by the existence of pairs of adjoint functors between appropriate categories. An adjunction between two categories gives rise to a family of universal morphisms, one for each object in the first category and one for each object in the second. In this way, each object in the first category induces a certain property in the second category and the universal morphism caries the object to the universal for that property, that is it transports it to an agent of objectification in the second category exemplifying this property in a paradigmatic way. Most important, every adjunction extends to an *adjoint equivalence of certain subcategories* of the initially correlated categories. As we have explained previously this is described by the notion of *functorial duality*, when both the unit and counit natural transformations of the formed adjunction become natural isomorphisms.

0.5.3   *Probes and Adjoints to Realization Functors*

A pertinent problem that arises frequently is the determination of *not directly accessible* objects of some kind L by means of *controllable probes* Y. The categorical formulation of this problem involves two categories, namely the *small category of probes* $\mathcal{A}$ and the category of the not directly accessible objects $\mathcal{E}$, which is considered to be a *cocomplete* category. Moreover, we assume the existence of a *shaping or coordinatization functor* $\mathbb{M} : \mathcal{A} \to \mathcal{E}$, viz. a functor assigning to each probe Y in $\mathcal{A}$ the underlying object L in $\mathcal{E}$ and analogously for the arrows. In some cases the category of probes $\mathcal{A}$ is a subcategory, or even a generating subcategory of $\mathcal{E}$, such that a complete localization of $\mathcal{E}$ in terms of $\mathcal{A}$ is feasible. In the general case, the aim is to approximate an indirectly accessible object L in $\mathcal{E}$ inductively by means of appropriate structured families of probes Y. The important issue is if there exists a universal solution to this problem, which can be thus obtained in terms of a categorical adjunction.

It is rarely the case that the shaping functor $\mathbb{M} : \mathcal{A} \to \mathcal{E}$ has an adjoint such that an adjunction can be formed in a straightforward manner. For this reason, it is necessary to extend the probes from the categorical level of $\mathcal{A}$ to the categorical level of diagrams in $\mathcal{A}$, such that the initial probes can be embedded in the latter extended category. This process is accomplished by means of the Yoneda embedding $\curvearrowright$: $\mathcal{A} \longrightarrow Sets^{\mathcal{A}^{op}}$, which constitutes the free completion of $\mathcal{A}$ under taking colimits of diagrams of probes.

The *functor category of presheaves of sets on probes*, denoted by $Sets^{\mathcal{A}^{op}}$, has objects all functors $\mathbb{P} : \mathcal{A}^{op} \to Sets$, and morphisms all natural transformations between such functors, where $\mathcal{A}^{op}$ is the opposite category of $\mathcal{A}$. Each object $\mathbb{P}$ in the category of presheaves $Sets^{\mathcal{A}^{op}}$ is a contravariant set-valued functor on $\mathcal{A}$, called a *presheaf* on $\mathcal{A}$, defined as follows: For each probe Y of $\mathcal{A}$, $\mathbb{P}(Y)$ is a set, and for each arrow $f : C \to Y$, $\mathbb{P}(f) : \mathbb{P}(Y) \to \mathbb{P}(C)$ is a set-theoretic function such that if $p \in \mathbb{P}(Y)$, the value $\mathbb{P}(f)(p)$ for an arrow $f : C \to Y$ in $\mathcal{A}$ is called the restriction of p along f and is denoted by $\mathbb{P}(f)(p) = p \cdot f$.

We notice that each probe Y of $\mathcal{A}$ gives rise to a contravariant Hom-functor $\curvearrowright[Y] := Hom_{\mathcal{A}}(-, Y)$. This functor defines a presheaf on $\mathcal{A}$ for each Y in $\mathcal{A}$. Concomitantly, the functor $\curvearrowright$ is a full and faithful functor from $\mathcal{A}$ to the contravariant functors on $\mathcal{A}$, viz.:

$$(0.5.28) \qquad \curvearrowright : \mathcal{A} \longrightarrow Sets^{\mathcal{A}^{op}},$$

defining the Yoneda embedding $\mathcal{A} \hookrightarrow Sets^{\mathcal{A}^{op}}$.

According to the *Yoneda Lemma*, there exists an injective correspondence between elements of the set $\mathbb{P}(Y)$ and natural transformations in $Sets^{\mathcal{A}^{op}}$ from $\curvearrowright[Y]$ to $\mathbb{P}$ and this correspondence is natural in both $\mathbb{P}$ and Y, for every presheaf of sets $\mathbb{P}$ in $Sets^{\mathcal{A}^{op}}$ and probe Y in $\mathcal{A}$. The functor category of presheaves of sets on probes $Sets^{\mathcal{A}^{op}}$ is a complete and cocomplete category. Thus, the *Yoneda embedding* $\curvearrowright : \mathcal{A} \longrightarrow Sets^{\mathcal{A}^{op}}$ constitutes the sought *free completion of $\mathcal{A}$ under colimits of diagrams of probes*.

The deep meaning of this fact is that if we consider a shaping or coordinatization functor $\mathrm{M} : \mathcal{A} \to \mathcal{E}$ there exists *precisely one* corresponding uniquely defined, up to isomorphism, *colimit-preserving* functor $\widehat{\mathrm{M}} : Sets^{\mathcal{A}^{op}} \to \mathcal{E}$, such that the following diagram commutes:

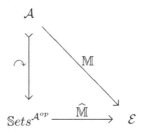

Consequently, every morphism from a probe Y in $\mathcal{A}$ to an indirectly accessible object L in $\mathcal{E}$ factors *uniquely* through the functor category $Sets^{\mathcal{A}^{op}}$ and the computation of the colimit-preserving functor $\widehat{\mathrm{M}} : Sets^{\mathcal{A}^{op}} \to \mathcal{E}$ is crucial for understanding the underlying structure of an object L in $\mathcal{E}$ in terms of diagrams of probes.

The physical significance of the functor $\widehat{\mathrm{M}}$ is unraveled by the fact that it plays the role of a left adjoint $\mathbb{L}$, and thus colimit-preserving functor, from $Sets^{\mathcal{A}^{op}}$ to $\mathcal{E}$. More precisely, the functor $\widehat{\mathrm{M}} := \mathbb{L}$ is the left adjoint of the *categorical adjunction* between the categories $Sets^{\mathcal{A}^{op}}$ and $\mathcal{E}$, where the right adjoint $\mathbb{R} : \mathcal{E} \to \mathbf{Sets}^{\mathcal{A}^{op}}$, is physically interpreted as the realization functor of $\mathcal{E}$ in terms of probes.

More precisely, the *probes-induced realization functor of $\mathcal{E}$ in $Sets^{\mathcal{A}^{op}}$* is defined as follows:

$$(0.5.29) \qquad\qquad \mathbb{R} : \mathcal{E} \rightarrow Sets^{\mathcal{A}^{op}},$$

where the action on a probe Y in $\mathcal{A}$ is given by:

$$(0.5.30) \qquad\qquad \mathbb{R}(L)(Y) := \mathbb{R}_L(Y) = Hom_{\mathcal{E}}(\mathbb{M}(Y), L).$$

The functor $\mathbb{R}(L)(-) := \mathbb{R}_L(-) = Hom_{\mathcal{E}}(\mathbb{M}(-), L)$ is called the *functor of probing frames* of L, where $\mathbb{M} : \mathcal{A} \rightarrow \mathcal{E}$ is a *shaping or coordinatization functor* of $\mathcal{E}$. Since the physical interpretation of the functor $\mathbb{R}(L)(-)$ refers to the functorial realization of the indirectly accessible object L in $\mathcal{E}$ in terms of probes Y in $\mathcal{A}$, we think of $\mathbb{R}(L)(-)$ as the *spectrum functor* of L through the probes Y.

We notice that the probing frames of L, denoted by $\psi_Y : \mathbb{M}(Y) \rightarrow L$, instantiated by the evaluation of the functor $\mathbb{R}(L)(-)$ at each probe Y in $\mathcal{A}$, are not ad hoc but they are inter-related by the operation of presheaf restriction. Explicitly, this means that for each arrow $f : C \rightarrow Y$, $\mathbb{R}(L)(f) : \mathbb{R}(L)(Y) \rightarrow \mathbb{R}(L)(C)$ is a function between sets of probing frames of L in the opposite direction, such that if $\psi_Y \in \mathbb{R}(L)(Y)$ is a probing frame, the value of $\mathbb{R}(L)(f)(\psi_Y)$, or equivalently the corresponding probing frame $\psi_C : \mathbb{M}(C) \rightarrow L$ is given by the restriction or pullback of ψ_Y along f, denoted by $\mathbb{R}(L)(f)(\psi_Y) = \psi_Y \cdot f = \psi_C$.

In this setting, the problem of *approximating* an indirectly accessible object L in $\mathcal{E}$ by means of diagrams of probes Y has a *universal solution*, which is provided by the left adjoint functor $\mathbb{L} : Sets^{\mathcal{A}^{op}} \rightarrow \mathcal{E}$ to the realization functor $\mathbb{R} : \mathcal{E} \rightarrow Sets^{\mathcal{A}^{op}}$. In other words, the existence of the left adjoint functor $\mathbb{L}$ paves the way for an explicit inductive synthesis of an object L in $\mathcal{E}$ by means of appropriate diagrams of probes in a functorial manner.

Theorem 0.25. *There exists a* categorical adjunction *between the categories $Sets^{\mathcal{A}^{op}}$ and $\mathcal{E}$. More precisely, there exists a pair of adjoint functors $\mathbb{L} \dashv \mathbb{R}$ as follows:*

$$(0.5.31) \qquad\qquad \mathbb{L} : Sets^{\mathcal{A}^{op}} \xleftarrow{}\xrightarrow{} \mathcal{E} : \mathbb{R}.$$

First, it is crucial to notice that every presheaf $\mathbb{P}$ in $Sets^{\mathcal{A}^{op}}$ gives rise to a category. *The category of elements of a presheaf* $\mathbb{P}$, denoted by $\int(\mathbb{P}, \mathcal{A})$, has objects all pairs (Y, p), and morphisms $(\acute{Y}, \acute{p}) \rightarrow (Y, p)$ are those morphisms $u : \acute{Y} \rightarrow Y$ of $\mathcal{A}$ for which $p \cdot u = \acute{p}$, that is, the restriction or pullback of p along u is $\acute{p}$. Projection on the second coordinate of $\int(\mathbb{P}, \mathcal{A})$, defines a functor $\int_{\mathbb{P}} : \int(\mathbb{P}, \mathcal{A}) \rightarrow \mathcal{Y}$ called the *split discrete uniform fibration* induced by $\mathbb{P}$, where $\mathcal{A}$ is the base category of the fibration as in the diagram below. We note that the fibers are categories in which the only arrows are identity arrows. If Y is a probe in $\mathcal{A}$, the inverse image under $\int_{\mathbb{P}}$ of Y is simply the set $\mathbb{P}(Y)$, although its elements are written as pairs so as to form a disjoint union.

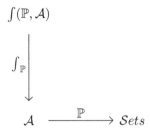

Second, a natural transformation τ between the presheaves $\mathbb{P}$ and $\mathbb{R}(L)$ on the category of probes $\mathcal{A}$, $\tau : \mathbb{P} \longrightarrow \mathbb{R}(L)$ is a family τ_Y indexed by probes Y of $\mathcal{A}$ for which each τ_Y is a map of sets,

$$(0.5.32) \qquad \tau_Y : \mathbb{P}(Y) \rightarrow Hom_{\mathcal{E}}(\mathbb{M}(Y), L),$$

such that the diagram of sets below commutes for each arrow $u : \acute{Y} \rightarrow Y$ of $\mathcal{A}$.

$$\begin{array}{ccc} \mathbb{P}(Y) & \xrightarrow{\tau_Y} & Hom_{\mathcal{E}}(\mathbb{M}(Y), L) \\ {\scriptstyle \mathbb{P}(u)} \downarrow & & \downarrow {\scriptstyle \mathbb{M}(u)} \\ \mathbb{P}(\acute{Y}) & \xrightarrow{\tau_Y} & Hom_{\mathcal{E}}(\mathbb{M}(\acute{Y}), L) \end{array}$$

Third, from the perspective of the category of elements of the presheaf P the map τ_Y, defined above, is identical with the map:

$$(0.5.33) \qquad \tau_Y : (Y, p) \rightarrow Hom_{\mathcal{L}}\left(\mathbb{M} \circ \int_{\mathbb{P}} (Y, p), L \right).$$

Therefore, such a τ may be represented as a family of arrows of $\mathcal{E}$ which is being indexed by objects (Y, p) of the category of elements of the presheaf $\mathbb{P}$, namely

$$(0.5.34) \qquad \{\tau_Y(p) : \mathbb{M}(Y) \rightarrow L\}_{(Y,p)}.$$

Thus, the condition of the commutativity of the above diagram, is equivalent to the condition that for each arrow u the following diagram commutes:

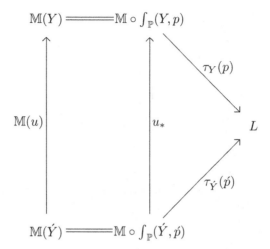

Consequently, according to the above diagram, the arrows $\tau_Y(p)$ form a cocone from the functor $\mathbb{M} \circ \int_{\mathbb{P}}$ to L. The categorical definition of a colimit, points to the conclusion that each such cocone emerges by the composition of the colimiting cocone with a unique arrow from the colimit $\mathbb{LP}$ to L. Equivalently, we conclude that there exists a bijection, which is natural in $\mathbb{P}$ and L:

$$(0.5.35) \qquad Hom_{\mathcal{S}ets^{\mathcal{A}^{op}}}(\mathbb{P}, Hom_{\mathcal{E}}(\mathbb{M}(-), L)) \cong Hom_{\mathcal{L}}(\mathbb{LP}, L),$$

$$(0.5.36) \qquad Hom_{\mathcal{S}ets^{\mathcal{A}^{op}}}(\mathbb{P}, \mathbb{R}(L)) \cong Hom_{\mathcal{L}}(\mathbb{LP}, L)$$

abbreviated as follows:

$$(0.5.37) \qquad Nat(\mathbb{P}, \mathbb{R}(L)) \cong Hom_{\mathcal{L}}(\mathbb{LP}, L).$$

Hence, the probes-induced realization functor of $\mathcal{E}$, realized for each L in $\mathcal{E}$ by the presheaf of probing frames $\mathbb{R}(L) = Hom_{\mathcal{E}}(\mathbb{M}(-), L)$ in $\mathcal{S}ets^{\mathcal{A}^{op}}$, has a left adjoint functor $\mathbb{L} : \mathcal{S}ets^{\mathcal{A}^{op}} \to \mathcal{E}$, which is defined for each presheaf $\mathbb{P}$ in $\mathcal{S}ets^{\mathcal{A}^{op}}$ as the colimit $\mathbb{L}(\mathbb{P})$. Thus, the following diagram, where the Yoneda embedding is denoted by $\curvearrowright$, commutes:

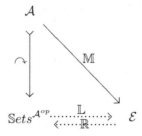

The pair of adjoint functors $\mathbb{L} \dashv \mathbb{R}$ formalizes category-theoretically the functorial process of *encoding and decoding* information between diagrams of probes Y and not directly accessible objects L via the action of probing frames $\psi_Y : \mathbb{M}(Y) \to L$.

Furthermore, the existence of an adjunction between two categories always gives rise to a family of *universal morphisms*, called unit and counit of the adjunction, one for each object in the first category and one for each object in the second. Most importantly, every adjunction gives rise to an *adjoint equivalence* of certain subcategories of the initial functorially correlated categories. It is precisely this category-theoretic fact which determines the necessary and sufficient conditions for the isomorphic representation of an object L in $\mathcal{E}$ by means of suitably restricted functors of probing frames.

For any presheaf $\mathbb{P}$ in the functor category $\mathcal{S}ets^{Y^{op}}$, the *unit natural transformation* of the adjunction is defined as follows:

$$(0.5.38) \qquad\qquad \delta_{\mathbb{P}} : \mathbb{P} \longrightarrow \mathbb{R}\mathbb{L}\mathbb{P}.$$

On the other side, for each object L in $\mathcal{E}$ the *counit natural transformation* of the adjunction is defined as follows:

$$(0.5.39) \qquad\qquad \epsilon_L : \mathbb{L}\mathbb{R}(L) \longrightarrow L.$$

The representation of an object L in $\mathcal{E}$, in terms of the functor of probing frames $\mathbb{R}(L)$ of L, is *full and faithful*, if and only if the counit of the adjunction is an isomorphism, that is structure-preserving, injective and surjective. In turn, the counit of the adjunction is an isomorphism, if and only if the right adjoint functor is full and faithful. In the latter case we characterize the shaping or coordinatization functor $\mathbb{M} : \mathcal{A} \to \mathcal{E}$ as a *proper* or *dense* functor.

0.5.4 *Hom-Tensor Adjunction*

We have shown that formally the left adjoint functor of the adjunction, viz. $\mathbb{L} : \mathcal{S}ets^{\mathcal{A}^{op}} \to \mathcal{E}$, is defined for each presheaf $\mathbb{P}$ in $\mathcal{S}ets^{\mathcal{A}^{op}}$ as the colimit $\mathbb{L}(\mathbb{P})$. The functorial representation of an object L in $\mathcal{E}$ through the category $\mathcal{S}ets^{\mathcal{A}^{op}}$ requires an explicit calculation of this colimit.

For this purpose, it is instructive to explain the general categorical method of calculating the colimit of any functor $\mathbb{X} : \mathcal{I} \to \mathcal{E}$ from some index category $\mathcal{I}$ to $\mathcal{E}$. Let $\mu_i : \mathbb{X}(i) \to \sum_i \mathbb{X}(i)$, i in $\mathcal{I}$, be the injections of $\mathbb{X}(i)$ into their coproduct $\sum_i \mathbb{X}(i)$. A morphism from this coproduct, $\chi : \sum_i \mathbb{X}(i) \to \mathcal{E}$, is determined uniquely by the set of its components $\chi_i = \chi\mu_i$. These components χ_i are going to form a cocone over $\mathbb{X}$ to the vertex L only when for all arrows $v : i \to j$ of the index category $\mathcal{I}$ the following conditions are satisfied:

$$(0.5.40) \qquad\qquad (\chi\mu_j)\mathbb{X}(v) = \chi\mu_i,$$

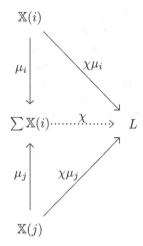

So we consider all $\mathbb{X}(dom\, v)$ for all arrows v with its injections μ_v and obtain their coproduct $\sum_{v:i\to j}\mathbb{X}(dom\, v)$. Next we construct two arrows ζ and η, defined in terms of the injections μ_v and μ_i, for each $v : i \to j$, by the conditions:

(0.5.41) $\zeta\mu_v = \mu_i,$

(0.5.42) $\eta\mu_v = \mu_j\mathbb{X}(v),$

together with their coequalizer χ:

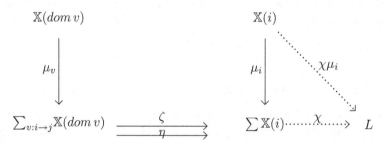

The coequalizer condition $\chi\zeta = \chi\eta$ means that the arrows $\chi\mu_i$ form a cocone over $\mathbb{X}$ to the vertex L. We further note that since χ is the coequalizer of the arrows ζ and η this cocone is the colimiting cocone for the functor $\mathbb{X} : \mathcal{I} \to \mathcal{E}$ from some index category $\mathcal{I}$ to $\mathcal{E}$. Hence, the colimit of the functor $\mathbb{X}$ can be constructed as a coequalizer of coproduct according to the following diagram:

$$\sum_{v:i\to j}\mathbb{X}(dom\, v) \; \underset{\eta}{\overset{\zeta}{\rightrightarrows}} \; \sum\mathbb{X}(i) \overset{\chi}{\longrightarrow} Colim\,\mathbb{X}$$

In our case, we need to calculate the colimit $\mathbb{L}(\mathbb{P})$ for a functor $\mathbb{P}$ in $Sets^{\mathcal{A}^{op}}$. According to the above general scheme, it is necessary to specify the index category corresponding to the presheaf functor $\mathbb{P}$ defined on the base category of probes $\mathcal{A}$.

The index category corresponding to the presheaf functor $\mathbb{P}$ is the category of its elements $\mathcal{I} \equiv \int(\mathbb{P}, \mathcal{A})$, whence the functor $[\mathbb{M} \circ \int_\mathbb{P}]$ plays the role of the functor $\mathbb{X} : \mathcal{I} \to \mathcal{E}$. Hence, we obtain:

$$(0.5.43) \qquad \mathbb{L}(\mathbb{P}) = \mathbb{L}_\mathbb{M}(\mathbb{P}) = Colim \left\{ \int(\mathbb{P}, \mathcal{A}) \xrightarrow{\int_\mathbb{P}} \mathcal{A} \xrightarrow{\mathbb{M}} \mathcal{E} \right\}.$$

In this manner, referring to the coequalizer of coproduct diagram of the general scheme, in the present case, the second coproduct is over all the objects (Y, p) with $p \in \mathbb{P}(Y)$ of the category of elements, while the first coproduct is over all the arrows $v : (\acute{Y}, \acute{p}) \to (Y, p)$ of that category, so that $v : \acute{Y} \to Y$ and the condition $p \cdot v = \acute{p}$ is satisfied. We conclude that the colimit $\mathbb{L}_\mathbb{M}(\mathbb{P})$ can be equivalently presented as the following coequalizer of coproduct:

$$\sum_{v : \acute{Y} \to Y} \mathbb{M}(\acute{Y}) \quad \underset{\eta}{\overset{\zeta}{\rightrightarrows}} \quad \sum_{(Y,p)} \mathbb{M}(Y) \xrightarrow{\chi} \mathbb{L}_\mathbb{M}(\mathbb{P}).$$

In order to analyze in more detail the colimit in the category of elements of $\mathbb{P}$ induced by the shaping functor $\mathbb{M}$, we consider the standard case where there exists a colimit-preserving functor from the category $\mathcal{E}$ to $\mathcal{S}ets$, such that the calculation can be performed set-theoretically. Then the coproduct $\sum_{(Y,p)} \mathbb{M}(Y)$ is a coproduct of sets, which is equivalent to the product $\mathbb{P}(Y) \times \mathbb{M}(Y)$ for $Y \in \mathcal{A}$. The coequalizer is thus equivalent to the definition of the tensor product $\mathbb{P} \otimes_\mathcal{A} \mathbb{M}$ of the set valued factors:

$$(0.5.44) \qquad\qquad \mathbb{P} : \mathcal{A}^{op} \to \mathcal{S}ets, \qquad \mathbb{M} : \mathcal{A} \to \mathcal{S}ets,$$

where the contravariant functor $\mathbb{P}$ is considered as a right $\mathcal{A}$-module and the covariant functor $\mathbb{M}$ as a left $\mathcal{A}$-module, in complete analogy with the algebraic definition of the tensor product of a right $\mathcal{A}$-module with a left $\mathcal{A}$-module over a ring of coefficients $\mathcal{A}$. We call this construction the *functorial tensor product decomposition of the colimit in the category of elements of $\mathbb{P}$ induced by the shaping functor* $\mathbb{M} : \mathcal{A} \to \mathcal{E}$ of $\mathcal{E}$:

$$\sum_{Y, \acute{Y}} \mathbb{P}(Y) \times Hom(\acute{Y}, Y) \times \mathbb{M}(\acute{Y}) \quad \underset{\eta}{\overset{\zeta}{\rightrightarrows}} \quad \sum_Y \mathbb{P}(Y) \times \mathbb{M}(Y) \xrightarrow{\chi} \mathbb{P} \otimes_\mathcal{A} \mathbb{M}.$$

Therefore, we formulate the following theorem:

Theorem 0.26. *The pair of adjoint functors* $\mathbb{L} \dashv \mathbb{R}$ *constitute a* Hom-Tensor *adjunction defined by the natural bijection:*

$$(0.5.45) \qquad Hom_{\mathcal{S}ets^{\mathcal{A}^{op}}} (\mathbb{P}, Hom_\mathcal{E}(\mathbb{M}(-), L)) \cong Hom_\mathcal{E}(\mathbb{P} \otimes_\mathcal{A} \mathbb{M}, L)$$

abbreviated as follows:

$$(0.5.46) \qquad\qquad Nat(\mathbb{P}, \mathbb{R}(L)) \cong Hom_\mathcal{E}(\mathbb{P} \otimes_\mathcal{A} \mathbb{M}, L).$$

According to the above diagram for elements $p \in \mathbb{P}(Y)$, $v : \acute{Y} \to Y$ and $\acute{q} \in \mathbb{M}(\acute{Y})$ the following equations hold:

$$(0.5.47) \qquad \zeta(p, v, \acute{q}) = (p \cdot v, \acute{q}), \qquad \eta(p, v, \acute{q}) = (p, v(\acute{q}))$$

symmetric in $\mathbb{P}$ and $\mathbb{M}$. Hence the elements of the set $\mathbb{P} \otimes_{\mathcal{A}} \mathbb{M}$ are all of the form $\chi(p, q)$. This element can be written as:

$$(0.5.48) \qquad \chi(p, q) = p \otimes q, \quad p \in \mathbb{P}(Y), q \in \mathbb{M}(Y).$$

Thus, if we take into account the definitions of ζ and η above, we obtain:

$$(0.5.49) \qquad p \cdot v \otimes \acute{q} = p \otimes v(\acute{q}), \quad p \in \mathbb{P}(Y), \acute{q} \in \mathbb{M}(\acute{Y}), v : \acute{Y} \to Y.$$

We conclude that the set $\mathbb{P} \otimes_{\mathcal{A}} \mathbb{M}$ is actually the quotient of the set $\sum_Y \mathbb{P}(Y) \times \mathbb{M}(Y)$ by the smallest equivalence relation generated by the above equations, whence the elements $p \otimes q$ of the quotient set $\mathbb{P} \otimes_{\mathcal{A}} \mathbb{M}$ are the equivalence classes of this relation.

Furthermore, if we define the arrows:

$$(0.5.50) \qquad k_Y : \mathbb{P} \otimes_{\mathcal{A}} \mathbb{M} \to L, \qquad l_Y : \mathbb{P}(Y) \to Hom_{\mathcal{E}}(\mathbb{M}(Y), L),$$

they are related under the adjunction by:

$$(0.5.51) \qquad k_Y(p, q) = l_Y(p)(q), \qquad Y \in \mathcal{A}, p \in \mathbb{P}(Y), q \in \mathbb{M}(Y).$$

Here we consider k as a function on $\sum_Y \mathbb{P}(Y) \times \mathbb{M}(Y)$ with components $k_Y : \mathbb{P}(Y) \times \mathbb{M}(Y) \to L$ satisfying:

$$(0.5.52) \qquad k_{\acute{Y}}(p \cdot v, \acute{q}) = k_Y(p, v(\acute{q})),$$

in agreement with the equivalence relation defined above.

Now, we replace back the category $\mathcal{S}ets$ by the category $\mathcal{E}$ under study. The element q in the set $\mathbb{M}(Y)$ is replaced by a generalized element $q : \mathbb{M}(C) \to \mathbb{M}(Y)$ from some object $\mathbb{M}(C)$ of $\mathcal{E}$. Then we consider k as a function $\sum_{(Y,p)} \mathbb{M}(Y) \to L$ with components $k_{(Y,p)} : \mathbb{M}(Y) \to L$ for each $p \in \mathbb{P}(Y)$, which for all arrows $v : \acute{Y} \to Y$ satisfy:

$$(0.5.53) \qquad k_{(\acute{Y}, p \cdot v)} = k_{(Y,p)} \circ \mathbb{M}(v).$$

Then the condition defining the bijection holding by virtue of the Hom-Tensor adjunction is given by:

$$(0.5.54) \qquad k_{(Y,p)} \circ q = l_Y(p) \circ q : \mathbb{M}(C) \to L.$$

This argument, being natural in the object $\mathbb{M}(C)$, is determined by setting $\mathbb{M}(C) = \mathbb{M}(Y)$ with q being the identity map. Hence the bijection takes the simple form:

$$(0.5.55) \qquad k_{(Y,p)} = l_Y(p).$$

As a first application of the above method, we consider the case $\mathbb{P} = \curvearrowright[Y]$, viz. the representable Hom-functor $Hom_{\mathcal{A}}(-, Y)$, represented by the probe Y in $\mathcal{A}$. Then, we have:

$$(0.5.56) \quad \mathbb{L}(\curvearrowright[Y]) = Colim \left\{ \int (\curvearrowright[Y], \mathcal{A}) \xrightarrow{\int \curvearrowright[Y]} \mathcal{A} \xrightarrow{\mathbb{M}} \mathcal{E} \right\} = \curvearrowright[Y] \otimes_{\mathcal{A}} \mathbb{M} \cong \mathbb{M}(Y).$$

Thus, we express the evaluation of the shaping functor $\mathbb{M}$ at each probe Y in $\mathcal{A}$ as the colimit taken in the category of elements of the representable Hom-functor $\frown[Y]$ according to the above.

The next case of interest is the one where $\mathbb{P} = \mathbb{R}(L)$, viz. the functor of probing frames of L, $Hom_{\mathcal{E}}(\mathbb{M}(-), L)$. The significance of the calculation of the colimit $\mathbb{L}(\mathbb{R}(L))$:

$$(0.5.57) \qquad \mathbb{L}(\mathbb{R}(L)) = Colim\{\int (\mathbb{R}(L), \mathcal{A}) \xrightarrow{\int_{\mathbb{R}(L)}} \mathcal{A} \xrightarrow{\mathbb{M}} \mathcal{E}\} = \mathbb{R}(L) \otimes_{\mathcal{A}} \mathbb{M},$$

stems from the fact that if the counit of the Hom-Tensor adjunction:

$$(0.5.58) \qquad\qquad \epsilon_L : \mathbb{L}\mathbb{R}(L) \longrightarrow L,$$

is an isomorphism, that is structure-preserving, injective and surjective, then we obtain a full and faithful isomorphic representation of an object L in $\mathcal{E}$, in terms of the functor of probing frames $\mathbb{R}(L)$ of L. Thus, it is important to specify the conditions which force the counit ϵ_L to be an isomorphism.

The counit natural transformation ϵ_L elucidates the notion of *spectral observation via probes* of an indirectly accessible object L in $\mathcal{E}$ with a precise operational interpretation. More concretely, spectral observation of L constitutes a natural transformation of the identity of L. This natural transformation is expressed concretely by the counit ϵ_L for every L in $\mathcal{E}$. In this sense, spectral observation of an object L, via the functor of probing frames $\mathbb{R}(L)$, where the probes Y in $\mathcal{A}$ play the role of partial or local information carriers classifying the global information content of L, is interpreted as the operational implementation of the counit natural transformation ϵ_L. Thus, the composite endofunctor:

$$(0.5.59) \qquad\qquad \mathbb{G} := \mathbb{L}\mathbb{R}(L) : \mathcal{E} \to \mathcal{S}ets^{\mathcal{A}^{op}} \to \mathcal{E},$$

may be called the *global spectral observation functor* of L via the functor of probing frames $\mathbb{R}(L)$. Notice, that if the counit is an isomorphism, then L can be considered as a fixed point of the corresponding global spectral observation functor $\mathbb{G}$. The counit natural transformation ϵ_L is a natural isomorphism, if and only if the right adjoint functor is full and faithful, or equivalently if and only if the cocone from the functor $\mathbb{M} \circ \int_{\mathbb{R}(L)}$ to L is universal for each L in $\mathcal{E}$. In the latter case, we characterize the functor $\mathbb{M} : \mathcal{A} \to \mathcal{E}$ as a dense shaping or coordinatization functor. In this way, it is important to specify the necessary and sufficient conditions which force the counit ϵ_L to be an isomorphism. This requirement leads to the notion of sheaf-theoretic localization of L through the probing frames $\psi_Y : \mathbb{M}(Y) \to L$.

0.6 Grothendieck Topos Interpretation of the Hom-Tensor Adjunction

We introduce the notion of a *functor of prelocalizations* $\mathbb{T}(L)$ for an object L in $\mathcal{E}$ as follows: $\mathbb{T}(L)$ is defined as a subfunctor of the functor of probing frames $\mathbb{R}(L)$ of L, viz. $\mathbb{T}(L) \hookrightarrow \mathbb{R}(L)$, or equivalently $\mathbb{T}(L)(Y) \subseteq [\mathbb{R}(L)](Y)$ for each probe Y in $\mathcal{A}$. A functor of prelocalizations for an object L evaluated at a probe Y can be expressed in the form of a right ideal $\mathbb{R}(L)(Y) \triangleleft T(L)(Y)$ consisting of prelocalizing probing frames $\psi_Y : \mathbb{M}(Y) \to L$. In more detail, this means that *prelocalizing probing frames of L are characterized by the following property fitting them into* right ideals: If $[\psi_Y : \mathbb{M}(Y) \to L] \in \mathbb{T}(L)(Y)$, and $\mathbb{M}(v) : \mathbb{M}(\acute{Y}) \to \mathbb{M}(Y)$ in $\mathcal{E}$, for $v : \acute{Y} \to Y$ in $\mathcal{A}$, then $[\psi_Y \circ \mathbb{M}(v) : \mathbb{M}(\acute{Y}) \to \mathcal{E}] \in \mathbb{T}(L)(Y)$. A functor of prelocalizations $\mathbb{T}(L)$ of L is equivalently called a *(spectral) sieve* of L.

A family of probing frames $\psi_Y : \mathbb{M}(Y) \longrightarrow L$, Y in $\mathcal{A}$, is the *generator of a right ideal of prelocalizations* $\mathbb{T}(L)(Y)$, if and only if, this ideal is the smallest among all that contains that family. The right ideals of prelocalizations for an L in $\mathcal{E}$ constitute a partially ordered set under inclusion. The minimal right ideal is the empty one, namely $\mathbb{T}(L)(Y) = \emptyset$ for all Y in $\mathcal{A}$, whereas the maximal right ideal is the functor of probing frames $\mathbb{R}(L)$ of L itself.

In order to demonstrate explicitly the implications of the notion of a sieve $\mathbb{T}(L)$, it is essential to calculate explicitly the colimit $\mathbb{L}(\mathbb{T}(L))$:

$$(0.6.1) \qquad \mathbb{L}(\mathbb{T}(L)) = Colim \left\{ \int (\mathbb{T}(L), \mathcal{A}) \xrightarrow{\int_{\mathbb{T}(L)}} \mathcal{A} \xrightarrow{\mathbb{M}} \mathcal{E} \right\} = \mathbb{T}(L) \otimes_{\mathcal{A}} \mathbb{M},$$

for any sieve $\mathbb{T}(L)$. According to the preceding, the sought colimit is expressed as the tensor product:

$$\sum_{Y, \acute{Y}} \mathbb{T}(L)(Y) \times Hom(\acute{Y}, Y) \times \mathbb{M}(\acute{Y}) \quad \xrightarrow[\eta]{\zeta} \quad \sum_Y \mathbb{T}(L)(Y) \times \mathbb{M}(Y) \xrightarrow{\chi}$$

$$\xrightarrow{\chi} \mathbb{T}(L) \otimes_{\mathcal{A}} \mathbb{M}.$$

Thus, for prelocalizing probes $\psi_Y \in \mathbb{T}(L)(Y)$, $v : \acute{Y} \to Y$ and $\acute{q} \in \mathbb{M}(\acute{Y})$ we obtain the following identification equations:

$$(0.6.2) \qquad\qquad \psi_Y \cdot v \otimes \acute{q} = \psi_Y \otimes v(\acute{q}).$$

We conclude that the set $\mathbb{T}(L) \otimes_{\mathcal{A}} \mathbb{M}$ is the quotient of the set $\sum_Y \mathbb{T}(L)(Y) \times \mathbb{M}(Y)$ by the smallest equivalence relation generated by the above equations. Clearly, this equivalence relation generates a right ideal of prelocalizations, where the elements $\psi_Y \otimes q$ of the $\mathbb{T}(L) \otimes_{\mathcal{A}} \mathbb{M}$ are the equivalence classes of this relation. An equivalence class of the form $\psi_Y \otimes q$ demarcates a *maximal connected component* in the category of elements $\int (\mathbb{T}(L), \mathcal{A})$ of the sieve $\mathbb{T}(L)$, where the identification is forced by means of admissible underlying transition arrows according to the above description. Notice that the elements in $\int (\mathbb{T}(L), \mathcal{A})$ are expressed in the form of pairs (ψ_Y, q), and thus, they should be thought of as *pointed probing frames* of L.

The *pullback of the prelocalizing probing frames* $\psi_Y : \mathbb{M}(Y) \longrightarrow L$, Y in $\mathcal{A}$, and $\psi_{\acute{Y}} : \mathbb{M}(\acute{Y}) \longrightarrow L$, $\acute{Y}$ in $\mathcal{A}$, with common codomain L, consists of the common refinement $\mathbb{M}(Y) \times_L \mathbb{M}(\acute{Y})$ and two arrows $\psi_{Y\acute{Y}}$ and $\psi_{\acute{Y}Y}$, called projections, as shown in the following diagram. The square commutes and for any K and arrows h and g that make the outer square commute, there exists a unique $u : \mathbb{M}(\acute{Y}) \longrightarrow \mathbb{M}(Y) \times_L \mathbb{M}(\acute{Y})$ that makes the whole diagram commute.

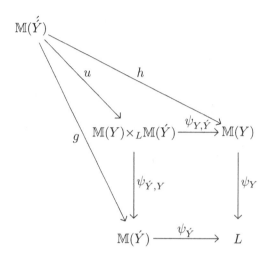

We notice that if the prelocalizing probing frames ψ_Y and $\psi_{\acute{Y}}$ are injective, then their pullback is given by their intersection. Consequently, we define the *gluing isomorphism* of the prelocalizing probing frames ψ_Y and $\psi_{\acute{Y}}$, as follows:

$$(0.6.3) \qquad \Omega_{Y,\acute{Y}} : \psi_{\acute{Y}Y}(\mathbb{M}(Y) \times_L \mathbb{M}(\acute{Y})) \longrightarrow \psi_{Y\acute{Y}}(\mathbb{M}(Y) \times_L \mathbb{M}(\acute{Y})),$$

$$(0.6.4) \qquad \Omega_{Y,\acute{Y}} = \psi_{Y\acute{Y}} \circ \psi_{\acute{Y}Y}^{-1}.$$

From the previous definition, we deduce the satisfaction of the following *cocycle conditions*:

$$(0.6.5) \qquad \Omega_{Y,Y} = id_Y,$$

$$(0.6.6) \qquad \Omega_{Y,\acute{Y}} \circ \Omega_{\acute{Y},\acute{\acute{Y}}} = \Omega_{Y,\acute{\acute{Y}}},$$

$$(0.6.7) \qquad \Omega_{Y,\acute{Y}} = \Omega^{-1}_{\acute{Y},Y}$$

where, in the first condition id_Y denotes the identity of $\mathbb{M}(Y)$, in the second $\psi_Y \times_L \psi_{\acute{Y}} \times_L \psi_{\acute{\acute{Y}}} \neq 0$, and in the third $\psi_Y \times_L \psi_{\acute{Y}} \neq 0$.

Thus, the *gluing isomorphism* between any two prelocalizing frames of a sieve $\mathbb{T}(L)$ assures that $\psi_{\acute{Y}Y}(\mathbb{M}(Y) \times_L \mathbb{M}(\acute{Y}))$ and $\psi_{Y\acute{Y}}(\mathbb{M}(Y) \times_L \mathbb{M}(\acute{Y}))$ probe L on their common refinement in a compatible way. This provides the sought criterion of qualifying prelocalizing frames of $\mathbb{T}(L)$ to localizing ones.

Before we formulate the notion of localizing sieves of L, based on the above criterion characterizing localizing probing frames of L, we need to introduce the notion of *covering sieves* of L. In other words, it is necessary to distinguish among the set of all sieves of L the ones that we will think of as covering sieves of L by means of the following constitutive properties:

[i]. The maximal L-sieve $\mathbb{S}_m(L)$ is a covering L-sieve for any L in $\mathcal{E}$.

[ii]. The covering L-sieves are stable under pullback operations in $\mathcal{E}$, and in particular, the intersection of L-covering sieves is also a L-covering sieve.

[iii]. The covering L-sieves are transitive, such that any two covering L-sieves in $\mathcal{E}$ admit a common refinement.

The above conditions characterizing the properties of covering L-sieves may be equivalently formulated in terms of their *generating families* of probing frames $\{\psi_{Y_j} : \mathbb{M}(Y_j) \to L\}$, Y_j in $\mathcal{A}$, as follows:

[i]. The family of all probing frames of L is a L-covering family.

[ii]. If $\{\mathbb{M}(Y_j) \to L\}$ is a L-covering family and $D \to L$ is any morphism in $\mathcal{E}$, then for every index j in this family, the following diagram is a pullback diagram over L:

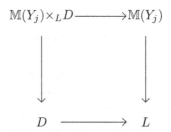

such that $\{\mathbb{M}(Y_j) \times_L D \to D\}$ is a D-covering family.

[iii]. If $\{\mathbb{M}(Y_j) \to L\}$ is a L-covering family and for every index k such that $\{\mathbb{M}(Y_{jk}) \to \mathbb{M}(Y_j)\}$ is an $\mathbb{M}(Y_j)$-covering family, then the collection of composite probing frames $\{\mathbb{M}(Y_{jk}) \to \mathbb{M}(Y_j) \to L\}$ is a L-covering family.

We call any probing frame $\{\psi_{Y_j} : \mathbb{M}(Y_j) \to L\}$, Y_j in $\mathcal{A}$, of a L-covering sieve $\mathbb{S}(L) := \mathbb{S}L$, an $\mathbb{M}(Y_j)$-spectrum *cover* of L belonging to some L-covering generating family of $\mathbb{S}L$. As a consequence of conditions [ii] and [iii], we notice that if $\{\mathbb{M}(Y_j) \to L\}$ is a L-covering family and $\{\mathbb{M}(Y_k) \to L\}$ is another L-covering family, then for each index j in the first family and for each index k in the second, the following diagram is a *pullback diagram* over L:

$$\mathbb{M}(Y_j) \times_L \mathbb{M}(Y_k) \xrightarrow{\psi_{jk}} \mathbb{M}(Y_j)$$

$$\psi_{kj} \downarrow \qquad\qquad\qquad \downarrow \psi_j$$

$$\mathbb{M}(Y_k) \xrightarrow{\quad\psi_k\quad} L$$

and $\{[\mathbb{M}(Y_j)\times_L\mathbb{M}(Y_k)] \to \mathbb{M}(Y_j) \to L\}$, $\{[\mathbb{M}(Y_j)\times_L\mathbb{M}(Y_k)] \to \mathbb{M}(Y_k) \to L\}$ are both L-covering families. Moreover, it is clear that the L-covering family $\{[\mathbb{M}(Y_j)\times_L\mathbb{M}(Y_k)] \to L\}$ is a common refinement of the L-covering families $C_j = \{\mathbb{M}(Y_j) \to L\}$ and $C_k = \{\mathbb{M}(Y_k) \to L\}$. In particular, if both of the L-covering families C_j and C_k consist of injective covers of L, then the L-covering family $C_{jk} = \{[\mathbb{M}(Y_j)\times_L\mathbb{M}(Y_k)] \to L\}$, viz. their common refinement is actually the same as their intersection. Then for each index j in the first family and for each index k in the second one, we define the following morphism:

$$(0.6.8) \qquad \Omega_{jk} : \psi_{kj}(\mathbb{M}(Y_j)\times_L\mathbb{M}(Y_k)) \to \psi_{jk}(\mathbb{M}(Y_j)\times_L\mathbb{M}(Y_k))$$

by the assignment:

$$(0.6.9) \qquad \Omega_{jk} = \psi_{jk} \circ \psi_{kj}^{-1}.$$

If for each index j in the first family and for each index k in the second one the morphism Ω_{jk} is an isomorphism, then we say that the $\mathbb{M}(Y_j)$-covering family $\{[\mathbb{M}(Y_j)\times_L\mathbb{M}(Y_k)] \to \mathbb{M}(Y_j)\}$ is compatible with the $\mathbb{M}(Y_k)$-covering family $\{[\mathbb{M}(Y_j)\times_L\mathbb{M}(Y_k)] \to \mathbb{M}(Y_k)\}$ with respect to L. Now, since the L-covering family $C_{jk} = \{[\mathbb{M}(Y_j)\times_L\mathbb{M}(Y_k)] \to L\}$ is a common refinement of the L-covering families $C_j = \{\mathbb{M}(Y_j) \to L\}$ and $C_k = \{\mathbb{M}(Y_k) \to L\}$, if Ω_{jk} is an isomorphism for each j, k, it is called a *gluing isomorphism*, because $\psi_{kj}(\mathbb{M}(Y_j)\times_L\mathbb{M}(Y_k))$ and $\psi_{jk}(\mathbb{M}(Y_j)\times_L\mathbb{M}(Y_k))$ cover the same part of L in a compatible way. Equivalently, L has compatible $[\mathbb{M}(Y_j)\times_L\mathbb{M}(Y_k)]$-Spectrum covers belonging to the common refinement L-covering family C_{jk} of the L-covering families C_j and C_k.

It is straightforward to show that the gluing isomorphism Ω_{jk} satisfies the following *cocycle compatibility conditions* (whenever they are defined), where j, k refer to the common refinement L-covering family C_{jk}:

$$(0.6.10) \qquad \Omega_{jj} = id_{\mathbb{M}(Y_j)},$$

$$(0.6.11) \qquad \Omega_{jk} \circ \Omega_{km} = \Omega_{jm},$$

$$(0.6.12) \qquad \Omega_{jk} = \Omega^{-1}{}_{kj}.$$

Consequently, we conclude as follows:

Theorem 0.27. *A $\mathbb{M}(-)$-spectrum L-sieve $\mathbb{S}(L)$ is a localizing L-sieve, or equivalently a functor of localizations of L, if and only if it is closed with respect to covers of L and the above cocycle compatibility conditions are satisfied.*

For our purposes, the conceptual significance of a localizing $\mathbb{M}(-)$-spectrum L-sieve, lies on the fact that the functor $\mathbb{R}(L)$ becomes a *sheaf* when restricted to it.

Technically, in order to be able to transform the functor of probing frames $\mathbb{R}(L)$ into a sheaf it is necessary to define a notion of topology on the base category of probes $\mathcal{A}$. This is made possible by restricting the description of covering sieves on the category $\mathcal{A}$ as follows:

We define that a Y-sieve $\mathbb{S}(Y)$ on a probe Y in $\mathcal{A}$ is a covering Y-sieve, if the images of all the arrows $s_j : Y_j \to Y$ belonging to the Y-sieve $\mathbb{S}(Y)$, under the action of the shaping functor $\mathbb{M} : \mathcal{A} \to \mathcal{E}$, together form a covering $\mathbb{M}(Y)$-sieve in $\mathcal{E}$, generated by the $\mathbb{M}(Y)$-covering family $\{\mathbb{M}(Y_j) \to \mathbb{M}(Y)\}$ in $\mathcal{E}$.

The function $\mathbb{J}$ which assigns to each probe Y in $\mathcal{A}$ a set $\mathbb{J}(Y)$ of covering Y-sieves defines a *Grothendieck topology* on the base category $\mathcal{A}$. The category $\mathcal{A}$ together with a topology $\mathbb{J}$ is called a *site*, denoted by $(\mathcal{A}, \mathbb{J})$.

Then, the functor of probing frames $\mathbb{R}(L)$ on $\mathcal{A}$ is a sheaf for $\mathbb{J}$, if and only if for any Y-covering family $\{s_j : Y_j \to Y\}$ belonging to $\mathbb{J}(Y)$, any compatible family of probing frames $\{\psi_j : \mathbb{M}(Y_j) \to L\}$ can be glued together uniquely. Equivalently, any compatible family of frames $\{\psi_j\}$ has a unique amalgamation, in the sense that there exists a unique frame $\psi : \mathbb{M}(Y) \to L$ in the set $\mathbb{R}(L)(Y)$, such that the restriction of ψ at s_j gives ψ_j, that is $\psi \cdot s_j = \psi_j$. If the above holds for a particular covering sieve in the topology, we obtain that the functor of probing frames $\mathbb{R}(L)$ satisfies the sheaf condition, that it becomes a sheaf with respect to this covering sieve.

A technically equivalent formulation of the sheaf condition, see Mac Lane - I. Moerdijk [1], referring in the present context of enquiry to the functor of probing frames $\mathbb{R}(L)$, is described in terms of an equalizer condition, expressed in terms of covering Y-sieves $\mathbb{S}(Y)$, as in the following diagram in $\mathcal{S}ets$:

$$\prod_{f \circ g \in \mathbb{S}(Y)} \mathbb{R}(L)(dom(g)) \;\rightleftarrows\; \prod_{f \in \mathbb{S}(Y)} \mathbb{R}(L)(dom(f)) \;\xleftarrow{\;e\;}\; \mathbb{R}(L)(Y).$$

If the above diagram is an equalizer for a particular covering Y-sieve $\mathbb{S}(Y)$, we obtain that the functor of probing frames $\mathbb{R}(L)$ satisfies the sheaf condition with respect to the covering Y-sieve $\mathbb{S}(Y)$.

A *Grothendieck topos* over the small category of probes $\mathcal{A}$ is a category which is equivalent to the category of sheaves $\mathcal{S}h(\mathcal{A}, \mathbb{J})$ on a site $(\mathcal{A}, \mathbb{J})$. The site can be thought of as a system of generators and relations for the topos. We note that a category of sheaves $\mathcal{S}h(\mathcal{A}, \mathbb{J})$ on a site $(\mathcal{A}, \mathbb{J})$ is a full subcategory of the functor category of presheaves $\mathcal{S}ets^{\mathcal{A}^{op}}$.

The basic properties of a Grothendieck topos are the following:

(1). It admits finite projective limits; in particular, it has a terminal object, and it admits fibered products.

(2). If $(Y_i)_{i \in I}$ is a family of objects of the topos, then the sum $\coprod_{i \in I} Y_i$ exists and is disjoint.

(3). There exist quotients by equivalence relations and have the same good properties as in the category of sets.

0.7 The Grothendieck Topology of Epimorphic Families

As a concrete application of this framework, we define a covering $\mathbb{M}(-)$-spectrum L-sieve $\mathbb{S}(L)$ consisting of a generating family of coordinatizing probing frames $\{\psi_{Y_j} : \mathbb{M}(Y_j) \to L\}$, Y_j in $\mathcal{A}$, which are stable under pullback operations and jointly define an epimorphism in $\mathcal{E}$:

(0.7.1)
$$T : \coprod_{(Y_j \in \mathcal{Y},\, \psi_j : \mathbb{M}(Y_j) \to L)} \mathbb{M}(Y_j) \to L.$$

Concomitantly, we define that a Y-sieve $\mathbb{S}(Y)$ on a probe Y in $\mathcal{A}$ is a covering Y-sieve, if the images of all the arrows $s_j : Y_j \to Y$ belonging to the Y-sieve $\mathbb{S}Y$, under the action of the shaping functor $\mathbb{M} : \mathcal{A} \to \mathcal{E}$, together form an *epimorphic family* in $\mathcal{E}$.

Theorem 0.28. *The specification of covering sieves on the probes Y in $\mathcal{A}$, in terms of epimorphic families of coordinatizing probing frames in $\mathcal{E}$, expressed in terms of the morphism*

(0.7.2)
$$G : \coprod_{(s_j : Y_j \to Y) \in \mathbb{S}Y} Y_j \to Y,$$

such that G is an epimorphism in $\mathcal{E}$, is a Grothendieck topology $\mathbb{J}$ on $\mathcal{A}$, called the Grothendieck topology of epimorphic families.

First of all, we notice that the maximal sieve on each probe Y in $\mathcal{A}$ includes the identity $Y \to Y$, and thus it is a covering sieve. The satisfaction of the transitivity property of the defined covering sieves is obvious. It remains to demonstrate that the covering sieves remain stable under pullback. For this purpose we consider the pullback of such a covering sieve $\mathbb{S}(Y)$ on Y along any morphism $h : Y' \to Y$ in $\mathcal{A}$, according to the following diagram:

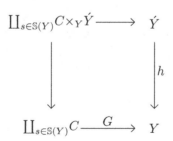

For each morphism $s : C \to Y$ in $\mathbb{S}(Y)$ there exists a morphism $\coprod_{s \in \mathbb{S}Y} [D]^s \to C \times_Y \acute{Y}$ being an epimorphism in $\mathcal{E}$ under the action of the shaping functor $\mathbb{M}$. Equivalently, under $\mathbb{M}$ there exists a jointly epimorphic family of morphisms $\{[D]^s_j \to C \times_Y \acute{Y}\}_j$, where each domain $[D]^s$ is a probe. Consequently, the collection of all the composites:

(0.7.3)
$$[D]^s_j \to C \times_Y \acute{Y} \to \acute{Y}$$

for all $s : D \to Y$ in $\mathbb{S}(Y)$, and all indices j jointly form an epimorphic family in $\mathcal{E}$, which is contained in the sieve $h^*(\mathbb{S}(Y))$, being the pullback of $\mathbb{S}(Y)$ along the morphism $h : \acute{Y} \to Y$. Therefore, the sieve $h^*(\mathbb{S}(Y))$ is a $\acute{Y}$-covering sieve. Thus, the operation $\mathbb{J}$, which assigns to each probe Y in $\mathcal{A}$, a collection $\mathbb{J}(Y)$ of covering Y-sieves, being epimorphic families of arrows in $\mathcal{E}$ under $\mathbb{M}$, constitutes a Grothendieck topology $\mathbb{J}$ on $\mathcal{A}$.

Now, we may express a covering Y-sieve, that is

(0.7.4)
$$G : \coprod_{(s_j : Y_j \to Y) \in \mathbb{S}Y} Y_j \to Y$$

being an epimorphism in $\mathcal{E}$ (under the action of $\mathbb{M}$), equivalently as the pullback of G along itself:

$$
\begin{array}{ccc}
\coprod_{\acute{s}} \acute{C} \times_Y \coprod_s C & \longrightarrow & \coprod_s C \\
\downarrow & & \downarrow {\scriptstyle G} \\
\coprod_{\acute{s}} \acute{C} & \xrightarrow{\ \ G\ \ } & Y
\end{array}
$$

Given that pullbacks in $\mathcal{E}$ preserve coproducts, the epimorphic family associated with a covering sieve on Y, admits the following coequalizer presentation:

$$
\coprod_{\acute{s},s} \acute{C} \times_Y C \ \underset{q_2}{\overset{q_1}{\rightrightarrows}}\ \coprod_s C \ \xrightarrow{\ G\ }\ Y
$$

Further, we notice that for each pair of morphisms $s : C \to Y$ and $\acute{s} : \acute{C} \to Y$ in a covering Y-sieve $\mathbb{S}Y$, there exists (under the action of $\mathbb{M}$) an epimorphic family $\{D \to \acute{C} \times_Y C\}$, such that each domain D is a probe in $\mathcal{A}$. Consequently, each epimorphic family of morphisms associated with a covering Y-sieve $\mathbb{S}(Y)$ may be represented by a commutative diagram of the following form:

$$
\begin{array}{ccc}
D & \xrightarrow{\ \ l\ \ } & C \\
\downarrow {\scriptstyle k} & & \downarrow {\scriptstyle s} \\
\acute{C} & \xrightarrow{\ \ \acute{s}\ \ } & Y
\end{array}
$$

Now we may combine the representation of epimorphic families by commutative squares, together with the coequalizer presentation of the same epimorphic families as follows:

$$\coprod_D D \quad \underset{y_2}{\overset{y_1}{\rightrightarrows}} \quad \coprod_s C \xrightarrow{\ G\ } Y$$

where, the first coproduct is indexed by all D in the commutative squares representing epimorphic families.

Then, for each L in $\mathcal{E}$, we consider the functor of probing frames $\mathbb{R}(L)$ in $\mathcal{S}ets^{\mathcal{A}^{op}}$. If we apply the functor $\mathbb{R}(L)$ to the above coequalizer diagram we obtain an equalizer diagram in $\mathcal{S}ets$ as follows:

$$\prod_D \mathbb{R}(L)(D) \quad \overset{\longleftarrow}{\underset{\longleftarrow}{=\!=\!=}} \quad \prod_{s \in \mathbb{S}Y} \mathbb{R}(L)(C) \quad \longleftarrow \quad \mathbb{R}(L)(Y)$$

where the first product is indexed by all D in the commutative squares representing epimorphic families. Consequently, since the above diagram is an equalizer in $\mathcal{S}ets$ we conclude that the functor of probing frames $\mathbb{R}(L)$ in $\mathcal{S}ets^{\mathcal{A}^{op}}$, satisfies the sheaf condition for the covering Y-sieve $\mathbb{S}(Y)$. Moreover, the equalizer condition holds for every covering sieve in the Grothendieck topology of epimorphic families. By rephrasing the above, we conclude as follows:

Theorem 0.29. *The functor of probing frames $\mathbb{R}(L)$ is a sheaf for the Grothendieck topology of epimorphic families on the base category $\mathcal{A}$ of probes.*

0.8 Unit and Counit of the Hom-Tensor Adjunction

We focus again our attention in the Hom-Tensor adjunction and investigate the unit and the counit of it. For any presheaf $\mathbb{P} \in \mathcal{S}ets^{\mathcal{A}^{op}}$, we deduce that the unit $\delta_{\mathbb{P}} : \mathbb{P} \longrightarrow Hom_{\mathcal{L}}(\mathbb{M}(\_), \mathbb{P} \otimes_{\mathcal{A}} \mathbb{M})$ has components:

(0.8.1) $$\delta_{\mathbb{P}}(Y) : \mathbb{P}(Y) \longrightarrow Hom_{\mathcal{E}}(\mathbb{M}(Y), \mathbb{P} \otimes_{\mathcal{A}} \mathbb{M})$$

for each probe Y of $\mathcal{A}$. If we make use of the representable presheaf $y[Y]$, we obtain:

(0.8.2) $$\delta_{\curvearrowright[Y]} : \curvearrowright[Y] \to Hom_{\mathcal{L}}(\mathbb{M}(\_), \curvearrowright[Y] \otimes_{\mathcal{A}} \mathbb{M}).$$

Hence, for each probe Y of $\mathcal{A}$ the unit, in the case considered, corresponds to a morphism:

(0.8.3) $$\mathbb{M}(Y) \to \curvearrowright[Y] \otimes_{\mathcal{A}} \mathbb{M}.$$

But, since

(0.8.4) $$\curvearrowright[Y] \otimes_{\mathcal{A}} \mathbb{M} \cong \mathbb{M}(Y),$$

the unit for the representable presheaf of probes, which is a sheaf for the Grothendieck topology of epimorphic families, is clearly an isomorphism. By the preceding discussion we can see that the diagram commutes:

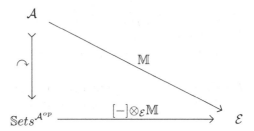

Theorem 0.30. *The unit of the Hom-Tensor adjunction, referring to the representable sheaf $\curvearrowright[Y]$ of the category of probes, $\delta_{\curvearrowright[Y]} : \mathbb{M}(Y) \longrightarrow \curvearrowright[Y] \otimes_{\mathcal{A}} \mathbb{M}$, is an isomorphism.*

Next, we show that the counit natural transformation

$$(0.8.5) \qquad \epsilon_L : \mathbb{L}\mathbb{R}(L) \to L$$

of the Hom-Tensor adjunction is a natural isomorphism if the domain is restricted to the sheaf of probing frames $\mathbb{R}(L)$ (for the Grothendieck topology $\mathbb{J}$ of epimorphic families on $\mathcal{A}$, or equivalently for each covering Y-sieve $\mathbb{S}(Y)$ in $\mathbb{J}$).

For each L in $\mathcal{E}$, the counit is defined as follows:

$$(0.8.6) \qquad \epsilon_L : Hom_{\mathcal{L}}(\mathbb{M}(\_), L) \otimes_{\mathcal{A}} \mathbb{M} \longrightarrow L.$$

The counit corresponds to the vertical map in the following coequalizer diagram:

$$\coprod_{v:Y \to E} \mathbb{M}(Y) \underset{\eta}{\overset{\zeta}{\rightrightarrows}} \coprod_{(E,\psi)} \mathbb{M}(E) \longrightarrow [\mathbb{R}(L)](-) \otimes_{\mathcal{A}} \mathbb{M}$$

$$\searrow \qquad \Big\downarrow \epsilon_L$$

$$L$$

where the first coproduct is indexed by all arrows $v : Y \to E$, with Y, E probes in $\mathcal{A}$, whereas the second coproduct is indexed by all probes Y in $\mathcal{A}$ and probing frames $\psi : \mathbb{M}(E) \to L$, belonging to a covering sieve of L by objects of the category of probes $\mathcal{A}$.

We remind that a covering $\mathbb{M}(Y)$-spectrum L-sieve $\mathbb{S}(L)$ consists of a generating family of coordinatizing probing frames $\{\psi_{Y_j} : \mathbb{M}(Y_j) \to L\}$, Y_j in $\mathcal{A}$, which are stable under pullback operations and jointly define an epimorphism in $\mathcal{E}$:

$$(0.8.7) \qquad T : \coprod_{(Y_j \in \mathcal{Y}, \psi_j : \mathbb{M}(Y_j) \to L)} \mathbb{M}(Y_j) \to L.$$

Equivalently, a covering L-sieve $\mathbb{S}(L)$ admits the following coequalizer diagram presentation in $\mathcal{E}$:

$$\coprod_\nu \mathbb{M}(D) \;\;\underset{w_2}{\overset{w_1}{\rightrightarrows}}\;\; \coprod_\psi \mathbb{M}(C) \xrightarrow{\;\;T\;\;} L$$

where, the first coproduct is indexed by all ν, representing commutative squares in $\mathcal{E}$ corresponding to epimorphic families of coordinatizing probing frames:

$$
\begin{array}{ccc}
\mathbb{M}(D) & \xrightarrow{\;\mathbb{M}(l)\;} & \mathbb{M}(C) \\
{\scriptstyle \mathbb{M}(k)}\big\downarrow & & \big\downarrow{\scriptstyle \psi_{\mathbb{M}(C)}} \\
\mathbb{M}(\acute{C}) & \xrightarrow{\;\psi_{\mathbb{M}(\acute{C})}\;} & L
\end{array}
$$

where D, C, $\acute{C}$ are probes in $\mathcal{A}$.

Next, we show that a covering $\mathbb{M}(Y)$-spectrum L-sieve $\mathbb{S}(L)$ is the coequalizer of (w_1, w_2) if and only if it is the coequalizer of (ζ, η), under the proviso that $\mathbb{R}(L)$ is a sheaf of probing frames.

We notice that the coequalizer condition

$$(0.8.8) \qquad T \circ w_1 = T \circ w_2,$$

is equivalent to the condition:

$$(0.8.9) \qquad T_{(\mathbb{M}(C), \psi_{\mathbb{M}(C)})} \circ \mathbb{M}(l) = T_{(\mathbb{M}(\acute{C}), \psi_{\mathbb{M}(\acute{C})})} \circ \mathbb{M}(k)$$

for each commutative square ν of the form above.

Correspondingly, the coequalizer condition $T \circ \zeta = T \circ \eta$ is equivalent to the condition:

$$(0.8.10) \qquad T_{(\mathbb{M}(C), \psi_{\mathbb{M}(C)})} \circ \mathbb{M}(u) = T_{(\mathbb{M}(\acute{C}), \psi_{\mathbb{M}(C)} \circ \mathbb{M}(u))}$$

for every morphism of probes $u : \acute{C} \to C$, with $\acute{C}$, C probes in $\mathcal{A}$ and $\psi : \mathbb{M}(C) \to L$, a coordinatizing probing belonging to a covering sieve of L. Therefore, our claim is proved if we show that the above conditions are equivalent.

On the one hand, $T \circ \zeta = T \circ \eta$, implies for every commutative diagram of the form ν:

$$
\begin{array}{ccc}
\mathbb{M}(D) & \xrightarrow{\;\mathbb{M}(l)\;} & \mathbb{M}(C) \\
{\scriptstyle \mathbb{M}(k)}\big\downarrow & & \big\downarrow{\scriptstyle \psi_{\mathbb{M}(C)}} \\
\mathbb{M}(\acute{C}) & \xrightarrow{\;\psi_{\mathbb{M}(\acute{C})}\;} & L
\end{array}
$$

where D, C, $\acute{C}$ are probes in $\mathcal{A}$, the following relations hold:

$$(0.8.11) \qquad T_{(\mathbb{M}(C),\psi_{\mathbb{M}(C)})} \circ \mathbb{M}(l) = T_{(\mathbb{M}(\acute{C}),\psi_{\mathbb{M}(\acute{C})})} \circ \mathbb{M}(k).$$

On the other hand, the coequalizer condition $T \circ w_1 = T \circ w_2$, implies that for every morphism $\mathbb{M}(u) : \mathbb{M}(\acute{C}) \to \mathbb{M}(C)$, with C, $\acute{C}$ probes in $\mathcal{A}$ and $\psi_{\mathbb{M}(C)} : \mathbb{M}(C) \to L$, the diagram of the form ν below:

$$
\begin{array}{ccc}
\mathbb{M}(\acute{C}) & \xrightarrow{\;\mathbb{M}(u)\;} & \mathbb{M}(C) \\[2pt]
{\scriptstyle id}\Big\downarrow & & \Big\downarrow{\scriptstyle \psi_{\mathbb{M}(C)}} \\[2pt]
\mathbb{M}(\acute{C}) & \xrightarrow{\hspace{2cm}} & L
\end{array}
$$

commutes and provides the condition:

$$(0.8.12) \qquad T_{(\mathbb{M}(C),\psi_{\mathbb{M}(C)})} \circ \mathbb{M}(u) = T_{(\mathbb{M}(\acute{C}),\psi_{\mathbb{M}(C)} \circ \mathbb{M}(u))}.$$

Consequently, the pairs of arrows (ζ, η) and (w_1, w_2) have isomorphic coequalizers, proving that the counit of the Hom-Tensor adjunction restricted to sheaves for the Grothendieck topology of epimorphic families on $\mathcal{A}$ is an isomorphism.

Theorem 0.31. *The epimorphism $T : \coprod_{(Y_j \in \mathcal{Y}, \psi_j : \mathbb{M}(Y_j) \to L)} \mathbb{M}(Y_j) \to L$ that defines a covering L-sieve, induces an isomorphism in $\mathcal{E}$: $\mathbb{LR}(L) \cong L$ if the functor of probing frames $\mathbb{R}(L)$ is a sheaf for the Grothendieck topology $\mathbb{J}$ of epimorphic families. Equivalently, the counit natural transformation $\epsilon_L : \mathbb{LR}(L) \to L$ of the Hom-Tensor adjunction is a natural isomorphism if restricted to the sheaf $\mathbb{R}(L)$ for this topology $\mathbb{J}$.*

Hence, the global information content of an object L in $\mathcal{E}$ is completely described up to isomorphism in terms of the colimit $\mathbb{LR}(L)$ taken in the fibred category $\int(\mathbb{R}, \mathcal{A})$ of the sheaf of probing frames $\mathbb{R}(L)$, because the counit ϵ_L of the Hom-Tensor adjunction for every covering L-sieve $\mathbb{S}(L)$ is an isomorphism.

As a consequence of the unit and counit isomorphism, we obtain the following:

Theorem 0.32. *The Hom-Tensor adjunction can be restricted to a dual functorial equivalence of categories:*

$$(0.8.13) \qquad \mathbb{L} : \mathcal{S}h(\mathcal{A}, \mathbb{J}) \underset{\longrightarrow}{\overset{\longleftarrow}{\rule{1.5cm}{0pt}}} \mathcal{E} : \mathbb{R},$$

$$(0.8.14) \qquad \mathbb{L} : \mathcal{S}h(\mathcal{A}, \mathbb{J}) \cong \mathcal{E} : \mathbb{R}.$$

Thus, the category $\mathcal{E}$ can be made equivalent to the category of sheaves $\mathcal{S}h(\mathcal{A}, \mathbb{J})$ on the site $(\mathcal{Y}, \mathbb{J})$.

We note that the Grothendieck topos extension of the Hom-Tensor adjunction according to the above general functorial framework has been applied for the sheaf-theoretic localization and representation of quantum logical event algebras via local Boolean probing measurement frames, see Zafiris [1, 2, 5, 6, 10] and Zafiris-Karakostas [1], quantum measure algebras via local Boolean probabilistic frames, see Zafiris [9], and quantum observables algebras via local commutative observables frames, see Zafiris [3, 4, 7, 8] and Epperson-Zafiris [1].

It is instructive to summarize the previous discussion as follows: The problem of approximating an indirectly accessible object L in $\mathcal{E}$ by means of diagrams of probes Y has a *universal solution*, which is obtained by the Hom-Tensor adjunction. More precisely, the universal solution is given by the left adjoint functor $\mathbb{L} : \mathcal{S}ets^{\mathcal{A}^{op}} \to \mathcal{E}$ to the realization functor $\mathbb{R} : \mathcal{E} \to \mathcal{S}ets^{\mathcal{A}^{op}}$. If the functor of probing frames $\mathbb{R}(L)$ of L is restricted to a localization functor, or equivalently to a sheaf for the Grothendieck topology $\mathbb{J}$ of epimorphic families, then the counit natural transformation ϵ_L of the Hom-Tensor adjunction is a natural isomorphism if restricted to the sheaf $\mathbb{R}(L)$ for this topology $\mathbb{J}$ and the right adjoint is a full and faithful functor. It is useful to recapitulate the role of localization functors by means of the following diagram displaying the function of the counit of the Hom-Tensor adjunction:

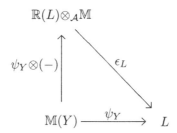

We conclude that the process of *spectral observation via probes* of an indirectly accessible object L in $\mathcal{E}$ by means of a localization functor $\mathbb{R}(L)$ constitutes a natural isomorphism of the identity of L. Therefore, under the hypothesis that there exists a colimit-preserving functor from the category $\mathcal{E}$ to $\mathcal{S}ets$, the elements of an object L can be represented isomorphically in terms of equivalence classes or germs $\psi_Y \otimes q$, $q \in \mathbb{M}(Y)$, of pointed probing frames qualified as local covers of L. Thus, the action of localization functors on L makes the category $\mathcal{E}$ a reflection of $\mathcal{S}ets^{\mathcal{Y}^{op}}$, which can be further identified with the category of sheaves $\mathcal{S}h(\mathcal{A}, \mathbb{J})$ on the site $(\mathcal{Y}, \mathbb{J})$.

Chapter 1

General Theory

"... God always geometrizes."
(Plato)*

"... find a purely algebraic theory for the description of reality."
(A. Einstein)

1.0 General Introduction

The *principle in the title** of this work refers to the fact that:

(0.1) though our *descriptions of physical quantities* may be given *in terms of non-commutative algebraic structures* [see e.g. $\mathcal{E}nd\,\mathcal{E}$ below], our *measurements are* always *effectuated through commutative ones* (!). [Cf. $\mathcal{E}nd\,\mathcal{E} \cong M_n(\mathcal{A})$, *locally*, with $\mathcal{A}$ *commutative*; see (9.25) in the sequel].

We note that this happened already from the early years of classical quantum mechanics, see e.g. *von Neumann/C^*-algebras* and the corresponding *spectra* therein, even *smooth $\mathbb{C}$-valued functions*, of the operators involved. This fact has still been promptly emphasized by Niels Bohr, therefore the denomination applied above. [Now, just for a historical perspective, in that context, I still wanted to mention here that I about used in effect the previous remarks, as in (0.1), pertaining to the principle at issue, in a meeting on Differential Geometry in Patras (Greece), in 1999 already, without actually knowing (!) its relevance with Bohr's dictum; so it was Yannis Raptis who, sometime later on, kindly drew my attention to that fact. Yet, it is still good to say here, in connection also with the ensuing discussion, that the scholium, as in (0.1), was actually my response to a relevant question

---

*One might say here that Plato, at the time, spoke actually of *"geometry"* in, what we may rather construe as, a *"Pythagorean perspective"* (viz. "cosmos (: universe)"-geometry-arithmetic/ number (thus still e.g., *Weil conjectures*); see also, for instance, in that context, a relevant (terminology employed in the) correspondence of Pascal-Fermat (!).

*The original title of Mallios's handwritten notes has been *"Bohr's Correspondence Principle (: the Commutative Substance of the Quantum), Abstract (: Axiomatic) Quantum Gravity, and Functor Categories"*.

of a physicist, pertaining to *"Non-commutative Geometry"*, during my talk in the aforementioned meeting]. For convenience of reference, we also cite below *Bohr's adage*:

(0.2)   *". . . the description of our own measurements of a quantum system must use classical, commutative c-numbers . . ."*

(emphasis above is ours); see e.g. A. Mallios [GT, Vol. II: p. 143]. In this context, we still note that an analogous aspect has been formulated, in a way of post-anticipating N. Bohr, as before, by G. Ludwig [2: p. vii], in that (emphasis below is ours);

(0.3)   *". . . it is necessary to use the classical mode of description in order to describe . . . the measurement process in quantum mechanics."*

Indeed, as we shall see, the said principle permeates essentially all that follows, concerning the two other items in the title of the present account. That is, that which we might call *"abstract quantum gravity"*, together with the rôle, within that context, of a basic result, regarding the notion of *"colimits"* in the theory of *"functor categories"* (see [MM: p. 13]). So by referring to the first of the above two issues, as in the title, by the term *"abstract"*, we mean the *absence of any notion of "space"*, in the classical sense of this issue. We thus treat the subject of *quantum gravity* quite *axiomatically*, and as we shall see, *in* pure *categorical terms*, viz. quite *algebraically*; thus in accordance with the *classical demand*, as well: see, for instance, the relevant epigraph in the RHS, under the title of this study. On the other hand, the present discussion can still be construed, as a *natural continuation of our* previous similar *treatise in* A. Mallios [18]. Thus, we come, straight away, by the ensuing account to analyze and also add further commentary on that perspective. The initiative to this was the profound relevant study of E. Zafiris [7]; yet, see also A. Mallios-E. Zafiris [3], as well as the recent detailed monograph by Zafiris [8], while the same work in A. Mallios-E. Zafiris [1], [2] was still our initial motive for relevant thoughts in our treatise in Mallios [18], as this has been already mentioned therein.

Now, as a prelude to what follows, we also discuss, by the next Section, certain basic consequences of the point of view, in general, that we follow, so far, throughout our study. On this we still return later on, via the subsequent discussion (see, for instance, Section 1.10 in the sequel), indicating thus further potential applications of the present general perspective, vis-à-vis to fundamental standard results of the *classical theory* (CDG: *Classical Differential Geometry* on smooth (C^∞-)manifolds). Yet, within the same vein of ideas, one might still recall herewith the relevant *Feynman's aphorism*,

(0.4)   *". . . the greatest discoveries abstract away from the model* and the model never does any good."

(The emphasis above is ours). See R. P. Feynman [1: p. 57].

1.1 Basic Assumptions of ADG (: Abstract Differential Geometry)

The *fundamental principle/axiom* of the whole perspective *of ADG* is that the *Nature* (alias, by using the Greek word, *Physis*) *is given*, together with *its function;* thus, equivalently, the *natural laws* are given. So to state it in one word,

(1.1)   the *Physis*/Nature *and* (its function, viz.) the *physical laws are given.*

Therefore, as a *fundamental conclusion*, and in effect, as an *equivalent way of saying* (1.1), we note that,

(1.2)   *Physis*, hence *the physical laws*, as well, that is its function, *is indepen-dent of us!*

Now, the phrase *"independent of us"*, as above, refers of course to the particular *manner we observe*/understand *the physical laws* (i.e., as already said, the function of Physis); hence, the *Nature* itself, according to the *physical theories*, we afford at each particular period of time, based as usual on the so-called *"mathematical-physical" theories* we dispose/display (but, how else (!)). Thus, we are led to say that we usually refer to

(1.3)   *the physical laws, as we* actually *understand*/detect *them*, on the basis of *our own* (physical) *theories*. That is, *theories* aiming always at *describing*, not (!) explaining *Nature*; to recall here too, for instance, *N. Bohr* in that (by paraphrasing him),

(1.3.1)   the task of *physics* is to see *what we can tell about Nature*, *not* (!) to say *how Nature is.*

Or even, L. Wittgenstein [2: p. 17],

(1.3′)   *"Physics does not explain anything, it simply describes concomitant cases."*
(Emphasis above is ours).

So, here comes the symbol "$\mathcal{A}$", collectively indicating in the sense of ADG, our *"arithmetic $\mathcal{A}$"*, yet, the (*spaceless* (!)) *"Calculus"*, in the broadest sense of this term,
that we employ throughout ADG. Thus, we get at the connotation,

(1.4)   "$\mathcal{A}$" ⟵——————⟶ *we*

(alias our *"antenna"* (!), so to say), often used all along the treatise of ADG (see also e.g. A. Mallios [18: p. 1932, (1.12)]).

As a result, in view of the preceding, one can now conclude further, still the celebrated classical,

(1.5)   *principle of general covariance*, hence *that of General Relativity*, as well, just, as a *spin-off of* (1.2).

In that context, see also, for instance, R. Torretti [1: p. 153], or even M. Nakahara [1: p. 28]. Therefore, in other words, the same principles, as above, are but

(1.6)   an outcome of our ability to *properly participate Nature* in its function, *via a suitably chosen*, at each particular case, *arithmetic/Calculus* $\mathcal{A}$.

Thus, our experience so far is that,

(1.7)   *our* conclusions/*results* (based on *observation*/experiments, *technical* or, even "*Gedanken*" – ones), therefore, still the so-called "*differential equations*" emanated therefrom, should be "$\mathcal{A}$-*invariant*"; yet, "*tensorial*", hence, at the very end, "*functorial*" in character! [The previous three adjectives, which are associated with the word "*results*", as before, are characteristic of each other, *explaining* thus *the meaning of* the term "$\mathcal{A}$-*invariant*". In that respect, see also (1.12)/(1.13) in the sequel]. Consequently, the *universal validity* of the same (results), being thus, in that way, the description/"*representation*" of the physical laws at issue.

Of course, the important item here is *the way* we do the aforesaid procedure; namely, that of *choosing/concocting*, in effect, our "*arithmetic*" $\mathcal{A}$. Thus, based again on our experience thus far, we are very tempted, indeed, to say, in that context, that;

(1.8)   the more our *choice/concoction of* $\mathcal{A}$, as above, is *functorial*, the *more akin* our description of Nature is *to the real essence of things* (: the "*reality*", viz. that, which we are trying to describe; cf. also (1.3′)).

So in this context, the intervention here of ADG is proved, thus far, to be able to provide us with the appropriate *mechanism*, the same being *absolutely functorial* (!), in nature, hence, with what we might also characterize, as a "*relational*" one: In this respect, see thus A. Mallios [17: p. 267, (2.6)], along with [20]. Consequently, as it concerns the whole perspective of ADG, we can also remark that,

(1.9)   "*analysis*" therein, to the extent that we understand it, classically speaking, does *not* arrive at all from any notion of "*space*", *whatsoever*!

In this context, cf. also previous relevant remarks, yet, in conjunction with potential applications of this *general point of view of ADG* in problems of *quantum gravity*, in A. Mallios [17: p. 270, (3.5)-(3.7)], along with [16: p. 1583, (4.32.1), (4.33)].

On the other hand, in connection with our previous remarks in (1.5), and in anticipating our later discussion in Section 1.10 below, we can still note here that:

(1.10)   the same *principle of general covariance* (see comments following (1.5)) seems that it can be expressed, *equivalently*, via the so-called *transformation law of potentials* (see (1.11) below).

Now, classically, the aforesaid law characterizes the existence of an $(\mathcal{A}-)$ *connection*, hence, within the context of ADG, that of a *Yang-Mills field*, in general (see (4.6), (8.15) and (10.3) in the sequel). On the other hand, technically speaking the same law concerns, in that context, the relation,

$$(1.11) \qquad \qquad \delta(\omega) = \tilde{\vartheta}(g).$$

For the notational details, concerning the last relation, we refer to (10.1), (10.2) and (10.8) below; see also (8.5). Yet, full details for the same relation can be found in A. Mallios [VS, Vol. II: p. 116, Theorem 3.1, and p. 119, Theorem 3.2, in particular, (3.17) therein], along with [GT, Vol. I: p. 17, (2.56), and p. 20, (2.71)]. It is worth noticing here that the said law is still applied in what we may call a "*cohomological classification*" of elementary particles (here, precisely, of *Maxwell fields*; ibid., p. 219, Lemma 4.1. Yet, an analogous classification, pertaining to *Yang-Mills fields*, in general, can also be found in the latter treatise, as above: Vol. II, Chapt. I; p. 70, Section 9).

All told, it is the *principle of general covariance* (acronym, PGC), which thus appears to express the way, the *physical laws/"dynamics"* (see also below, together with A. Mallios [17: p. 280, (6.16.1), and 270, (3.3)]) are revealed to us, the same being also a very *characterization of their existence*; cf. , for instance, the afore-mentioned *transformation law of potentials* (acronym, tlp), as in (1.11), concerning the existence of $\mathcal{A}$-*connections* (: "*dynamics*". See also A. Mallios [16: p. 1569, (2.23.1), along with (2.22.1)]). In this context, cf. further A. Mallios [18: p. 1931, (1.7), and the comments following it], together with [17: p. 270, (3.5)-(3.7)]. Yet, cf. also Section 1.10 in the sequel.

On the other hand, as already said (see (1.1)),

(1.12)   *Physis/Nature is given*, which *Physis*, hence *the physical laws*, as well (: viz. *its function*, ibid.), are also "*functorial*"! (In this context, see also (1.7)/(1.8), as above, together with A. Mallios [18: p. 1930, (1.5)]).

Now, by further commenting on the previous statement, we should note here that, *strictly speaking*,

(1.13)   it is *not* actually *Physis, which is functorial* (cf., just, (1.1)/(1.2)); *in point of fact*, this is *the way we realize* its function, viz. *the physical laws*; namely, *as* "$\mathcal{A}$-*invariant*", this being still *the* (only) *way* (so far) *we detect them*! (See also ibid. p. 1930, (1.3), (1.3′)). So, it is *essentially we*, who should behave *accordingly*; i.e., in a "*functorial* (yet, *discrete* (!), toward its function) *way*", by suitably choosing "$\mathcal{A}$" (: *our presence* (!)). Cf. also (1.4)/(1.6), as above, together with the same Ref., as before, p. 1930, (1.5)/(1.6).

Now, it is good to comment here, a bit more, on our previous remark, pertaining to *our behavior* (see also (1.20) below) with respect to Nature: So participate Nature, thus *its function*, that is, finally the *physical laws*, means literally, that

(i)   *we respect Her/them.*

(1.13.1)   Therefore, technically speaking, this amounts to the fact that we do satisfy (i), as before, whenever we are able to

(ii)   *be "A-invariant".*
Precisely, the latter refers, of course, to our *equations* that are based thus on our theories, we afford at each particular time (see also (9.3) in the sequel).

Therefore, it is in that sense, as above, that one can say (or even, this is actually the *meaning of saying* that,

(1.14)   *Physis is "functorial"*, hence, the *physical laws*, as well.

Now, as a result of (1.12)-(1.14), we thus arrive at the remark that,

(1.15)   (1.12) is still another *equivalent* formulation of the *principle of general covariance*. The latter still follows from saying that

(1.15.1)   *"Physis is A-invariant".*

See also ibid., p. 1931, (1.7). Thus, in anticipation of our discussion in Section 1.9 below, one can still remark here that,

(1.16)   *fundamental equations* in nowadays Physics, as e.g. *Einstein, Yang-Mills*, or even, more generally, *Utiyama's equation*, are but, particular (*"technical"*) *realizations of the same principle*, as before.

See thus (9.51)/(9.42) in the sequel. Yet, regarding the aforementioned two first equations, as above, within the framework of ADG, cf. also A. Mallios [GT, Vol. I-II]; moreover, see [5: p. 167, (2.3)/(2.3′)] for the *Yang-Mills equation(s)*, within the same context, as before.

In this context, it is also instructive to recall here too, Einstein's motto, in that (emphasis below is ours);

(1.17)   "... *Nature* is the realization of the *mathematically simplest*".

See, for instance, R. Torretti [1: p. 283]. Furthermore, within this same vein of ideas, see also our later discussion, pertaining to the famous so-called *"principle of least action"* : cf. thus, for instance, (6.3), (9.10) in the sequel, along with Section 1.10, in general.

Thus, by paraphrasing the preceding, one can still say that,

(1.18)   any time we are able to be *mathematically simple*, then *we mimic*, in effect, in that concern, *Nature herself*! Indeed, then, we *"best approximate"* Her! Yet, the *level of simplicity*, we achieve, characterizes that of our being *more efficient in* our *description* (always, *of course* (!), *not explanation* (see thus (1.3.1) in the preceding) *of Nature*/Physis.

Therefore, still, to tell it in other words, we can also say that,

(1.19) the *"mathematically simplest"* corresponds to a *most appropriate description/"realization"* of Nature.

On the other hand, the preceding can be further associated, appropriately related, with another *fundamental notion*, in connection with the above, that we also discuss in the sequel (see thus Section 1.2 below); namely, that one of an *"adjunction"*: technical details will be provided later on. Yet, in this context, the following telling apostroph/definition of J. A. Goguen (1971) is quite relevant; that is, the

(1.20) *"minimal realization"* is a *functor, left adjoint to* (the *functor* of [*"natural"* (*:physical*)]) *"behavior"*. Or even, *"realization is universal"*. We call this,

(1.20.1) *Goguen's principle*; or even, *Goguen's adjunction*.

The above is to be related with the aforementioned *"least action principle"* (cf. the remarks following (1.17), along with (1.13)), the same being, otherwise, a kind of an *adjunction*. In this regard, see also S. Mac Lane [1: p. 87].

1.2 Basic Framework

Continuing now as in the previous Section, we start by pointing out our *basic set-up*: thus, we first consider an arbitrary, in principle, category,

$$(2.1) \qquad\qquad\qquad \mathcal{E},$$

which, as we shall see in the sequel, we are going to describe/*"approximate"* by another, more concrete, *"small"* category,

$$(2.2) \qquad\qquad\qquad \mathcal{A}.$$

Concerning undefined categorical notions, we refer unless otherwise indicated, to S. Mac Lane [1]. So the latter category, as in (2.2), is by assumption a *small subcategory of* $\mathcal{E}$ (hence, the objects of $\mathcal{A}$ yield a *set*), in such a manner that, we first have the (functor) inclusion,

$$(2.3) \qquad\qquad\qquad \mathcal{A} \underset{\to_i}{\subseteq} \mathcal{E}.$$

Now, the above *inclusion functor i* will also be, at occasions, the so-called (Zafiris), *"coordinatization functor"* of the objects of $\mathcal{A}$, with respect to those (in effect, their images through i), which underlie/*"coordinatize"* them in $\mathcal{E}$ (see also (2.13) below). On the other hand, *we* further *stipulate* for $\mathcal{A}$ that, the (coordinatization) *functor i*, as in (2.3), is *full* and *faithful*, while the small (sub)category $\mathcal{A}$ is still *separating* (alias, *generating*; cf. also (2.6) below). Thus,

(2.4)   (2.4.1) $\mathcal{A}$ is a *full separating* (small) *subcategory* of $\mathcal{E}$. Yet, we still note, in anticipation, that, furthermore

(2.4.2) $\mathcal{A}$ is a "*dense*" subcategory of $\mathcal{E}$.

See thus (3.14) in the sequel.

Therefore, the previous assumptions/properties of $\mathcal{A}$, indeed of the pair $(\mathcal{A}, i)$, entail that,

(2.5)   *one knows the* (sub)*category* $\mathcal{A}$ by simply *knowing* the set of *its objects*, $\mathcal{O}b(\mathcal{A})$; the *arrows in* $\mathcal{A}$ are the "*same*" (: in bijective correspondence) *as* (their images through *i*) *in* $\mathcal{E}$.

See also S. Mac Lane [1: p. 15]. Of course, the *faithfulness* (of the functor *i*) *is* already *redundant*, in view of (2.3), i.e. $\mathcal{A}$ is a subcategory of $\mathcal{E}$ (ibid.). On the other hand, the "*separation*" property of $\mathcal{A}$ (cf. (2.4)) means, by definition, that;

(2.6)   any two different and "*parallel*" *arrows* in $\mathcal{E}$, can be discerned by an arrow of $\mathcal{A}$.

For the technical aspect of the previous terminology, see e.g. S. Mac Lane-I. Moerdijk [1: p. 576]. Now, given that $\mathcal{A}$ is, by assumption (cf. (2.2)), a *small category*, the above *condition* (2.6) *still defines* (equivalently) the *set of objects of* $\mathcal{A}$, $\mathcal{O}b(\mathcal{A})$, therefore,

(2.7)   *the* same *category* $\mathcal{A}$ (see (2.5)), as a "*generating*" *subcategory of* $\mathcal{E}$ (ibid.).

Now, the latter notion is, as we shall see, really important for what follows. Indeed, what *one can* just *define through the arrows of* $\mathcal{A}$, is *the following* (family of) *sets*; namely,

for every object E *of* $\mathcal{E}$, one defines the *set* (cf. (2.2)/(2.3));

(2.8)

$$epi(E) := \bigcup_{A \in \mathcal{O}b(\mathcal{A})} (Hom_{\mathcal{E}}(i(A), E) \cong Hom_{\mathcal{E}}(A, E))$$

(2.8.1)

$$= [\mathcal{E} \downarrow_E; i(\mathcal{A})] \cong [\mathcal{E} \downarrow_E; \mathcal{A}].$$

Concerning the "*comma category*" appeared in the last two terms of (2.8.1), see also e.g. S. Mac Lane [1: p. 46ff]. On the other hand, by referring to the non-technical/ "*physical*" perspective (:*physical significance*), one gets through the same relation (2.8.1), we note that;

(2.9) for every object E of $\mathcal{E}$ (: "*quantum domain*", see also below), we collect, via $Hom_{\mathcal{E}}(A, E) \cong Hom_{\mathcal{E}}(i(A), E)$, all the "*information*" we can have *for it*, by resorting to (our *commutative* "*arithmetic*" (cf. (0.2)/(0.3), still (1.4)) $\mathcal{A}$; that is, *for any A, object of $\mathcal{A}$.*

Equivalently, we can still express the above, by saying that:

(2.9′) *we observe* the object $E \in \mathcal{O}b(\mathcal{E})$, at issue (we employed, just before, for convenience, an obvious "abuse of notation", in general), *through* (the objects of) $\mathcal{A}$ [*as it* actually *were* (!), in effect]; precisely, in terms of their images in $\mathcal{E}$, via the functor i, as in (2.3). Thus, in applying physical parlance; in terms of "*experiments*", through the objects of our basic "*small*" category, $\mathcal{A} \underset{\rightarrow i}{\subset} \mathcal{E}$ (see (2.2), (2.3)).

In this context, cf. also (2.4.2) in the preceding, or even, more precisely, (3.9) in the sequel, still pertaining to the rôle of $\mathcal{A}$ in $\mathcal{E}$.

Now, we call the *set epi(E)*, with $E \in \mathcal{O}b(\mathcal{E})$, as in (2.8.1), an *epimorphic family* (of arrows) on (or even, a *covering* of) E. Indeed, it is via such "*epimorphic families*", that one can endow the given (small) category $\mathcal{A}$, as in (2.3), with a so-called *Grothendieck topology*, say $\mathcal{J}$. Thus, the pair,

(2.10) $$(\mathcal{A}, \mathcal{J}),$$

constitutes what we denominate, as a *site*; see e.g. [MM: p. 110]. So one can further consider the *category of sheaves on it*, denoted by

(2.11) $$Sh(\mathcal{A}, \mathcal{J}).$$

We thus obtain the prototype for what we call, a *Grothendieck topos*, associated with the site (2.10), as above (ibid., p. 127 Definition 3). In this context, it is also instructive to recall here that the same category,

(2.12) $Sh(\mathcal{A}, \mathcal{J})$ *is a full subcategory* of the *functor category*,

(2.12.1) $$(Sets)^{\mathcal{A}^{op}}$$

that is, the category of *set-valued presheaves on $\mathcal{A}$*. Therefore, one has the *inclusion functor*,

(2.12.2) $$Sh(\mathcal{A}, \mathcal{J}) \underset{\rightarrow i}{\subset} (Sets)^{\mathcal{A}^{op}},$$

which is thus *full* and *faithful*.

See also [CWM: p. 88]. Indeed, it is essentially the above "*functor category*" (cf. ibid., p. 40), or even [MM: p. 25, (viii)]), as in (2.12.1), that we are mostly interested in, throughout the sequel.

1.2.1 *Adjoint Functors*

> *"Adjoint functors arise everywhere"*.
> (S. Mac Lane)

It is a basic result, see e.g. [MM: p. 128, Theorem 1] that, *the* above *inclusion functor i*, as in (2.12.2), *has a left-adjoint*; i.e., the so-called *"associated sheaf functor"*,

$$(2.13) \qquad \qquad \boldsymbol{a} : (\mathcal{S}ets)^{\mathcal{A}^{op}} \longrightarrow \mathcal{S}h(\mathcal{A}, \mathcal{J}),$$

the definition of which is based on the standard *"plus-construction"* (ibid., p. 129ff). Thus, we are led here to the fundamental notion of the existence of an *adjunction*, which still explains our assertion, concerning the functor *i*, as in (2.12.2). Namely, one has the relation,

$$(2.14) \qquad \qquad \mathcal{S}h(\mathcal{A}, \mathcal{J}) \xrightarrow[\ \ \boldsymbol{a}\ \ ]{\ \ i\ \ } ((\mathcal{S}ets)^{\mathcal{A}^{op}}).$$

More precisely, one gets at the following triple

$$(2.15) \qquad \qquad (\mathcal{S}h(\mathcal{A}, \mathcal{J}), (\mathcal{S}ets)^{\mathcal{A}^{op}}, \phi)$$

(see [CWM: p. 78, Definition]), where the *bijection* ϕ is defined by the following *"natural isomorphism"*,

$$(2.16) \qquad \qquad \boldsymbol{a} \cdot i \cong \mathbf{1}_{\mathcal{S}h(\mathcal{A}, \mathcal{J})}$$

(cf. also [MM: p. 133, Corollary 6]).

Indeed, by considering the *"coordinatization functor"*,

$$(2.17) \qquad \qquad i \equiv \mathbb{M},$$

(see (2.3), change of notation is made here, just for convenience, see thus still (3.35) in the sequel), one defines the next functor (viz. the so-called *"functor of points"* of $\mathcal{E}$; cf. also (2.33) below),

$$(2.18) \qquad \qquad \mathbb{F} : \mathcal{E} \longrightarrow (\mathcal{S}ets)^{\mathcal{A}^{op}},$$

according to the relation,

$$(2.19) \qquad \qquad \mathbb{F}(\boldsymbol{z}) : a \longmapsto Hom_{\mathcal{E}}(\mathbb{M}(a), \boldsymbol{z}),$$

for any $a \in \mathcal{O}b(\mathcal{A})$ and $\boldsymbol{z} \in \mathcal{O}b(\mathcal{E})$, the latter correspondence being also *"natural"* in a, $\boldsymbol{z}$ (cf. (2.22) below). Therefore, by further applying the standard *"$\mathcal{H}om$-tensor adjunction"* (see also Section 1.2.2, along with the next Section 1.3), one defines a *left-adjoint* for $\mathbb{F}$, say,

$$(2.20) \qquad \qquad \mathbb{L} : (\mathcal{S}ets)^{\mathcal{A}^{op}} \longrightarrow \mathcal{E},$$

such that one sets,

$$(2.21) \qquad \qquad \mathbb{L}(P) := P \otimes_{\mathcal{A}} \mathbb{M}, \ \ \text{or yet,} \ \ \mathbb{L}(\cdot) := (\cdot) \otimes_{\mathcal{A}} \mathbb{M},$$

for any P a (*Sets*)-*valued presheaf on* $\mathcal{A}$ (cf. (2.12.1)). Hence, one has the *natural isomorphism*,

$$(2.22) \qquad Hom_{\mathcal{E}}(\mathbb{L}(P), \boldsymbol{z}) \cong Nat(P, \mathbb{F}(\boldsymbol{z})),$$

where in the second member of the last relation one sets

$$(2.22') \qquad Nat(.\,,.) \equiv Hom_{\mathcal{A}}(.\,,.).$$

See also, for instance, S. Mac Lane [1: p. 40, (1)]; yet, all the "*hom-sets*" in (2.22) are supposed to be "*small*". More on the material of this subsection will be supplied throughout the ensuing discussion.

Scholium 2.1. — As a moral of the preceding, and still based on Zafiris' work [7] (we also occasionally refer therein for undefined terminology in the sequel), we remark the following:

(2.23)

> for every *commutative* "*observables algebra*", there exists an underlying "*quantum observables algebra*" E. Zafiris [1], [2]; that is, being also in accordance with the preceding terminology, one gets at the following association,
>
> $$(2.23.1) \qquad \mathbb{M} : \mathcal{A}_c \equiv \mathcal{A} \longrightarrow \mathcal{E} \equiv \mathcal{A}_Q,$$
>
> realized thus by what we called in the foregoing the "*coordinatization functor*", $i \equiv \mathbb{M}$ (see (2.17)/(2.3)).

Now, the point of view, pertaining to the functor $\mathbb{M}$, as above, is that one gets in effect an *equivalent version of (2.23)*, in the sense that,

(2.24)

> to every "*commutative object*", thus, element of $\mathcal{O}b(\mathcal{A})$, there corresponds, via $\mathbb{M}$, an "*underlying object*", element of $\mathcal{O}b(\mathcal{E})$; that is, *its image through* $\mathbb{M}$. Hence, one obtains the relation,
>
> $$(2.24.1) \qquad x \longmapsto \mathbb{M}(x) \in \mathcal{O}b(\mathcal{E}),$$
>
> for every element $x \in \mathcal{O}b(\mathcal{A})$.

Accordingly, through the above considerations, one gets at an analogous situation with that one concerning the *natural correspondence*,

$$(2.25) \qquad \text{"reality"} \longrightarrow quantum.$$

Now, within the previous framework, one can think, of course, by looking at (2.25), as something that goes back, even to *Democritus* (!), certainly by an obvious *extension of terminology*; that is, in other words, to the "*atomic*" *substance of the matter*. Thus, one might consider it, as an aspect of a so-called,

$$(2.26) \qquad \text{"Democritian principle"},$$

realized here, just, through the "*coordinatization functor*"; either in its *abstract* form (see ADG), or the more *concrete* one, as before (cf. (2.26)). On the other hand, our previous argument, pertaining at least to its *"applicable"* (!) counterpart, as above, can still be rooted on a more recent and known adage, due at least to D. R. Finkelstein, in that

(2.27)   *"all is quantum"*.

See thus the said author [1: p. 477]; the same is further specified by saying,

(2.27')   *"all is quantum, and topology is all"*.

Here, Finkelstein's *"topology is all"* can still be associated, within our present context (viz. ADG), with the point of view that,

(2.28)   *"all" is relations* (!).

Therefore, *sheaf theory*, or even *topos theory*, in the sense this is applied herewith. See also the subsequent discussion, along, for instance, with A. Mallios [17: p. 268; (2.9)], [18: p. 1940; (4.2)] and [16: pp. 1557-1558; (1.1) and (1.5)], as well as, [20]. So to repeat it, here too,

(2.29)   it is actually the *physical laws*, that make what we may call, or even theoretically construed, as (*"physical"*) *geometry*, hence, *topology* as well, in the previous sense.

Furthermore, it is also worthwhile to remark, in this context, that the aforementioned aspect of *"topology"* (viz. the, so to say, *"relational"* one) is *better understood*, even in the case that this (technical) term, in the usual sense, is referred, for instance, still to the case of a *"differential manifold"*; of course, when this *"structural ingredient"* of the latter notion is scrutinized, relative always to its *"physical"* (: esoteric) *significance* (!): indeed, to recall here H. Weyl [3: p. 86],

(2.30)   "While topology has succeeded ... in mastering continuity, [however] *we do not* yet *understand the inner meaning of the restriction to differentiable manifolds ... one day physics will be able to discard it.* At present it seems indispensable since *the laws of transformation of most physical quantities are intimately connected with* that of the *differentials ...*"

Concerning, in particular, the last part of the above utterance of H. Weyl, we come to it again in the sequel; it follows that, still there, one has something essential to say, when *looking at it from the point of view of* ADG.

Yet, within the same vein of ideas, in conjunction also with (2.30), as above, and further relevant comments of H. Weyl (ibid., p. 86ff), still in relation with an analogous argument of D. R. Finkelstein (see e.g. (2.27'), as before), one should also notice here the quite relevant work of R. D. Sorkin, concerning *quantum gravity*, via his *"finitary"* ideas, pertaining to *"continuous topology"* R. D. Sorkin [1]; in this context, see also the relevant work of I. Raptis [1], and still of A. Mallios-I. Raptis [1], [2], within the framework of ADG.

1.2.2 *Natural Adjunction*

Now to come back to the aforementioned *inclusion/"coordinatization" functor*,

$$(2.31) \qquad\qquad \mathcal{A} \xrightarrow[i\equiv\mathbb{M}]{\subset} \mathcal{E},$$

as in (2.23.1), we remark that the same gives rise, in effect, to a hint for a *"natural adjunction"* (see below) of the categories involved, in terms also of the so-called *"Yoneda embedding"* (cf. Scholium 3.1 in the sequel). That is, in other words, to an (adjoint) *pair of functors*, between the categories at issue (viz. $\mathcal{A}$ and $\mathcal{E}$), together with a *"natural" isomorphism* for the corresponding *sets of arrows* (see also e.g. (2.15), and scholia after (2.22) in the preceding. Yet, cf. S. Mac Lane [1: p. V] and [2: p. 269]): Indeed, as already remarked in the previous Section 1.2.1, the aforesaid *adjunction* is, in effect, a sort of the standard *"Hom-Tensor adjunction"*, realized in particular as a pair of functors,

$$(2.32) \qquad\qquad (\mathbb{L}, \mathbb{F}),$$

given by (2.20) and (2.18). Furthermore, one gets at the relations (cf. also the next Section 1.3; (3.1)),

$$(2.33) \qquad Hom_{\mathcal{E}}(\mathbb{L}(P), z) \cong Hom_{\mathcal{A}}(P, Hom_{\mathcal{E}}(\mathbb{M}(a), z)) \equiv Nat(P, \mathbb{F}(z)),$$

for any $a \in \mathcal{O}b(\mathcal{A})$ and $z \in \mathcal{O}b(\mathcal{E})$, as also *explained by* (2.20), (2.22) *and* (2.23).

In toto, (2.33) realizes, within the present framework, that, which one might still denominate, as

$$(2.34) \qquad \text{``\textit{principle of adjoint transformations}/(: \textit{functors})''},$$

in general; its *"natural" character* is, of course, expressed through the last term of (2.33): namely,

$$(2.34') \quad \begin{array}{l} \textit{the aforesaid principle, as in (2.33), concerns, in effect, a ``natural trans-} \\ \textit{formation (alias, ``morphism'') of (adjoint) functors''.} \end{array}$$

Precisely, by referring in particular to the case at hand, one actually gets here at a

$$(2.35) \qquad \text{``\textit{natural equivalence}'' (thus, \textit{isomorphism}) of (\textit{adjoint}) \textit{functors}.}$$

See S. Mac Lane [2: p. 269, *Note on Adjoints*]; furthermore it is also worthwhile to point out here, once more, the *"relational character"* of the preceding material: Indeed, to paraphrase in that respect S. Mac Lane again (ibid., p. 8; beginning of Chapter one), we remark that,

$$(2.36) \quad \begin{array}{l} \textit{Homology is a ``relational'' theory (!), dealing with formal properties of} \\ \textit{functions.} \end{array}$$

In this context, see also (2.28) in the foregoing. Thus, in sum, as a result of the previous discussion, one can further formulate the aspect that,

$$(2.37) \qquad \text{``\textit{all}'' is a ``\textit{natural equivalence}'' of (adjoint) \textit{functors}.}$$

As we are going to realize later on, the above still *generalizes the* well-known *"symplectic creed"* (A. Weinstein), pertaining to Symplectic Differential Geometry; cf. A. Weinstein [1].

Now, within the same vein of ideas, and by anticipating, in a sense, what we are going to consider throughout the subsequent discussion, we still remark that:

(2.38)
> an *"adjunction"*-situation/context *transforms* *"geometry"*, thus, *metric/"Hom-functor"*, or even a *quadratic form*; however, this always with respect to $\mathcal{A}$ (viz. *"we"*, cf. (1.4)), therefore, *in the sense of* ADG. In that way, we are thus led to the concept of *topology* à la Finkelstein, at least (cf. (2.27′). That is, in other words, we finally arrive in a

(2.38.1)
> *"relational affaire"*; therefore, *"connection"* $\longleftrightarrow$ *physical law* $\longleftrightarrow$ $\otimes$-*functor* (: *Kähler*, hence ADG, again).

See also, for instance, (4.6), (8.13)/(8.14), and (10.2), (10.14.1) in the sequel.

As already said above, we are going to be more precise, concerning the previous account, as in (2.38), straightforwardly through the ensuing discussion. Yet, the same material, as above, is going to be appropriately related, as we shall see, with another *fundamental principle* in Nature, namely, the so-called, *"least action principle"*: See thus Sections 1.6 and 1.9 in the sequel.

On the other hand, in connection with our previous terminology in (2.18), (2.19), we further note that,

(2.39)
> $\mathbb{F}(z)$, with $z \in \mathcal{O}b(\mathcal{E})$, *represents*, by its very definition (see e.g. (2.39.2) below), *all the information* we can have/detect, *about the* particular *element* (viz. *"quantum object"*) $z \in \mathcal{O}b(\mathcal{E})$, at issue, through (our *"arithmetic"*) $\mathcal{A}$ (see also (2.23)-(2.25): precisely, in the form of a (*Sets*)-*valued presheaf on* $\mathcal{A}$ (!); that is one has,
>
> (2.39.1) $\mathbb{F}(z) \in \mathcal{O}b((\mathcal{S}ets)^{\mathcal{A}^{op}})$,
>
> for every $z \in \mathcal{O}b(\mathcal{E})$ (cf. (2.18)/(2.19)), such that one sets,
>
> (2.39.2) $\mathbb{F}(z)(a) := Hom_{\mathcal{E}}(\mathbb{M}(a), z)$,
>
> with $a \in \mathcal{O}b(\mathcal{A})$ (see (2.24.21)).

Consequently, one can still assert here that,

the previous *"functor of points"*, as given by (2.39.2) (see also (2.18)/(2.10)), constitutes in effect the

(2.40)

(2.40.1) *"materialization"*, or else *realization/effectuation of relations* (viz. "arrows")/yet, of ideas (: details) and the like,

which are (: can be) associated with the particular (*quantum*) z, as before; therefore, in other words, even a

(2.40.2) *"realization of the physical law"* itself, associated with z, as above.

As a result, one can think here, for instance, of the *gravity*, as this is expressed, by means of the *"curvature"*; the same is actually, technically speaking, just, a particular case of the *Hom-functor* (see e.g. A. Mallios [VS, Vol. II: Chapt. VIII, p. 192, Lemma 2.1; (2.10)]). Yet, via a similar vein of ideas, we are led, in conjunction with (2.39.2) and its physical analogue/interpretation, through (2.40.1), to a telling phrase, pertaining to relevant situations (S. A. Seleснick), referring to the *"meat on the bones"* ...!

On the other hand, as a further result of (2.40), one still realizes a

(2.41) *change of* the *"geometry"*/product (: $\otimes$-*functor*, hence, H. Grassmann), through the underling *"adjunction"*, as actually alluded to in (2.38) above; yet, see the following comments.

Indeed, the effectuation/transformation of the aforesaid *change of geometry/product* (Grassmann-Kähler, see also below) is achieved, *through the function of* what we called in the preceding,

(2.42) *"Hom - $\otimes$ adjunction"*

(cf. also the next Section 1.3, along with Sections 1.6, 1.10 in the sequel). Now, *the intervention* here *of these two* basic *functors*, as above, *is* indeed *characteristic* of the whole procedure, as it concerns our previous claim in (2.41). Yet, the same function of the two (adjoint) functors at issue has been still realized, as a *"natural transformation"*, indeed, *"equivalence"* (: *isomorphism*) *of functors*; cf. (2.35), along with S. Mac Lane [1: pp. 80, 81; Theorems 1, 2]. In particular, cf. (3.11) below.

Now, within the same context, as before, we can still remark that,

(2.43) the function of *differentiation* (: *dynamics*, hence *kinematics*, as well) herewith, amounts actually to an appropriate *enlargement/extension of* our initially employed *"arithmetic"* (: $\mathcal{A}$). That was indeed the contribution first by H. Grassmann (: the $\otimes$-*functor*), and then by E. Kähler (: differential).

More comments thereon will be also supplied in the sequel. Yet, in that context see A. Mallios [VS, Vol. II: p. 321ff], along with A. Mallios-E. Zafiris [1], [2], [3].

The preceding lead to a telling diagram connecting the above with the *fundamental pair* $(\mathcal{E}, \mathcal{D})$ (cf. (6.1) in the sequel). It refers to what we may call a

(2.44)   *quantum-interactions field,*

as expressed via the following diagram:

(2.45) $(\mathbb{F}(z) := Hom_{\mathcal{E}}(\mathbb{M}(\cdot), z),\ \nabla_{\mathbb{F}}(z)).$

That is, a

(2.46)   (quantal) *"transliteration"* of the fundamental pair $(\mathcal{E}, \mathcal{D})$.

See also (the terminological) *Note* 6.1 in the sequel for the notation employed, herewith, along with (8.20)/(9.5), and (10.2)/(10.3), and (10.18), as well as, *Note* 10.1, in general.

1.3   Bohr's Correspondence

The present Section might be viewed as a *"triumph of the Bohr's correspondence principle"*, in the sense the latter was stated by (0.2). Now, as a result of the preceding, one gets at the following *"adjunction"*,

(3.1) $\mathbb{L} \dashv \mathbb{F};$

namely, one defines, the relation,

(3.2) $(\mathcal{S}ets)^{\mathcal{A}^{op}} \underset{\mathbb{L}}{\overset{\mathbb{F}}{\longleftarrow\!\!\!\longrightarrow}} \mathcal{E}.$

The *counit* of the aforesaid adjunction *determines the objects of* $\mathcal{E}$ (see also (3.17)/(3.18) below). Precisely, one has the next *fundamental result* (yet, see (3.8), (3.9) in the sequel); this may still be perceived, on the basis of (0.2), as an echo of a sort of *"Bohr's Theorem"*. That is, one gets at the following.

   Theorem 3.1. (*N. Bohr*). Every object of the *category* $\mathcal{E}$ (cf. (2.1)/(2.23)) can be construed as an object of the *site* $(\mathcal{A}, \mathcal{J})$ (cf. (2.10)), in such a manner that, one gets at the relation,

(3.3) $\mathcal{O}b(\mathcal{E}) \cong \varinjlim \mathcal{O}b((\mathcal{S}ets)^{\mathcal{A}^{op}}),$

within a *category equivalence*. Furthermore, one can, *equivalently, still look at* (3.3) as an *equivalence of the initial categories involved* (see (2.2)/(2.23) and (2.12.1)).

   In this context, see also e.g. A. Mallios [18: p. 1943, (4.17)]. Now, the justification of our previous claim, contained in Theorem 3.1, is based on a crucial result, within the present context, known as, the *"converse part of Giraud's theorem"*. So we clarify/highlight presently below the relevant parts of that theorem; see also [MM: p. 580, Section 3]:

Now, basically the adjunction (3.2) is a particular instance of a "*Hom-tensor adjunction*" of the categories at issue, defined by the *inclusion functor*,

$$(3.4) \qquad\qquad i \equiv \mathbb{M} : \mathcal{A} \longrightarrow \mathcal{E},$$

cf. also (2.31), such that one posits,

$$(3.5) \qquad - \underset{\mathcal{A}}{\otimes} \mathbb{M} : (\mathcal{S}ets)^{\mathcal{A}^{op}} \overset{\mathbb{F}}{\underset{\mathbb{L}}{\longleftarrow}} \mathcal{E} : \mathcal{H}om_{\mathcal{E}}(\mathbb{M}, .),$$

while we have further set in the preceding,

$$(3.6) \qquad\qquad \mathbb{L} \equiv - \underset{\mathcal{A}}{\otimes} \mathbb{M} \ \text{ and } \ \mathbb{F} \equiv \mathcal{H}om_{\mathcal{E}}(\mathbb{M}, .)$$

see (2.21) and (2.19). On the other hand, we first have, in that context, the following *basic result*.

Lemma 3.1. *The functor* $\mathbb{F}$, *as in* (3.6)/(3.5), *sends the objects of* $\mathcal{E}$, not just into set-valued presheaves on $\mathcal{A}$ (cf. (3.5)/(2.18)), *but indeed into sheaves for the site* $(\mathcal{A}, \mathcal{J})$ (cf. (2.10)). *That is, into objects of the Grothendieck topos* (of sheaves on $(\mathcal{A}, \mathcal{J})$),

$$(3.7) \qquad\qquad \mathcal{S}h(\mathcal{A}, \mathcal{J}).$$

Precisely, this in such a manner that, one finally gets at the following *equivalence of categories*,

$$(3.8) \qquad\qquad \mathcal{E} \cong \mathcal{S}h(\mathcal{A}, \mathcal{J}).$$

In this context, see also [MM: p. 580ff], along with A. Mallios [18: p. 1942, (4.12)]. Consequently, as a result of (3.8), one thus obtains that,

(3.9)   *every object **z** of the category $\mathcal{E}$ can be represented by a complete presheaf, hence a sheaf, on the category $\mathcal{A}$, relative to a suitable Grothendieck topology on it.*

Before we proceed further, it would be worthwhile to make here some remarks, pertaining to the framework we applied in the preceding, just, as it concerns the *sheaf-theoretic* one. Thus, we should notice, of course, that

(3.10)   "... *the very notion of sheaf is ... central to topos theory* ..."

(emphasis is ours: See [MM: p.2]. However, the *topos-theoretic context*, applied herewith, has a *particular*

(3.11)   *physical significance/importance*, providing thus an *increased* (yet, *exclusive*) *relational character/nature of the* technical (: mathematical) *background* supplied thereupon. The latter sort of the framework thus obtained appears, indeed, to be *more akin*, even *to the type of "topology"*, we really need/should eventually apply *in the "quantum deep"*.

In this context, cf. also our previous remarks in (2.28), along the subsequent comments therein. Yet, more generally, see for instance A. Mallios [20].

Now, by further commenting on (3.9), concerning the image of the functor $\mathbb{F}$, as given by (2.18)/(2.19), we will conclude that,

(3.12)   *any presheaf* (object of the category (2.12.1)) *is a colimit* (of a diagram) *of "representables"*, realized through the (associated with the presheaf at issue) *"functor of elements"* along with the *Yoneda embedding*. (See also Scholium 3.1 in the sequel).

See [MM: p. 41, Proposition 1, and p. 42, Corollary, together with the relevant comments on p. 139]. Thus, one obtains (see also (3.45) below).

Lemma 3.2. *Every object z of the category $\mathcal{E}$ (in effect, its image under $\mathbb{F}$, cf. (2.19)/(2.39.2)) is of the form* (see also Scholium 3.1 below),

$$(3.13) \qquad z \cong \varinjlim(\boldsymbol{ay}(x_k)),$$

for some *"diagram of objects"* (x_k) in $\mathcal{A}$.

Furthermore, in other words, one still concludes that,

(3.14)   the subcategory $\mathcal{A}$ of $\mathcal{E}$ (cf. (2.3)) is *dense* in $\mathcal{E}$.

In this context, see also [CWM: p. 241, §6]. Now, *(3.13) proves* indeed *the rel. (3.3)*, hence, *Theorem 3.1*, as well.   □

On the other hand, we further remark here that,

(3.15)   *the set of all sheaves*,

    (3.15.1) $\boldsymbol{ay}(x)$, *with* $x \in \mathcal{O}b(\mathcal{A})$

(see also (2.14), along with (3.23)/(3.31) in the sequel), *"generate"* the *category* $\mathcal{S}h(\mathcal{A}, \mathcal{J})$. Therefore, in view of (3.8), *the category $\mathcal{E}$, as well*.

For the terminology applied above, see [CWM: p. 123], or even [MM: p. 139]; yet, cf. also (2.4) in the preceding. So the term *"generate"*, as before, referring particularly to the set (3.15.1), means *equivalently* that

(3.16)   *the category $\mathcal{A}$ is a "generating"* (alias, *"separating"*) *subcategory of* $\mathcal{E} \cong \mathcal{S}h(\mathcal{A}, \mathcal{J})$.

More on this, still in conjunction with (3.13), will be given in Scholium 3.1 below.

Now, before we continue further, we should recall here, yet for later use, the *fundamental rôle* played, in connection with the *Hom-tensor adjunction*, as above (see (3.5)), *by the* relevant notions, *"unit"* and *"counit"*. See thus also e.g. [CWM: pp. 79-81], and/or [MM: p. 18ff]: So looking at the *natural bijective correspondence*

(2.22), one defines, for any object $z \in Ob(\mathcal{E})$, the corresponding *counit isomorphism* factor (see also (2.19),

(3.17) $$\epsilon_z : \mathbb{F}(z) \otimes_\mathcal{A} \mathbb{M}(\cdot) \cong z.$$

In particular, the same defines the following *isomorphism* (in fact, "*natural transformation*"; see also comments after (2.34) in the foregoing),

(3.18) $$\epsilon_z : \mathcal{H}om_\mathcal{E}(\mathbb{M}(\cdot), z) \otimes_\mathcal{A} \mathbb{M} \equiv (\mathbb{L}\mathbb{F})(z) \cong z,$$

for any object z of $\mathcal{E}$ (cf. also (3.2) and (2.39.2)). At this point, we still note that, as we shall see later on, the same principle, as in (2.34) (:"*principle of adjoint transformations*"), is going to be appropriately viewed as a sort, in effect, of a "*least action principle*". Now, the last two relations yield also the following specification:

(3.19)   the *sheaf-theoretic description of the category $\mathcal{E}$*, through (3.8), *is actually realized via the counit isomorphism*, as above, provided *by the adjunction (3.5)*. Moreover, in other words, *the* same *counit of the adjunction* considered, when restricted to sheaves of the site $(\mathcal{A}, \mathcal{J})$ (see (2.10)/(2.12.2)), *yields an equivalence of categories*, as in (3.8).

Thus, one actually obtains the following *functor-diagram*, where the counit isomorphism is displayed clearly:

(3.20)

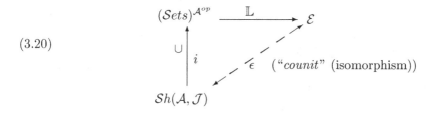

see also (2.10), along with [CWM: p. 88, Theorem 1]. Furthermore, to say it still otherwise, one thus realizes that, *the functor*

(3.21.1) $\mathbb{F}(\cdot) \equiv \mathcal{H}om_{\mathcal{E}}(\mathbb{M}, \cdot),$

as in (3.6) and (2.18), *sends the objects of* $\mathcal{E}$ not just into presheaves (ibid.), but actually *into sheaves*, with respect to a suitable topology $\mathcal{J}$ (Grothendieck) on the small subcategory $\mathcal{A}$ of $\mathcal{E}$, and *in such a manner that,*

(3.21)

> *the adjunction* (3.5) *is finally restricted* to an *equivalence of categories*, as in (3.8); that is, one gets,
>
> (3.21.2) $\mathcal{E} \cong \mathcal{S}h(\mathcal{A}, \mathcal{J}),$
>
> through the *natural transformation - counit -* ϵ, affording thus by definition, $\mathcal{E}$, as a *Grothendieck topos*.

Therefore, in other words,

(3.21.3)   the *counit* ϵ (see also (3.20)) becomes an *isomorphism*.

For a nice/telling *physical interpretation of (3.21.3)*, see also e.g. E. Zafiris [4: p. 113], yet, [6: p. 1513; *"Subobject Representation Theorem"*].

Now, we continue by further commenting, through the next Scholium, on the important notion of the so-called *"functor of points"*; see also e.g. remarks on (2.18) in the preceding.

Scholium 3.1. — *Category of elements (points) of a contravariant functor* (alias, *"Grothendieck construction"*). — We come to discuss below the classical construction, as in the title of this Scholium, in greater detail than that already employed in the preceding (see, for instance, (2.18)/(2.33)). Yet, the comments herewith are also in conjunction with our previous relevant account in Scholium 2.1. On the other hand, as we shall see, the same construction, at issue, acquires further a quite *"natural"* interpretation that is, after all, still of foremost concern all along the present treatise; so see, in particular, the discussion toward the end of this scholium.

Now, to start with, we first note, concerning our terminology as applied herewith (hence, having in mind applications in ADG), that a *contravariant functor* will be, in principle, just any *(Sets)-valued presheaf on* $\mathcal{A}$; that is, in other words, an element/object of the category (2.12.1), as above, say

(3.22) $\mathcal{P} \in \mathcal{O}b((\mathcal{S}ets)^{\mathcal{A}^{op}}).$

Yet, for convenience, we also set, occasionally, following here the French school (see also e.g. [MM: p. 25, (1)]),

(3.23) $(\mathcal{S}ets)^{\mathcal{A}^{op}} \equiv \widehat{\mathcal{A}}.$

Thus, by delineating/complementing the preceding discussion, what we first notice is that;

(3.24) *to every object $z \in \mathcal{O}b(\mathcal{E})$ (: quantum, see also (2.23)), one can associate a "representable functor/presheaf"; thus in effect, an element (presheaf) object of $(\mathcal{S}ets)^{\mathcal{A}^{op}}$, through the coordinatization functor $\mathbb{M} \equiv i$ (cf. (2.8.1), (2.17) along with (2.9)/(2.9')). Yet, see (3.28) below).*

Namely, one defines the association,

$$(3.25) \qquad z \longmapsto \mathcal{H}om_{\mathcal{E}}(\cdot, z), \quad z \in \mathcal{O}b(\mathcal{E}),$$

such that one gets at the *"representable functor"*/presheaf, as asserted in (3.24); that is, one has (see also e.g. (2.8.1)),

$$(3.26) \qquad \mathcal{H}om_{\mathcal{E}}(\cdot, z) \in (\mathcal{S}ets)^{\mathcal{A}^{op}} \equiv \widehat{\mathcal{A}}, \quad \text{with } z \in \mathcal{O}b(\mathcal{E}),$$

such that one sets,

$$(3.27) \qquad a \longmapsto \mathcal{H}om_{\mathcal{E}}(\mathbb{M}(a), z), \quad a \in \mathcal{O}b(\mathcal{A}).$$

Indeed, one actually proves that,

(3.28) *the above association, as in (3.25), yields, in effect, a sheaf for the "Grothendieck topology" on $\mathcal{A}$, viz. an object of the category ("Grothendieck topos"),*

$$(3.28.1) \qquad \mathcal{S}h(\mathcal{A}, \mathcal{J}).$$

See [MM: p. 581, Lemma 1], along with (3.9) in the preceding. Precisely, one thus obtains that,

(3.29) *all the "representable functors/presheaves", as above, are in fact sheaves, in the previous sense (see (3.28)).*

See also ibid., p. 583, proof of the quoted Lemma therein.

Now, to clarify things, within the same context, we further note that one sets (see also (3.13) in the preceding),

$$(3.30) \qquad i \equiv \mathbb{M} = \boldsymbol{y} \circ \delta,$$

where $\boldsymbol{y}$ denotes the standard *Yoneda embedding* of $\mathcal{A}$, viz. one has,

$$(3.31) \qquad \boldsymbol{y} : \mathcal{A} \longrightarrow \widehat{\mathcal{A}} \equiv (\mathcal{S}ets)^{\mathcal{A}^{op}},$$

such that one sets, by definition,

$$(3.32) \qquad \boldsymbol{y}(a) := \mathcal{H}om_{\mathcal{A}}(\cdot, a) \equiv \left[\mathcal{A}\big|_a\right] \in \mathcal{O}b(\widehat{\mathcal{A}}),$$

for every $a \in \mathcal{O}b(\mathcal{A})$ (see also (2.12.1) and (2.8.1), along with (3.43) below). Yet, we also recall here for later use that,

(3.33) *the Yoneda embedding is a full and faithful functor.*

This is a consequence of the Yoneda Lemma ; see [CWM: p. 61], or even [MM: p. 26]. Furthermore, concerning (3.30), one denotes by

$$(3.34) \qquad\qquad \delta : \mathcal{J} \longrightarrow \mathcal{A},$$

a so-called *"diagram in $\mathcal{A}$ of type $\mathcal{J}$"*, thus, a *functor*, as indicated, yet an element/object of (the *functor category*) $\mathcal{A}^{\mathcal{J}}$, with $\mathcal{J}$ a small *"indexing category"* (see also [MM: pp. 20, 41]); indeed, one defines/constructs $\mathcal{J}$ in a *"canonical"* way: ibid., p. 41, Proposition 1 (see also below).

On the other hand, by further employing the so-called *"Grothendieck construction"* on an element/functor $\mathcal{P}$, as in (3.22) (alias, *"diagram of $\mathcal{P}$"*, see also ibid., p. 386, for the terminology applied herewith), one is led to the relation,

$$(3.35) \qquad\qquad \mathcal{P} \cong \varinjlim_{\mathcal{J}} (\boldsymbol{y} \circ \delta)$$

(cf. also (3.41) below, in connection with (3.34) and (3.31)). Further details will be given straightforwardly in the sequel: So, by first looking at the definition of the *Yoneda embedding*, as in (3.32), we still elaborate on our previous assertion in (3.13); thus, for convenience of reference, we first recall the *definition of the functor $\mathcal{J}$*, as in (3.34): That is, for every object/presheaf on $\mathcal{A}$, say $\mathcal{P}$, in $(\mathcal{S}ets)^{\mathcal{A}^{op}}$, one considers the so-called *"category of elements/points of $\mathcal{P}$"*, denoted by

$$(3.36) \qquad\qquad \int_{\mathcal{A}} \mathcal{P} \equiv \int \mathcal{P}$$

(see also Note 3.1 below, concerning the terminology applied). Its objects are *pairs*,

$$(3.37) \qquad\qquad (x, p) : x \in \mathcal{O}b(\mathcal{A}) \ \text{ and } \ p \in \mathcal{P}(x)$$

(cf. also (3.22)), while the *morphisms* are defined by,

$$(3.38) \qquad\qquad (y, p') \longrightarrow (x, p),$$

in such a manner that one has,

$$(3.39) \qquad\qquad (y \xrightarrow{f} x) \in \mathcal{H}om_{\mathcal{A}}(y, x), \ \text{ with } \ p' = \bar{f}(p) \equiv p \cdot f,$$

where we set,

$$(3.39') \qquad\qquad \bar{f} \equiv \mathcal{P}(f), \ \text{ such that } \ \mathcal{P}(f)(p) \equiv \bar{f}(p) \equiv p\big|_f \equiv p \cdot f;$$

viz. what one may still view, as a *"restriction of $p \in \mathcal{P}(x)$, along f"*; here f is written after p, due to the *"contravariant character"* of (the functor) $\mathcal{P}$ (see also [MM: p. 25]). On the other hand, one further considers the *"projection functor"*,

$$(3.40) \qquad\qquad \pi_{\mathcal{P}} : \int_{\mathcal{A}} \mathcal{P} \longrightarrow \mathcal{A}, \ \text{ with } \ \pi_{\mathcal{P}}(x, p) := x.$$

Thus, $\int_{\mathcal{A}} \mathcal{P}$ is determined by the given *presheaf $\mathcal{P}$ on $\mathcal{A}$* (cf. (3.22)), so that one proves the following *isomorphism* (of functors),

$$(3.41) \qquad\qquad \mathcal{P} \cong \varinjlim_{\mathcal{J}} \left(\int_{\mathcal{A}} \mathcal{P} \xrightarrow{\pi_{\mathcal{P}}} \mathcal{A} \xrightarrow{\boldsymbol{y}} \widehat{\mathcal{A}} \right).$$

See also [MM: p. 41, Proposition 1, along with the ensuing comments therein]. As a result of the preceding, we actually get at the following (diagram of) functors;

(3.42)

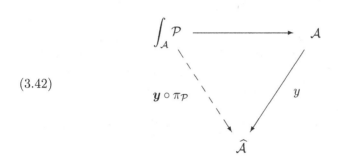

such that, *for any* $(x, p) \in \mathcal{O}b(\int \mathcal{P})$ *(see (3.37)), one obtains,*

$$(3.43) \qquad (\boldsymbol{y} \circ \pi_{\mathcal{P}})(x, p) = \boldsymbol{y}(\pi_{\mathcal{P}}(x, p)) = \boldsymbol{y}(x) \equiv \big[\mathcal{A}\big|_x\big] \in \mathcal{O}b(\widehat{\mathcal{A}})$$

(cf. also (3.40) and (3.32)), that also further explains (3.35), in conjunction with (3.34); in this context, still see relevant comments in [MM: p. 44, (14)].

> **Note 3.1.** — Looking at the previous relation (3.40), one could still consider (the category) $\int_{\mathcal{A}} \mathcal{P}$, yet, the *"diagram of $\mathcal{P}$"*, as
>
> (3.44)   a sort of a *"sum"* over $\mathcal{P}$ (: *"integral"*), à la *Feynman* (!);
>
> in view also of (3.41)/(3.42) and (3.43). Thus, as an
>
> (3.45)   "assemblage of *all the information one can have for x, in terms of our arithmetic A"* (: hence, *information about something, through arrows on it*: the (physical) meaning, in effect, of *"Grothendieck's construction"*.
>
> Therefore, the same as before, might be actually construed, still, as a type of a *"Feynman's principle"*.

On the other hand, *the* object/*presheaf* $\mathcal{P}$, as in (3.22), transformed, via the functor (: *"associated sheaf functor"*, cf. (2.13)),

$$(3.46) \qquad\qquad \boldsymbol{a} : (\mathcal{S}ets)^{\mathcal{A}^{op}} \longrightarrow \mathcal{S}h(\mathcal{A}, \mathcal{J}),$$

into an object/sheaf in the codomain of the latter, as indicated, *is further expressed through the adjunction* $(\mathbb{L}, \mathbb{F})$ *(see (3.1)/(3.2)) into the form, as* for instance *indicated* by the last term *in* (2.33). Indeed, we are actually led by the preceding to a *much simpler expression* of the same, in view of the following relations; cf. also (3.13) and (2.16). That is, one obtains,

$$(3.47) \qquad \boldsymbol{z} = \boldsymbol{a}i(\boldsymbol{z}) = \boldsymbol{a}(i(\boldsymbol{z})) = \boldsymbol{a}\left(\varinjlim_{k}(\boldsymbol{y}(x_k))\right) = \varinjlim_{k}(\boldsymbol{a}\boldsymbol{y}(x_k)),$$

with (x_k), a "*diagram of objects*" in $\mathcal{A}$ (see (3.34)), a "*dense*", as already said, *full subcategory* of $\mathcal{E}$ (cf. also (3.14)/(3.15), as well as, [CWM: p. 88, Theorem 1]). Yet, we also applied in (3.47) the fact that, *the functor $\boldsymbol{a}$* (see (3.46)), *as a left adjoint* (of $i \equiv \mathbb{M}$, cf. (2.16)/(2.17)), *commutes with/respects colimits* (ibid., p. 115, or even [MM: p. 22, and p. 139]).

Therefore, based on (3.18), (3.20), and also on (3.21.3), thus, in other words, on the actual meaning of the *adjunction* $(\mathbb{L}, \mathbb{F})$, as in (2.32)/(2.33), one comes to the conclusion that,

(3.48)
> the functor $\mathbb{L}$ (cf. (2.20)), *restricted to* $\mathcal{S}h(\mathcal{A}, \mathcal{J})$ (see also (2.12.2)), yields a *natural equivalence* of the categories concerned, viz. one gets (cf. (3.21.2)) at the relation,
>
> (3.48.1) $\qquad\qquad \mathcal{S}h(\mathcal{A}, \mathcal{J}) \cong \mathcal{E}$
>
> (see also (2.10), (2.12), and (2.33)).

In this context, as already noted, $\mathbb{L}(\mathcal{P})$, with $\mathcal{P} \in \mathcal{O}b(\widehat{\mathcal{A}})$, is given by (3.41), or else (3.35); see also [MM: p. 589, Corollary 1, and p. 590, (1)]. Thus, one gets at the following *diagram of functors*, supplementing that in (3.20):

(3.49)

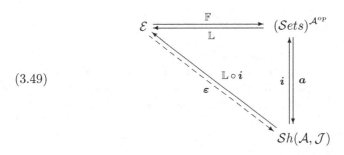

Now, terminating the preceding discussion, we further remark that it is, at least instructive, even from a *physical perspective* to point out, once more, the *significance of* our previous conclusion in *(3.28)*; thus, the fact that *the above procedure*, viz. *the functor $\mathbb{L}$* (cf. (3.49)), and even, strictly speaking,

(3.50)
> the adjunction $(\mathbb{L}, \mathbb{F})$ (cf. (2.32)/(2.33)) yields precisely a "*complete information*", yet, from a "*physical point of view*": Thus, otherwise said (however, still technically speaking) a *sheaf*! But, now in terms of the so-called, "*Grothendieck topology*", that is, through a *sheaf on a "site*".

On the other hand, the same *presheaf*, obtained through the functor $\mathbb{F}$, on the basis of the aforementioned adjunction, appears in the form of a *"representable functor/presheaf"*, that is, in terms of a (contravariant) *"$\mathcal{H}om$-functor"* (cf. e.g. (2.39.2), along with (3.25)/(3.27); however, see also (3.35)/(3.41)). So, in a *"realizable"* form (!), pertaining in effect to an element/object $z \in \mathcal{O}b(\mathcal{E})$ in the domain of $\mathbb{F}$ (cf. (2.18)). Thus, in other words, still speaking *physically*, (see (3.24), in conjunction with (2.23)/(2.39.1)), one arrives at the familiar (at least, in ADG) association,

$$(3.51) \qquad\qquad\qquad quantum \longleftrightarrow \text{``field''}$$

(see e.g. (10.2)/(10.3) in the sequel), which we are interested in/looking for. As an outcome, we thus realize, once more, that;

(3.52)   it is indeed *richly rewarding*, anytime we try/achieve to work in a, so to say, *"relational"* setting!

See also, for instance, A. Mallios [20]; thus, otherwise said, we come again here to the realization that,

(3.53)   *we do not* actually *study "objects"* (viz. *events*), *but*, in effect, the *relations* (:*"morphisms"*) *that characterize the events*.

Yet, this because,

(3.54)   *objects are*, indeed, still *"particular expressions"*/realizations *of relations* (viz. of *physical laws*).

This, to recall here Leibniz himself, as quoted by G. Lassner in A. Mallios [17: p. 267, (2.6), yet, p. 268, (2.9)]; cf. also A. Mallios [16: p. 1557, (1.1)]. The preceding *lead us naturally to the aspect* that,

(3.55)   the problem of *"quantum gravity"*, or even that of a *"unified field theory"*, might become still *quite simple*, when viewed from the perspective of an appropriate *"algebra of coefficients"* (: *"arithmetic"*); yet, through a suitable way of observing/measuring.

Of course, this also under the proviso that,

(3.56)   the same *"arithmetic"* supplies further *the means of detecting*/following *the* change/*"variation"* of the *information* we possess/have collected.

In this context, we can further say that, the *moral of* ADG so far is that,

(3.57)   *"adjunction functors"* indeed *respect the* ADG-*mechanism* (thus, see *Leibniz–Kähler–de Rham*). Therefore, one should always try to find out the inner/(esoteric) *"relational character"* of the (classical) *mechanism, we* usually *employ* at each particular case.

Concerning the last part of (3.57), see also relevant remarks in A. Mallios [16: p. 1559, (1.13)]. Furthermore, in connection with the same perspective, as before, relative to (3.57), cf. also the subsequent Sections 1.4, 1.5, together for instance with (9.30) in the sequel; yet, in connection also with (3.55) above, cf. as well, ibid., p. 1591, Subsection 6.(a). Therefore, here again, *one* thus *concludes the* so to say,

(3.58)   *"functorial character"*, in general, *of the entire mechanism of* ADG !

Cf. also the relevant discussion, throughout the next Section; see, for instance, (4.12.1) below.

1.4   Functorial, Topos-Theoretic Mechanism of ADG

Our task in what follows is to point out, even more (cf. also (3.58)), the behavior of the *mechanism* of what we called, so far, *Abstract Differential Geometry* (ADG), within the previous quite *algebraic*, thus *topos-theoretic* (hence, *sheaf-theoretic* is an important instance) framework. Indeed, as we shall see, *this mechanism* is entirely *"functorial"*, as already noted above; precisely, one realizes, in effect, that:

(4.1)   the *form* of the mechanism *of* ADG developed so far, in terms of $\mathcal{S}h(X)$, viz. within the *topos of sheaves* on a *topological space* X (a very particular case of a *"Grothendieck topos"*, cf. (2.11), and [MM: p. 111]), remains *quite untouched by* (viz. still *respects*) *the* present more general *topos-theoretic context*, as this has been employed hitherto in the foregoing.

In this context, pertaining in particular to what we may call, the *"dynamical"* aspect *of* ADG, the principal issue of the aforementioned mechanism/*function of* ADG is the *really fundamental* notion of an (𝒜-)*connection*, in the broadest possible sense of this term (cf. also e.g. (4.8)/(4.9), and (4.12.1)/(4.15) in the sequel): Thus, the latter concept appears to be extremely *"geometric"*, that is, *"natural"*, and, therefore, *"relational"* (!) in character; the preceding in the sense that

(4.2)   *"physical geometry" is the outcome of the physical laws.*

See thus A. Mallios [16: p. 1557, (1.1)], still, in accordance with P. Bergmann (see also below).

Note 4.1. — We should notice here that the previous term,

(4.3)   *"geometrical"*,

has the meaning of a *"physical"* one, as already explained above. There-fore, *not "geometrical"*, in the sense of *"analysis"*: in other words, *"arithmetical"* (!), in character; thus, still in the sense of *Functional Analysis*, viz. in effect, *topological vector space* structure and the like. Accordingly, the whole edifice of *Classical Differential Geometry* (CDG: smooth *manifolds*, finite dimensional or not), being thus *not* a *"physical"* (:*"relational"*) *one*, in the previous sense (still, see e.g. (4.2) above. Yet, cf. also A. Mallios [20]).

On the other hand, the latter comments, as before, regarding CDG, reminds us still the famous utterance that (emphasis below is ours).

(4.4)   *"The introduction of numbers as coordinates ... is an act of violence..."*

See H. Weyl [3: p. 90]. Furthermore, within the same vein of ideas, cf. also (2.28) in the preceding, together with the remarks following it, in particular, (2.29)/(2.30).

Yet, it is now the term,

(4.5)   *"relational"*, in character,

as employed above, that is, something being still *in accord with* (or even, repre-senting) *a physical law*, which essentially characterizes, within the previous context, even when referring to ADG, what we actually mean by talking about a *"geometric"* object/notion. Thus, we are looking, in effect, here for a *"physical geometry"* (P. Bergmann; see also (4.2) above), *not* the other way round, that is, for *"geometrical physics"*! The latter, of course, refers (cf. also Note 4.1) to our own *"cartesian-analytic"* one (ibid), thus to the so-called *"mathematical physics"*. Now, roughly speaking, by referring to the *point of view, advocated by the present study*, this means that

(4.6)   to confront with *physical problems* (viz. those pertaining to the *description of the function of the nature* - yet, to the *physical laws* - especially when this refers to the *"quantum"* (deep: the microscopic world), it is the perspective of this treatise to look always at the sort of mathemat-ics, that might be characterized, as *"relational"* ones (cf. also e.g. A. Mallios [20]); namely those that point out/underscore the *"relational* (: conceptual) *character"* that might be *inherent in our calculations*. Yet, in other words, those that work, for instance, *without any "spatial"* envi-ronment, in particular e.g., *no*- *"spacetime"*, concerning *"mathematical physics"*, as above. Instead, just, within a *"relational"* *set-up*, in the previous sense (cf. also below).

Of course, the above remind us of the phrase of S. Mac Lane,

(4.6.1)   *"... live without elements, using arrows instead"*,

from his classic [CWM: p. V]; or, even E. Galois, in that,

(4.6.2)   *"en un mot, les calculs sont impraticables"*,

as the latter is quoted by M. Artin in [1: p. 537].

In this context, we are thus led to the aspect of

(4.7)   *"doing without elements"* (: category theory), therefore, still *"doing without points"* (hence, *quantum theory*, see e.g. A. Mallios [17: p. 268, (2.9)], along with subsequent comments herewith).

Thus, consequently, we are also led, *by extension*, to the idea of

(4.8)   *"doing without manifolds"*, as well (!),

consenting now, even to H. Weyl himself (cf., for instance, (2.30) in the preceding). He postanticipated in that *A. Einstein*, who *already in 1916* declared that,

(4.9)   *"... continuum space-time"* [hence, 4-dimensional *manifold*] ... *corresponds to nothing real"* [(!)].

See e.g. A. Mallios [14: p. 59, (1.6)]. Yet, one can also quote here A. Einstein, in that;

(4.10)   *"... reality cannot at all be represented by a continuous field."*

Cf. A. Einstein [1: p. 165], or even that (ibid., p. 166);

(4.11)   *"... find a purely algebraic theory for the description of reality."*

Now, precisely, when looking at the aforementioned sort of mathematics, namely at that, which we may call/characterize, as already said, a *"relational"* one, *we are* thus *mostly interested in*/seeking for *the* eventual *mechanism* that might be hidden *beyond* any type of relevant *calculations* (cf. thus, for instance, (4.6.2)); let alone, if the latter appear to be especially effective. In particular, this happens exactly in the case of ADG, which interests us here. Thus,

(4.12)   the *differential-geometric mechanism* of the classical theory (CDG) of smooth manifolds *appears to be* just quite *"functorial"* in character; a *realization pointed out,* in effect, *mostly through* ADG. So, according to the foregoing,

(4.12.1)   the same mechanism may be construed/characterized, as a *"relational"* one (see also (4.5)/(4.6) above), alias *"functorial"*, which thus *functions without any surrounding "space"*; it is, therefore, *"spaceless"* (!), in *complete antithesis with that, which happens classically* (CDG), thus far.

Consequently, one should make it clear, whenever we refer to *the* aforesaid *mechanism of* ADG (viz., the *differential-geometric* one), that;

(4.13)   one can further deal with what we can denominate, as the *"dynamical part" of* ADG, in the same *"relational"* (i.e. the *"spaceless"* one) aspect, as before (see (4.12.1)). Of course, this gives to the latter part of ADG quite a broad gamma of *potential applications*: cf., for instance, *quantum gravity/"singularities"*, thus far; see e.g. A. Mallios [GT, Vol. II: Chapt. 4]. Yet, [17: p. 1946, last part of the Epilogue therein], for probable applications, within the same spirit, via ADG, in *"Geometric Measure Theory"*, and also in *"Geometrical Optics"*; now, concerning the last potential application, as above, we can still remark here that,

(4.13.1)   this is, in general, *true for any subject we apply* CDG, having that ADG proves to be, in that respect, *more "natural"*, hence, more *effective*, as well.

The above certainly *diversifies* ADG, *against the classical theory/("spatial")* perspective, pertaining to the well-known relevant obstacles of the latter.

In this context, by further referring to the aforementioned *"dynamical" part/aspect of* ADG, we remark that the fundamental notion of the so-called, within our framework, *A-connection*, stands, of course, for the *standard notion* of a *connection* (: "covariant derivative"), as it concerns the classical theory (CDG). So, within the present (*abstract*, hence, *non-"spatial"*) context, the notion at issue refers to a

(4.14)   *function* (of *"Leibniz type"*) between *sheaves of modules* (in particular, *"A-modules"*).

Cf. [VS, Vol. II: p. 11, Definition 3.1]. Hence, in other words, a *function between* (complete, à la Leray) *presheaves* of *A*-modules (ibid.), that is, a *function*/morphism *between contravariant functors* (appropriately valued), by applying category-theoretical terminology. Therefore, by further employing standard parlance, one concludes that an

$\mathcal{A}$-*connection*, in terms of ADG is just a *natural transformation of functors* (cf. [CWM: p. 16]), satisfying of course suitable conditions: Leibniz,

(4.15)   as before [VS, Vol. II]. For simplicity, we also refer in the sequel just to a *"connection"*, though we work within ADG, instead of the more technical/formal one, as above, unless it is otherwise specified.

Now, by the very definition of the *mechanism of ADG* (see [VS; Vol. I, II]), this essentially concerns a (co-)*homology theory of "differential complexes"*, the latter being, in particular, *complexes of "differential $\mathcal{A}$-modules"*; here $\mathcal{A}$ stands at the final end (cf. ADG, as above) for a *sheaf $\mathcal{A}$ of unital commutative algebras*, usually *over* $\mathbb{C}$, *or* even $\mathbb{R}$, based on an *arbitrary*, in principle, *topological space X*. So in the case of ADG, one finally considers *sheaves of $\mathcal{A}$-modules* (occasionally abbreviated, just, to *"$\mathcal{A}$-modules"*), that usually are *"locally free of finite rank"*, the so-called herewith *"vector sheaves"* (ibid.).

> **Note 4.2.** (*Terminological*) — The *algebra sheaf $\mathcal{A}$*, as before, refers in particular to what we call in the preceding our *"arithmetic"* (: *domain of coefficients* of the (sheaf) modules concerned; see e.g. (1.4) above). However, for convenience/*economy of the symbolism*, we also denote in the foregoing by $\mathcal{A}$ the whole *category of such sheaves of algebras* (*category of "observables"*, see e.g. (2.23), or even a *"site"*, cf. (2.10)). We wanted to expect the *distinction* to be always clear *from the context*!

On the other hand, *more generally*, one can work, of course, within a *topostheoretic context of ADG*, as this is also advocated, by the present study; yet, see relevant accounts in E. Zafiris [5], I. Raptis [1], and A. Mallios-E. Zafiris [1,2]. Furthermore, cf. also A. Mallios [18: p. 1940, Section 4]. In this context, we further notice that:

> one could also employ, concerning, in particular, the relevant *"differential-geometric"* mechanism, an *entirely categorical* (: homological algebra) *language*: thus, one can still consider, for instance, an appropriate *abelian category* endowed with a (sub)-*category of "cochain complexes"*, along with the corresponding *"differential"* morphisms, as e.g. in the standard case of ADG (see: [VS, Vol. I: Chapt. III]).

Within the same vein, see e.g. the classical account in R. Godement [1: in particular, p. 19, ftn. (1)]; yet, the more recent one of G. Kato [1]. *However*, in what follows, *for simplicity's sake*,

> *we restrict ourselves* to the standard *categories of sheaves of $\mathcal{A}$-modules*, as it was the case so far, *pertaining to* ADG; see A. Mallios [VS, Vol. I - II].

Now, the *basic moral*, thus far, of the entire development *of ADG* is that,

(4.18)   one realizes the *inherent/esoteric* therein *"differential-geometric mecha-nism"*, as an *outcome of an existing*, either *axiomatically* or *by construc-tion* (see also below),

(4.18.1)   *mechanism*/arrangement of the type of a *"complex"* à la *de Rham*, *"standard"* and/or *"covariant"*.

The preceding still clarifies the classical situation, one is confronted with, when dealing with the *standard differential geometry of smooth* (: $\mathcal{C}^\infty$-) *manifolds* (CDG): thus, here *the* aforementioned (cf. (4.18)) *mechanism is*, in effect, *constructed*; precisely, *by what we* essentially *assume*, in that context, namely, via the so-called therein, *"space-time"* (see also (4.6)). However, *this* particular *assumption* of the classical theory (CDG) appears, exactly, *in view of the preceding moral*, as in (4.18), to be *not necessary* (!). Indeed, *this* very *assumption*, viz. so to say this *"spa-cial"* component/*item*, yet, *characteristic*, of the standard theory, *is*, otherwise, the *"guilty"* issue for the usual *"pestilential impediments"* of that theory, in particular, *when confronting with quantum gravity*! The relevant situation herewith, in terms of ADG, has been already exposed elsewhere in several places; see e.g. [VS, Vol. I-II], [GT, Vol. I-II], and i n e x t e n s o in A. Mallios-I. Raptis [1-3], and, in particular, [4].

We come now straightaway by the next Section to clarify, *"technically"*, the phrase *"by construction"*, as in (4.18) above, referring to the so-called *"de Rham complex"*, in the classical context (CDG), or within that of ADG.

1.5   Kähler Construction

First, it is a *fundamental aspect* that in the *classical differential geometry* (CDG) on *smooth* (: $\mathcal{C}^\infty$-)*manifolds*,

(5.1)   the standard *de Rham complex* is an outcome of the *algebra of smooth* (: $\mathcal{C}^\infty$-)*functions* with values in $\mathbb{R}$ or $\mathbb{C}$ (*"numerical"* func-tions), defined on an a priori given smooth manifold X.

See e.g. [VS, Vol. I-II], and for a more recent account thereon Jet Nestruev [1]; cf. also, for instance, *"Fat Manifolds"*: A. De Paris-A. Vinogradov [1].

Indeed, the aforementioned *de Rham complex* is obtained by the standard *Kähler construction* of smooth (: $\mathcal{C}^\infty$-) *differential forms* on the manifold X. So, in other words, one considers here the classical *Kähler-Grassmann algebra* of X,

$$(5.2) \qquad\qquad {}^{\mathbb{C}}\Omega_X^* := \bigoplus_{p \in \mathbb{Z}_+} {}^{\mathbb{C}}\Omega_X^p,$$

in terms of the (*"exterior"*) ∧-*multiplication* of differential forms; namely, the *Grassmann algebra* of the $\mathcal{C}^\infty(X)$-*module* Ω_X^1 (:*differential* (smooth) *1-forms* on X). Here we denote by

$$(5.3) \qquad\qquad \mathcal{C}^\infty(X) \equiv \mathcal{C}^\infty(X, \mathbb{C}),$$

the *algebra* (: point-wise multiplication) *of* $\mathbb{C}$-*valued smooth functions* on X; hence, a *unital commutative* $\mathbb{C}$-*algebra*. Now, we already know from Homological/Commutative Algebra that,

(5.4)   *for any unital commutative algebra* (over $\mathbb{R}$ or $\mathbb{C}$) *one has the* corresponding "*Kähler construction*". So, to saying it in other words, and in effect, in a more general/"*physical*" perspective, one thus obtains a, so to say,

(5.4.1)   "*dynamics*" *of* commutative (unital) algebras *(à la Kähler*, which actually is also "*universal*"; see below). Yet, we may still refer to the above, as "*Kähler dynamics*".

In this context, see also e.g. N. Bourbaki [1], [2], E. Kunz [1], A. Mallios [VS,: Vol. II]. On the other hand, an entirely instrumental fact for the whole differential-geometric mechanism, within the framework of CDG, is certainly,

(5.5)   the *exactness of the* corresponding *de Rham (-Kähler) complex* (precisely, the "*de Rham's resolution of* (the *constant sheaf*) $\mathbb{C}$"; cf. [VS, Vol. I: p. 243; (8.45.1)]). An important consequence thereof is still the famous *Poincaré Lemma* (ibid., p. 244; (8.46)).

Cf., for instance, ibid. p. 241, (8.37), along with p. 243, Lemma 8.2 and comments following it]. On the other hand, the *Kähler construction* referred to in (5.4), is in effect, as already said, "*universal*". That is,

(5.6)   any construction "*similar*" *to/of the same character with it*, leads/ "*factorizes*", in fact, viz. within a *categorical isomorphism*, to *the same sort of construction*/procedure, *as the initial one*.

See e.g. N. Bourbaki [1], E. Kunz [1]. In the case of the algebra (5.3), the previous claim, as in (5.6), is also accomplished by still taking into account the *topology of the algebra* at issue! See A. Mallios [2: p. 131; (4.19)], along with D. Eisenbud [1: p. 389 ff, and p. 407, § 16.8]; yet, see also e.g. M. J. Pflaum [1: p. 207, B. 2].

Now, in the *abstract case* (ADG) we have, *by assumption*, at our disposal the situation (: the *differential-geometric mechanism*), afforded by the *exactness of the* corresponding here (*abstract*: always, "*spaceless*" (!)) *de Rham complex*; precisely speaking, still *in a generalized sense*, as the case may be: see, for instance, [VS, Vol. II: p. 254, Definition 3.1]. In this context, one should still notice here that,

	the whole study is *completely justified*, by a plethora of particular *non-standard cases*, namely, lying *outside of the classical differential geometry of* (smooth) *manifolds* (!), that actually happens.
(5.7)	

Yet, within this same vein of ideas, as before, one might refer to a relevant phraseology of S. A. Selesnick [1], pertaining to the aforementioned *abstract treatment*, quite *characteristic* of that general perspective, which we attain when working in terms of ADG. Namely, within the previous context, one is thus looking for

	"... a *structure sufficiently close to being a manifold* [so] that we may discuss the notion of a *connection* [:*"covariant derivative"*] on a bundle over it".
(5.8)	

Ibid., p. 390 (emphasis above is ours). Furthermore, one can still notice here that *sections* of the aforesaid bundle, as in (5.8) (hence, *"vector sheaves"*, in the terminology of ADG) are usually the *"coefficients"* or even the (unknown) *variables* of the *"differential"* equations, one is confronted with in problems e.g. of *quantum physics*.

Scholium 5.1. — The aforementioned *universal character of the "Kähler construction"* (see (5.4.1) and (5.6)), permeates in effect the whole presentation by means of ADG; namely, its *sheaf-theoretic mechanism*, the latter being in that sense entirely *"functorial"* (yet, *categorical in nature*). As a spin-off, one thus realizes that, indeed,

	the same mechanism (:*"Kähler–de Rham"*) can actually be appropriately transferred still *within a topos-theoretic context*.
(5.9)	

See, for instance, the relevant account in the basic work of E. Zafiris [6], [7], or even I. Raptis [3], [6] and also A. Mallios [18: p. 1940, Section 4], along with A. Mallios-E. Zafiris [1], [2].

Now, within the same vein of thoughts, it seems *very likely* that one might finally be led still to a general *"Representation Theorem"*, relative to *concrete instances of ADG*, in conjunction with appropriate *"generalized manifolds"*; yet, this even within a *topos-theoretic context*: See, for example, the notion of a *"newtonian spark"* in ADG; cf. A. Mallios [16: p. 1574, (4.3)], along with Refs therein, especially, A. Mallios-I. Raptis [1], [2]. Yet, relevant *more concrete* particular *cases of* ADG, as e.g. in J. A. Navarro González-J. B. Sancho de Salas [1]. Also A. Mallios [19] and M. Hakim [1], concerning, in particular, *topos-theoretic* aspects, as of the latter Ref.

1.6 Elementary Particles in the Jargon of ADG

Our aim throughout the following discussion is to point out the fundamental notion of an *"elementary particle"*, when applying the *jargon of ADG*; namely, by considering it, as a particular pair,

$$(6.1) \qquad\qquad\qquad (\mathcal{E}, D);$$

(see below, yet in relation with the ensuing (terminological) Note).

> **Note 6.1.** (*Terminological*) — The preceding notation in (6.1) is a
> familiar one in the jargon of ADG; cf., for instance, A. Mallios [VS, Vol.
> II: p. 244, (1.3), and also [GT: Vols I-II]. Yet, *Section 1.9 in the sequel*.
> On the other hand, for *economy of the symbolism*, we apply *similar
> notations* herewith, though we used them, *within a different context*,
> already in the foregoing (see also e.g. Note 4.2 above). We do *hope*,
> however, *and* still *ask* the reader for *understanding*, that this *will become
> clear*, at each particular case, *from the context*!

Now, the *important issue* herewith is that,

> due to the *functorial character of* ADG (viz. *of its mechanism*), as al-
> ready said (see e.g. (3.58)), one actually realizes that;

(6.2)

> (6.2.1)   more generally, *we can* further *employ analogous ter-
> minology*, referring to ADG, still *within a topos-theo-
> retic context*.

Indeed, as we shall see, the previous assertion is actually effectuated by appealing
to the *important* (categorical) *notion* of a "*fundamental adjunction*"; in our case
the latter is realized, through the following "*adjoint pair of functors*":

$$(6.3) \quad (\mathcal{A}\text{-})connection \; (:\text{``field''}; \; \text{``potential''}) \longleftrightarrow curvature \; (:\text{``field strength''}).$$

Technically speaking, the terms appeared in (6.3) are well defined already, within
the framework of ADG, in other places; see, for instance, A. Mallios [VS: Vol. II].
Concerning the categorical items, we refer to S. Mac Lane [1: Chapter IV]. Yet, on
our latter assertion, pertaining to the "*adjunction*", as in (6.3), we come again in
several places, throughout the subsequent discussion; see also Chapter 2. In this
context, we still note that the general notion of an "*adjunction*" has been considered
already in the foregoing (cf. Sections 1.2-1.4, in particular, 1.2.1).

On the other hand, it is now our next objective to show that the previous claimed
adjunction, as *in (6.3)*, is just another *consequence of*, in fact, another *equivalent*
version of *the fundamental*, in *Dynamics*,

$$(6.4) \qquad\qquad\qquad principle \; of \; least \; action.$$

See also e.g., "*duality*" *principle*: $\delta(\omega) = \tilde{\vartheta}(g)$, cf. (10.1). As we shall see by the
ensuing discussion,

(6.5)

> it is, in effect, *in the form of the above fundamental principle* that a
> *physical law appears to us.*

See e.g. (9.42), in particular, (9.44.1) in the sequel. Yet, in this regard, we also note that,

(6.6)   anytime we are able to detect/locate a *physical law*, we have then actually succeeded to behave *"functorially"* (!); viz. in other words, to *participate the function of physis*!

Cf. also e.g. (8.37)/(8.38) below. Therefore, we thus succeed *to be* essentially *part of it*, hence, of the *"force field"* (: the *"curvature"*, in technical terms) *itself*! We note that the latter is akin related with relevant thoughts of a recent ongoing study of I. Raptis in [8]. Now, within the same vein of ideas, we further remark that;

(6.7)   a *"variational problem"* might still be perceived, as a *pursuit* of the appropriate *physical law*. Hence (cf. (6.5)), of the appropriate version of *the* above *principle*, as *in* (6.4), as well; cf. also Section 1.10 in the sequel). The latter is actually *hidden behind the* (*"covariant"*)*variation*, at issue.

See also A. Cannas da Silva [1: p. 114]. On the other hand, and by still anticipating our later discussion (cf. Section 1.10 below), we also note, yet, in conjunction with the remarks following (6.6), as above, that; there exists, in effect, a *simultaneous variation of* the *field* (:the *physical law*) *and its effect* (:*"force field"*, alias *"curvature"*). This is expressed, for example, by the so-called,

(6.8)   *"transformation law of potentials"* (see (10.12) below); yet, more generally, by what one might call, in that context,

    (6.8.1)   *"Utiyama's principle"*.

See also (9.51) in the sequel and subsequent remarks therein.

More on this will be found in the subsequent discussion, Sections 1.9-1.11; we also consider therein the way the above may still be construed as *particular versions of* the fundamental principle, as in (6.4). Yet, this in conjunction with the fundamental *adjunction* (still, *"generalized duality"*) in (6.3). Furthermore, within the same point of view, as above, we may also express the corresponding situation therein, by saying that,

    under the previous circumstances,

(6.9)   (6.9.1)   the *physical law* (expressed through the so-called *"force field"*) *"disappears for us"*!

The *"technical"* counterpart of the above can be given, for instance, by *formulae of the type* (9.26) below. (See also the relevant remarks therein).

Yet, one should also remark here, *as it concerns* our previous note (6.9.1), that all the *fundamental equations in physics* (e.g. Einstein, Yang-Mills etc) are, just, *particular cases of* (6.9.1) (in the form, for instance, of (9.26), or similar formulas, ibid.).

We continue by further commenting on our claim, pertaining to the so-called "*adjunction*", as in (6.3): So, as already remarked in Scholium 5.1 above,

(6.10)
> the sheaf-theoretic mechanism of ADG is entirely *functorial*. Yet, apart from its *spaceless character*, it is still *independent of* the particular "*arithmetic*" (:"*domain of coefficients*"), viz. the *algebra sheaf* $\mathcal{A}$ (see [VS: Vol. II]), we occasionally employ. Furthermore, as another spin-off of its *sheaf-theoretic substance*, the same mechanism is, let alone, "*localization-universal*"; see e.g. *sheaf-theoretic Kähler construction*, as in (5.4)/(5.6). Of course, the latter is of a special significance, still, for a *topos-theoretic treatment of* the whole account of *ADG*, which as we shall see is, indeed, the case;

> yet, in that context, we also note that (6.3) *is*, just, *a particular important instance* thereof (cf. below).

Furthermore, the same *sheaf-theoretic substance of ADG*, being thus of an algebraic/*categorical nature*, makes it possible the *transference of its mechanism*, even, to a more general set-up, as for instance, *to a topos-theoretic one*; see e.g. relevant Refs after Note 4.2 in the preceding. Of course, one can still notice here, in that context, the fact that;

(6.11)   "... *the very notion of sheaf is ... central to topos theory.*"

Cf. [MM: p. 2] (emphasis above is ours). Therefore, within the present more general (abstract) setting, considered in Section 1.2 above, one has still to *explain the aforementioned ADG-mechanism in topos-theoretic terms*: As we shall see, this is intimately connected with the above idea of an appropriate *adjunction*, as hinted at already by (6.3); see also relevant remarks in (8.2) below. However, first we can further remark that;

(6.12)
> the main function of mathematics in physics must often be taken *locally*: cf. for instance, *differential equations*, therefore, also the "*arithmetic*" (alias, domain of "*coefficients*") we apply. Consequently, it should be at least appropriate to employ *sheaves of the pertinent structure* (see also below), getting thus also, in that way, the "*varying feature*" of the same due to the very character of the notion of a sheaf. Of course, this latter issue is always *of fundamental importance* pertaining in particular to "*relativistic*" *aspects* of the matter, hence let alone, to *quantum relativistic* ones.

All told, the inherent *sheaf-theoretic character of ADG* yields the possibility of looking at the *two fundamental items* of the notion *of sheaf*; that is, *simultaneously*, as well, at the

(6.13)   *local* and *varying* aspect of the matter (: the *differential-geometric mechanism* of ADG),

in a quite categorical manner, hence, still *functorial*. Yet, in such a way that, both items, as in (6.13), to be further *capable of admitting* even *a topos-theoretic formulation*; this is based, of course on the very nature of the notion of "*topos*". In this context, see also (6.11), as above, while within the same vein of ideas, we can further note that,

(6.14)   "*one of the basic reasons to considering sheaves*" was the problem of confronting with "*local*" (and also "*varying*") situations. Therefore, *general relativity*, or yet *quantum relativity/gravity*.

See also, for instance, [MM: p. 95], along with e.g. R. Haag [1: p. 326], concerning the physical counterpart of the above, in particular, of (6.12)/(6.14).

1.7   Relational Aspect of Space, Again

On the basis of the foregoing, one thus arrives at

(7.1)   an entirely *spaceless* framework, pertaining to the notions both of "*local*" and its *variation* (i.e., to the "*topological*", *not* the "*analytic*" one (!)).

So one gets at the above two notions, indeed, fundamental ones for the whole of our treatment; first, *sheaf-theoretically*, and then finally, within a more general setting, in terms of *topos theory*. In that way, we can further notice that, the *standard notion* of

(7.2)   "*space*" $\longleftrightarrow$ (smooth) *manifold* (: in general, CDG, and in particular, e.g. *general relativity*)

acquires a more "*natural*"/categorical, thus "*relational*" substance; cf. also A. Mallios [20]). Indeed, (physical) *geometry is the outcome of the physical laws*, cf. this author, as above, [16: p. 1557, (1.1)]. Whence, of "*relations*"; see also the same author [14: p. 60, (2.3)], along with (8.25) in the sequel. Yet, within the same vein of ideas, we can further notice (cf. also (2.29)/(2.30) in the preceding) that,

(7.3)   (physical) "*geometry*", being the *result of the physical laws*, as above, *is* always *based on an inherent algebraic*/relational *mechanism* (take e.g. the *physical field*. See also, ibid., p. 63, (3.3) and comments following it).

As a consequence, we thus come to the conclusion/*moral* that,

(7.4)   any time *"geometry"*, in the classical strict sense of the term, namely, *"analytic"*, viz.   *cartesian-newtonian*, even (classical) *differential-geometric* one (hence, *not "physical"* (!)), is *not available*, therefore, the notion of *"space"*, emanated therefrom, *does not match well* (cf., for instance, *quantum domain*), it is then, at least, good to *resort to* the more *"natural"* (see above) notion of *functors* (!). That is, in other words, to *categorical/"relational"* techniques (cf. also the comments, following (7.2), as above).

On the other hand, we can further remark, in that context, that, what we considered above, as

(7.5)   *"natural/physical"* geometry, thus a *"relational"* one, according to the very definitions (cf., for instance, A. Mallios [16: p. 1557, (1.1), and also p. 1558, (1.5)], along with [18: p. 1930, (1.5)]), *is*, indeed, *available*, always (!). Now, *its effectiveness*, concerning us, *depends* simply *on our* ability/*imagination* to provide *the* appropriate *mechanism to* use/*employing it*.

In this regard, see, for example, R. Feynman (*diagrams*; yet R. P. Feynman [1: p. 166], pertaining to his remarks on the *"geometry in the* (infinitely) *small"*. Cf. also, A. Mallios [14: p. 58, (1.3), and subsequent remarks therein]). Furthermore, within the same context, cf. L. Schwartz (*distributions*); yet, recall the famous *Plato's utterance*,

(7.6)   *"eternally God geometrizes"*;

in the sense, of course, of a *"relational"* (: natural) *geometry*, as above; therefore, the frontispiece of the present study. See also e.g. A. Mallios [16: p. 1557, epigraph, and starting scholia, therein, below it].

Thus, in other words, one speaks herewith of a

(7.7)   *geometry, referring directly to the "geometrical/physical" objects* themselves, *independently of* (their relative) *"position"*.

Anyway, within the previous perspective, the above reminds us of *G. W. Leibniz*, when asking for a *"geometrical calculus"*: See, for instance, ibid. p. 1560, (1.17), along with p. 1559, (1.11), or even, in that respect, A. Mallios [17: p. 267, (2.6)]. Yet *H. Grassmann* with his famous *"exterior algebra/calculus"*, or even *B. Riemann* in his celebrated *Habilitation* lecture, referring to the *"infinitesimals"* (thus, *"relational"*, in our sense, *objects*), as the *fundamental issues*. So we understand here *"geometry"*, as a *technique/mechanism of studying* "geometrical" objects, as before, or yet, *equivalently* (see also e.g. A. Mallios [12: (1.1), (1.4)]) the whole *"image"*, at least, *of the world* we live. Hence, *"geometry"*, as a *technique* to study, either the objects, as above, or yet the entire *"cosmos"*/image around us.

1.8 Dynamical Dressing, Extension: Kähler Construction (Contn'd)

We are thus finally in the position to confront now with the *"adjunction"*, set forth *in* (6.3) in the foregoing; this, after we have elaborated/explained the appropriate thereat, quite *abstract*, however, still *natural/"relational"* otherwise convenient framework, within the present discussion.

So our aim now in the sequel is to show the way, one can get at a, so to say,

(8.1)   *"dynamical dressing"* of that, which we talked about already in *the preceding.*

Indeed, as we shall presently see, this can be done by employing the *two fundamental "differential-geometric"* notions (of the classical theory, viz. of CDG), as appeared in (6.3); however, now within the present *functorial* (yet, *topos-theoretic*) *context!* The means to this end, appears, therefore, to be certainly *the* same *functorial character of ADG*, as this has been already explained in the foregoing (see e.g. (3.58), along with (4.12)/(4.13) and (4.16)). The *motivation to the* present perspective, which is, of course, an *axiomatic* (: abstract) *treatment*, a continuation in effect of our previous relevant discussion in A. Mallios [18: Section 4; cf. in particular, p. 1944, Scholium 4.1], *was* still *the* basic *work* on the subject by E. Zafiris in [2]; see also A. Mallios-E. Zafiris [1], [2] along with a related study by I. Raptis in [4], based also on ideas from ADG.

Now, in this context, what we actually have to do here is to look at the essential *message*/information we get *from the* well-known, indeed *fundamental "Hom-⊗ adjunction"*, by further considering the *intrinsic*/deeper *meaning* of its particular constituents, *when* these are *specialized, as in* (6.3): That is, in other words, we first notice that

(8.2)   *the* aforementioned *Hom-⊗ adjunction*, that concerns us herewith, *has been* already *considered* in the foregoing *by (3.5)/(3.6)*, so that *what we* further intend to *prove here is that,*

(8.2.1)   *the same adjunction*, as above, *can be expressed*, in particular, *by* (6.3).

However, before we proceed, let us still remark that, by strengthening the point of view, as stated by (8.2.1), and also *paraphrasing*, for instance, in that context, *H. Kleisli* (1965) (see e.g. S. Mac Lane [1: p. 143/251]), we can further note that,

(8.3)   *every physical (re)action is induced by a pair of adjoint functors*; thus, in other words,

(8.3.1)   *every physical (re)action is* (the outcome of) *a particular "adjunction".*

Yet, as we are going to see in the sequel, (8.3.1) *is* further *related with the principle of least action* (see also (6.4)/(6.5) in the preceding). Within the same context, see still A. Mallios [19: p. 65, (1.3)/(1.5)]. On the other hand, one might also relate (8.3.1), by still paraphrasing L. Wittgenstein [2], in that sense, with the adage (his own) that,

(8.4)   *"Physics ... describes concomitant cases"*,

associating thus *"concomitant"* with *"adjunction"*; in this regard, see also e.g. A. Mallios [16: p. 1571, (2.32)/(2.33)], and still ibid., p. 1570, (2.30), concerning another justification of the previous aspect. Thus, we may still say, by paraphrasing again Wittgenstein, in that context, that;

(8.5)   *physics describes/effectuates "adjunctions"*!

That is, to recall/paraphrase here further, S. Mac Lane (emphasis below is ours), one gets at an

(8.6)   *"... (adjoint) pair of functors* with a *natural isomorphism* between corresponding sets of arrows ..."*

See [CWM: p. V]; in this context, cf. also Section 1.3 in the preceding, in particular, (3.2)/(3.8) therein.

   The preceding is very tempting, indeed, so as to lead us actually to a *much more general perspective*; namely, to the idea related with the celebrated notion from Category Theory of

(8.7)   *"Kan Extension"*.

See, for instance, S. Mac Lane [1: Chapt. X, in particular, p. 244, along with Theorem 2 therein: *"Formal criteria for the existence of adjoints"* (J. Bénabu, 1965); yet, p. 230, Theorem 2]. So, within the previous context, one can now first remark that,

(8.8)   *even* the very notion of an (*A*-)*connection*, as exemplified for instance by the standard *"Kähler construction"* (cf. (5.4), viz. the so-called, *"Kähler dynamics"*, yet see also (5.6)), is, as we shall see, intimately related with that one of an *"extension of scalars"* functor.

Indeed, the same notion of *the aforementioned construction* (Kähler, *"universal"*, as it were, cf. (5.6) and the comments after it, therein) *is*, at the very end (see also below), *realized by* the intervention of *the ⊗-functor*. On the other hand, an *A-connection is* further *effectuated*, of course, even *physically speaking* (!), by the *"adjoint"* of the previous functor/procedure; that is, by that, which we may call, *"functor of curvature"*, as the latter can be *expressed by the (bi-) functor* $\mathcal{H}om(\cdot,\cdot)$.

Now, the above highlight, in brief and roughly speaking, the quintessence, in effect, of what we actually mean by the *"adjunction"* (6.3), as this could be *associated with* the adjoint functors, as in (3.5)/(3.6) in the preceding.

On the other hand, by the very definition of the standard *"Kähler construction"* (cf. also the ensuing discussion), an $\mathcal{A}$-connection is given, by defining/constructing, in effect, the domicile of the basic *"infinitesimal differentials"* (Leibniz, Riemann); thus, the space, and actually the $\mathcal{A}$-*module of* the so-called, *"1-forms,"*,

$$(8.9) \qquad\qquad \Omega^1.$$

Yet, this along with the definition of what we call, *basic ("flat")* $\mathcal{A}$-*connection*,

$$(8.10) \qquad\qquad \vartheta : \mathcal{A} \longrightarrow \Omega^1,$$

that is, in other words, *the* corresponding to our (abstract) case, *classical "dx"*, being thus now a suitably defined (see also below) $\mathbb{C}$-*linear morphism of* the $\mathcal{A}$-*modules* involved; alias, what we may still call a *"Leibniz map/*morphism". However, by looking at the very definition of Ω^1 (cf. below), one realizes that,

$$(8.11)$$

ϑ (hence Ω^1 itself) is the outcome of *an appropriate*

$$(8.11.1) \qquad \text{*"extension"* of the given } \mathcal{A}\text{-module (: our "arithme-} \\ \text{tic") } \mathcal{A}\text{, achieved actually, through the } \otimes\text{-}\textit{functor.}$$

The latter denomination (*in parenthesis*, as before) still reminds, of course, the *classical construction of Kähler*, viz. his manner of studying a *"field* (hence, in our terminology, that one of an *"arithmetic") extension"*.

Therefore, in other words, *"technically"* (: mathematically) *speaking*,

$$(8.12) \qquad \begin{array}{l} \text{the very definition/construction of } \Omega^1 \text{ entails already } \textit{one of the two} \\ \textit{parts of the adjunction } (3.5)/(6.3). \end{array}$$

Thus, now even, *"physically speaking"*, one obtains via

Ω^1, the *"world of infinitesimals"*, in effect, of *"infinitesimal variations"*, that represent the constituents of the *change of a* given situation/*"state"* $\mathcal{A}$, due to an *interaction*; therefore, the following diagram,

$$\textit{"adjunction"} \longleftrightarrow \textit{observer-state},$$

$$(8.13) \qquad (8.13.1) \quad \begin{array}{l} \text{realized, through the } \textit{change of the algebra} \text{ (of "observables")} \\ \mathcal{A}, \text{ viz. of the (initial) "geometry" (: yet, the "multiplica-} \\ \textit{tion")}. \end{array}$$

The same *"interaction"*, as before, is described by the application on $\mathcal{A}$ of the *fundamental* $\otimes$-*functor*, viz. by considering the new $\mathbb{C}$-*algebra* (sheaf), $\mathcal{A} \otimes_{\mathbb{C}} \mathcal{A}$, still viewed, in the sense of (8.13.1), as an *"extension"* of $\mathcal{A}$. (See also (8.14.1) and (8.19.1) below).

So, one thus arrives at the *quintessence of the Kähler construction,* viz. the *realization of the change* (of our *"arithmetic"*) $\mathcal{A}$, as above, yet, of what we already called, *"extension of $\mathcal{A}$"*. Indeed, this by considering the

(8.14)   *kernel,* say I, *of the* aforesaid, in technical terms, *interaction*; that is, the locus where, *the two multiplications* (: *"geometries"*) *coincide:* namely, the *initial multiplication on $\mathcal{A}$, and that* one defined, *through the $\otimes$-functor on $\mathcal{A} \otimes_{\mathbb{C}} \mathcal{A}$.* Thus, (classically) we set

(8.14.1) $$I := \ker\mu,$$

such that *one defines* (the so-called *"multiplication morphism"*),

(8.14.2) $$\mu(x \otimes y) := x \cdot y,$$

for any x, y in $\mathcal{A}$.

So we come now to the *definition of Ω^1,* through the previous kernel I, as in (8.14.1), in effect, *modulo* something, *"even smaller"* then I, viz. I^2. That is, *we set,*

(8.15) $$\Omega^1 := I/I^2,$$

while *one* further *defines,*

(8.16) $$d(x) := x \otimes 1 - 1 \otimes x, \ modulo \ I^2,$$

for any $x \in \mathcal{A}$; namely, *we set* (cf. (8.10)),

(8.17) $$\vartheta(x) := [d(x)] \in \Omega^1, \ with \ x \in \mathcal{A},$$

for the corresponding class of $d(x)$, as above. Therefore (still classically), one thus gets at the relation (:*"Leibniz formula"*),

(8.18) $$\vartheta(xy) = x(\vartheta y) + y(\vartheta x),$$

for any x, y in $\mathcal{A}$. In other words, one thus arrives at the desired (basic, *"flat"*) $\mathcal{A}$-*connection ϑ, as in (8.10);* that is, at a $\mathbb{C}$-*linear morphism* (of the corresponding $\mathbb{C}$-vector space sheaves), *obeying "Leibniz condition"*, as in (8.18). See also e.g. [VS: Vol. II].

In toto, *the* physical *field,* being here the *"causal agent"* of the *"interaction"*, hence, in effect, of the *"variation"* at issue (: the *change* we actually observe), appears as the *functor* of "*infinitesimal scalars-extension*" (cf. also (8.8) above); yet, in view of the following relation (: "*extension*" *of* $\mathcal{A}$, *via* I),

$$(8.19.1) \qquad\qquad \mathcal{A} \otimes_{\mathbb{C}} \mathcal{A} \cong \mathcal{A} \oplus I,$$

within an $\mathcal{A}$-*isomorphism*, the latter being an outcome of the *short exact* $\mathcal{A}$-*sequence*,

$$(8.19.2) \qquad 0 \longrightarrow I \equiv ker\mu \longrightarrow \mathcal{A} \otimes_{\mathbb{C}} \mathcal{A} \overset{\mu}{\longrightarrow} \mathcal{A} \longrightarrow 0$$

(8.19) (defining I, as in (8.14)). Thus, *the* same *field* is now still realized, "*technically speaking*", as an

$$(8.19.3) \qquad\qquad "(\mathcal{A}\text{-})connection\text{-}inducing\ functor".$$

Hence, the standard "*Kähler dynamics*", as a spin-off/ *description of the* "*field*": thus, another vindication of the still standard, throughout ADG, association,

$$(8.19.4) \qquad\qquad "field" \longleftrightarrow (\mathcal{A}\text{-})connection;$$

see e.g. A. Mallios [17: p. 278, (6.6)], and also (10.2)/(10.3) in the sequel.

Therefore, in other words, one gets again at the association,

$$(8.20) \qquad\qquad (\mathcal{E}, \mathcal{D}_{\mathcal{E}})(: "field") \longleftrightarrow \mathcal{D}_{\mathcal{E}} \equiv \mathcal{D};$$

cf. also (6.1) in the foregoing, along the following Note 6.1 therein, yet, the above quotations after (8.19.4). As an outcome of the preceding, *we do realize*, as it were, anyhow (see also the ensuing discussion, concerning "*curvature*"), *that*

(8.21) the *field* is known ("*technically speaking*"), in terms of its "*trace*", that is $\mathcal{D}$, *through* Ω^1; yet, cf. also (8.19.3), (the functor) $\mathcal{D}$, *via* its "*domain of arrival*" (: our "*antenna*") Ω^1, something, of course, that still reminds, R. Feynman's famous adage, in that,
(8.21.1) "*physics is number*" (!).

The same still refers to a similar, telling utterance of L. Wittgenstein (by para-phrasing it), that;

(8.22) "a *physical law* is *something* [we] *try to fathom* ["*gauge*"] *and make use of.*"

(Emphasis above is ours, as also in (8.21.1)); yet, see L. Wittgenstein [1: p. 70e].

On the other hand, the *curvature* of an ($\mathcal{A}$-)connection that *always exists*, any-time one has a connection (: physical law, cf. (8.19.4)/(8.20), along A. Mallios [11: p. 70, (5.9)/(5.10)], [16: p. 1558, (1.3)], *is* further *defined, through the* (*bi*−)*functor* $\mathcal{H}om_{\mathcal{A}}$: something "$\mathcal{A}$-*invariant*, viz., in general, "*covariant*" (alias, "*synvariant*" (I. Raptis)), according to the very definitions; cf., for instance, [VS, Vol. II: p. 192, Lemma 2.1]. Within the same context, see also the *functor* $\mathbb{F}$ in the foregoing, in particular, (2.39.2) pertaining to the *adjunction (6.3)* at issue, or yet to what we already called in the preceding "$\mathcal{H}$*om-*$\otimes$ *adjunction*", cf. also (2.42)/(8.16), as above, along with (2.38.1), concerning the $\otimes$-*functor*.

All told, we further notice here that, in any case, "*mathematically speaking*",

> (*physical*) *geometry* (cf. (7.3)/(7.5)), being the *spin-off of physical laws*, (ibid.), can be construed, as

(8.23)
> (8.23.1)   the "*solution set*" *of a* (partial nonlinear) *differential equation*; viz. the *kernel of a* "*differential operator*", describing the *physical law* ($\longleftrightarrow$ ($\mathcal{A}$-)*connection*) at issue. See also e.g. [GT, Vol. II: p. 72, (9.15)].

Therefore, one gets at the following conceptual diagram,

(8.24)
> "*fields*" (: *natural/physical law*) $\longleftrightarrow$ ($\mathcal{A}$-)*connection* $\longleftrightarrow$ *differential equation/operator*,

along with the "*solution set*" of the latter, as before. Indeed, this is, in fact,

(8.25)
> the *real* (!) "*space*", we actually have (based on, viz. control) through our equations/theory, on which the corresponding *natural law*, as above (cf. (8.23.1)), holds true. The same "*space*" is thus still the *result of* "*relations*", following the *jargon of ADG*; yet, also *in accord with Leibniz*: cf., for instance, A. Mallios [17: p. 267, (2.6)/(2.7)]. Therefore, "*space-less*"(!), in the classical sense(: CDG) of the latter term: so it is thus the *outcome* of the particular "*arithmetic*" we provide at a (given period of) time (see also ibid.).

Yet, within the same context, one should also remark that,

(8.26)
> it is actually the *curvature*, $R(\mathcal{D})$, *of the* given ($\mathcal{A}-$)*connection* $\mathcal{D}$, as above, that is always the *variable of* the *equations* that describe the *physical law*, under consideration (cf. (8.23)).

Take thus, for instance, *Einstein's equation* (in vacuo),

(8.27)
> $Ric(\mathcal{E}) = 0$, or in *abbreviated form*,
>
> (8.27.1) $\nabla \rho = 0.$

Here, it is actually the "*metric*" ρ, as above (viz. the appropriate "$\mathcal{A}$-*bilinear form*"), which finally *yields*/effectuates the *adjunction (6.3)*,

(8.28)
$$\mathcal{D} \xleftarrow{\hspace{2cm}} R(\mathcal{D}),$$

(cf. also (2.38), and (8.13) in the preceding). So one is finally led to an

(8.29)
> *adjunction between* "*algebra*" (: $\mathcal{D}$, $\otimes$-*functor*, see e.g. (8.16)) *and* "*geometry*" (: $R(\mathcal{D})$, $\mathcal{H}$*om-functor*).

In this context, see also [VS, Vol. II: p. 157, (8.28): the $(\mathcal{A})$-*connection* (:"*Christoffel symbols*"), in terms of the "$(\mathcal{A}\text{-})metric$"; "*2nd Christoffel identity*"]. Yet, the same *adjunction*, as in (8.29), viz.

$$(8.30) \qquad\qquad \text{``}algebra\text{''} \quad \longleftarrow\!\!\!\longrightarrow \quad \text{``}geometry\text{''},$$

is certainly still *in force*, through an appropriate "$(\mathcal{A}\text{-})bilinear\ form$", pertaining to what we may consider, as a "*Lagrangian perspective in Physics*", viz. to the so-called "*symplectic geometry*"; cf., for instance, the *algebraic part* of it, in the form of "*Geometric Algebra*": see e.g. E. Artin [1], R. Deheuvels [1]. On the other hand, the *ADG counterpart* of the latter has been already advocated throughout the ongoing recent work of A. Mallios-P. P. Ntumba [1-4] and that of Ntumba [1-3] and his collaborators P. P. Ntumba-A. C. Anyaegbunam [1], and P. P. Ntumba-B. Y. Yizengaw [1]. Furthermore, *the same* point of view, *as in (8.30)*, is actually of a *more general significance*, concerning *potential applications to Physics*: see, for instance, (9.22) below, along with Scholium 10.1 in the sequel.

In toto, and still in conjunction with the previous discussion, we can further remark, paraphrasing actually, e.g. in that context, S. A. Selesnick [1: p. 390] (cf. also (5.8) in the foregoing), in that: *we are* always *struggling to*

$$(8.31) \qquad \begin{array}{l} concoct\ a\ \text{(mathematical)}\ structure,\ so\ that\ be\ able\ to\ \text{provide/} \\ formulate\ \text{``}differential\ equations\text{''}\ (:\text{connections}),\ \text{with respect} \\ \text{to it.} \end{array}$$

Of course, that was also the whole story, even with the (classical) Calculus, since the time of Isaac Newton, already, and then still with the standard/(classical) Differential Geometry (CDG). On the other hand, to update and also paraphrase herewith, J.-M. Souriau [1: p. xvii], we can further remark that,

$$(8.32) \qquad \begin{array}{l} \text{only}\ theories\ \text{which are}\ \text{``}covariant\text{''},\ and,\ \text{yet, of course (based on our} \\ \text{experience, thus far, concerning the ``}quantum\text{''}),\ \text{``}localizable\text{''}\ \text{ones, to} \\ \text{a satisfactory degree,}\ can\ survive\,! \end{array}$$

Therefore, otherwise said, *theories* that are *able to* participate/*follow* the function of *Nature*, that is the (natural) *variation*, hence, the *dynamics of the physical laws*. We can thus still express (8.32), by saying that one should always look after

$$(8.33) \qquad theories,\ akin\ to\ the\ \text{``}relativistic\text{''}\ character\ \text{of the physical laws.}$$

Of course, by "*relativistic*", one means herewith, the point of view of

$$(8.34) \qquad \text{``}general\ relativity\text{''},\ \text{when viewed}\ as\ a\ varying\ dynamical\ theory.$$

Furthermore, and certainly, *not less important* (!),

(8.35) one has to *look at (8.32)/(8.33) not* in a *"spatial"* way (!), in the classical sense of the latter term (viz., in terms of CDG).

Of course, *such a theory is ADG*, by the very essence of its mechanism that is, *varying* (cf. (8.34), (8.35)); this has been already pointed out in the preceding (see e.g. (6.13)). Yet, see also A. Mallios [16: p. 1568, (2.18)], [18: p. 1931, (1.8)], [17: p. 271, (3.8)/(3.9)]. Therefore, the same

(8.36) functorial/*"covariant" character* of the whole machinery *of ADG* allows its transfer to a similar *categorical framework*, as e.g. within *topos theory*, cf. below; yet, along with the corresponding applications of the aforesaid *categorical perspective* to fundamental *physical notions*, as it actually were, of the same *functorial nature*.

See also, for instance, A. Mallios [18: p. 1930, (1.5.1)]. Furthermore, within the same framework, one should also remark here that, by the term,

(8.37) *"functorial"*, one still means something, that is *independent* of any *"gauge"* (: *measurement*), whatsoever! Precisely, one should further recall herewith the *Greek work*,

(8.37.1)   *"geometry"*,

and its etymology, where the second constituent of the word, viz. metrō : measure, refers exactly to *our* (: *"we"* ≡ *"𝒜"*, cf. also (1.4)) *intervention of observing/measuring*; viz. *describing Nature*, therefore, *"fathom the physical law"* (L. Wittgenstein, cf. (8.22)). Yet, within the same context, see also (1.2), along with (1.3.1) in the foregoing.

Thus, one arrives at one more realization that,

(8.38) what we usually name, *"mathematically speaking"*, *"geometry"*, *is not "physical"*, that is, *"functorial"*, as above, yet, *natural*! Thus, at the very end, *not a "relational"* one! It is, just, *our own description of the* real *"physical geometry"*; see also Note 4.1 in the foregoing, along with (4.5).

On the other hand, it is now still quite *worth noticing* here that,

(8.39) *ADG proves*, however, *to be* thus far *an appropriate* concrete *candidate* for that (see also below (8.42)/(8.43)).

Thus, within the same vein of ideas, we can further note with V. Guillemin and S. Sternberg [1: p. 1], that (emphasis below is ours)

(8.40) *"Not enough has been written* about the *philosophical problems* involved *in the application of mathematics ... to physics"*.

Furthermore, as another similar support of the above, we can still mention here G. 't Hooft's utterance, in that

(8.41)   *"... the problems of quantum gravity are much more than purely technical ones; they touch upon very essential philosophical issues"*.

See G. 't Hooft [1: p. 2]. Of course, the preceding under the proviso that *we are* still *able to recognize* the pertinent *"adjunctions"*, which are *involved* in the *physical notions*, we are interested in. But, see also our relevant remarks in the ensuing Section 1.9, in particular, (9.21) therein, and the comments following it.

As a result, one thus concludes that a theory, like the one which exhibited by (8.34)/(8.35), that is in other words,

(8.42)   a *varying dynamical theory*, being also a *"non-spatial"* one can still *be confronted with* problems of *the quantum*, as well. So it will be then *"quantum relativistic"* (!), in character.

Now, the standard case of ADG, as it concerns a quite general *topos-theoretic context*, as that one considered in the preceding, has been already treated, in detail, by E. Zafiris; see, for instance, [1], [2], along with [4]. Yet, a relevant account has still been given by I. Raptis [3], [5]. Thus, the preceding proves that one has, *by means of ADG*, what one may consider, as

(8.43)   *"functorial dynamics"*; that is, in other words, a

          (8.43.1)   functorial *"differential-geometric"* mechanism (!).

Yet, a *"functorial mechanism"*: thus, quite *categorical* in nature (cf., for instance, (4.6.1), or even (8.37)), *that* further *affords* a *"differential – geometric"* substance. That is, one can thus write down for example or even think (!) of a, *"differential equation*!

In this context, see also (3.58) in the preceding. On the other hand, as already pointed out in the foregoing, *the aforementioned mechanism* can be formulated, within a *topos-theoretic* framework. So, it might still be a *basic outcome* of the above, that

(8.44)   the whole *technique*/perspective *of ADG* might lead finally to a short, in effect, of a *Unified Field Theory*.

See also (9.28) in the sequel, along e.g. with A. Mallios [16: p. 1591, Subsection 6. (a)].

1.9   Adjunction, Least Action Principle

Our aim by the ensuing discussion is to correlate the preceding and, in particular, the *"adjunction"* (6.3), with the fundamental, so-called, *"least action principle"*. Indeed, as we shall see (cf., for instance, (9.10), (9.16)/(9.17)), these two basic perspectives, under consideration, are in effect *equivalent*. Yet, cf. also our previous remarks, following (6.3), *in conjunction with* (6.4)/(6.5): Thus, by seeking for to employing a *"physical-mathematical"* terminology, as it was essentially the case, so far, within the framework of ADG (see e.g. A. Mallios [VS, Vol. I-II], [GT, Vol. I-II]), one realizes that:

> *the physical law*, which is associated with/represented by the *"interac-tion"* between a couple of *"paired $\mathcal{A}$-modules"* (according to our theory, see e.g. (6.1), (1.19)/(1.20), or even A. Mallios [20], [11: p. 146, (1.1)] and [15], [21]), (seemingly) *disappears*, any time

(9.1)

> (9.1.1)   the aforementioned *pairing* is *"compatible"* with the law.

> Thus, technically speaking, *whenever one has the relation* (see right be-low for the terminology applied herewith),

> (9.1.2) $$\mathcal{D}_{\mathcal{H}om_{\mathcal{A}}(\mathcal{E},\mathcal{E}^*)}(\bar{\omega}) = 0.$$

> We may still remark here that the previous relation is actually the

> (9.1.3)   *point of view* pertaining *to any physical law-observation*.

> See also Section 2.1, *Utiyama's theme*.

On the other hand, one can further express the preceding, by also saying that;

(9.2)

> any time *we participate Nature*, viz. *we succeed our theory to be "co-variant"* with Her action (cf. also the ensuing discussion, concerning the meaning of the last expression), *we do not* in effect *feel anything*!

Furthermore, within the previous context, we should still notice here (see also A. Mallios [18: p. 1932, (1.13)/(1.17)]) that, what we require by (9.1.1), is essentially the

(9.3)

> *"compatibility"* of the pairing, at issue, *with the physical law* (as is ac-tually stated therein), and *not* (!), of course, *the other way round*, as is usually the case (cf. also (9.4) in the sequel).

Indeed, by an *abuse of* terminology/*language*, this is in fact the way we use to express the *situation*, as *pointed out through* (9.3), just by *simply misinterpreting phenomena*; see e.g. *"singularities"* (!), and the like.

Now, before we proceed further, we have to clarify, a bit more, the *terminology employed in (9.1.2)*: So $\mathcal{D}$ therein denotes the "*covariant derivative*" that corresponds, in view of our *setting in ADG*, to the $\mathcal{A}$-*connection*, representing the *physical law* at issue, according to the following diagram (see also (8.24) in the preceding),

$$(9.4) \qquad \mathcal{A} - connection(: \text{ "}potential\text{"}) \longleftrightarrow physical\ law.$$

In this context, cf. also, for instance, A. Mallios [14: p. 62, (3.2), and p. 63, (3.3) and Subsection iii), along with p. 70, (5.9)], and [16: p. 1558, (1.3)].

Furthermore, the $\mathcal{A}$-*connection* considered *in (9.1.2)*, corresponds not to the initially given *Yang-Mills field*,

$$(9.5) \qquad (\mathcal{E}, \mathcal{D} \equiv \mathcal{D}_\mathcal{E}),$$

(ibid.), but to what one might call, its "*Heisenberg representative*" (: "*matrix analogue*"),

$$(9.6) \qquad (\mathcal{E}nd\,\mathcal{E}, \mathcal{D}_{\mathcal{E}nd\,\mathcal{E}}).$$

Indeed, in a more general/distinct manner by considering, in particular, the (canonical) "*dual pair*",

$$(9.7) \qquad (\mathcal{E}, \mathcal{E}^*),$$

where we set,

$$(9.8) \qquad \mathcal{E}^* := \mathcal{H}om_\mathcal{A}(\mathcal{E}, \mathcal{A}),$$

viz. the "*dual*" ($\mathcal{A}$-module) *of* $\mathcal{E}$; cf. [VS, Vol. I: Chapt. IV; (5.1)]. In fact, the above (natural) *duality* (:"*self-$\mathcal{A}$-pairing*") can be attained, through an appropriate $\mathcal{A}$-*bilinear form*,

$$(9.9) \qquad \omega : \mathcal{E} \oplus \mathcal{E} \cong \mathcal{E} \times_X \mathcal{E} \longrightarrow \mathcal{A},$$

which, in turn, gives rise to an $\mathcal{A}$-*morphism*,

$$(9.10) \qquad \bar{\omega} : \mathcal{E} \longrightarrow \mathcal{E}^*,$$

of the $\mathcal{A}$-*modules* concerned; that is, *one sets*

$$(9.11) \qquad [\bar{\omega}(s)](t) := \omega(s, t),$$

for any *local* (continuous) *sections* s, t of $\mathcal{E}$ (viz. for any s, t in $\mathcal{E}(U)$ with U *open* in X, *base space* of the $\mathcal{A}$-*module* (sheaf) $\mathcal{E}$: cf. (9.5)/(9.9)). In particular, we further assume for the dual pair, as in (9.7), the basic (and still *useful*, see below) relation,

$$(9.12) \qquad \mathcal{E} \underset{\bar{\omega}}{\cong} \mathcal{E}^*;$$

viz., in other words, that $\bar{\omega}$ *is*, in effect, *an $\mathcal{A}$-isomorphism*.

Note 9.1. — Concerning the $\mathcal{A}$-*isomorphism* $\bar{\omega}$ in (9.12), this stands, in effect, for

(9.12′)   "*dynamics*", *realized by* $\mathcal{E}^*$.

Therefore, transformation of *matter* (: "geometry", that is $\mathcal{E}$) *into* "*energy*". So "*arithmetization*" leads to "*calculations*": cf. also Section 2.6.2. Yet,

$$\text{(9.12″)} \qquad \begin{array}{c} relativity, \text{ in the form of the diagram,} \\[4pt] \bar{\omega} \longleftrightarrow \rho. \end{array}$$

Furthermore, see subsequent remarks after (9.14), along with (9.18.2) and (10.9) in the sequel.

So we call then the given $\mathcal{A}$-*bilinear form* ω, as above (cf. (9.9)), "*strongly nondegenerate*". This happens e.g. in the case one has,

$$\text{(9.13)} \qquad \mathcal{E} \cong \mathcal{A}^n,$$

viz. when $\mathcal{E}$ is a *free* $\mathcal{A}$-*module of* (finite) *rank* $n \in \mathbb{N}$ (see A. Mallios [6: p. 298, (5.2.1)]). However, see also (9.23) below, when for the particular case considered therein, *viz.* $\mathcal{E}nd\,\mathcal{E}$, one gets at the basic (and *very useful*) relation,

$$\text{(9.14)} \qquad \mathcal{E} \cong \mathcal{E}^*,$$

within an $\mathcal{A}$-*isomorphism* (alias, we refer to (9.14), as a "*strong reflexivity*" of $\mathcal{E}$), *without the intervention of any* $\mathcal{A}$-*bilinear form* ω, as in (9.12), and *for any* "*vector sheaf*" $\mathcal{E}$ (: *locally free* $\mathcal{A}$-*module of finite rank*, ibid., p. 306, (6.19)).

On the other hand, (9.14) and, in particular, in the case under consideration herewith, (9.12) *still holds true*, by definition, in *important special cases*, as e.g. when $\mathcal{E}$ is a "*Riemannian* $\mathcal{A}$-*module*", with ω being then a so-called (Riemannian) "$\mathcal{A}$-*metric*" (ibid., p. 320, Definition 8.3); similarly, in the case of a "*Lorentzian* $\mathcal{A}$-*module*" (see A. Mallios [21: p. 161, Definition 2.1]). An analogous situation, as before, appears by still considering "*symplectic* $\mathcal{A}$-*modules*", viz. another *important special instance*, of what one may consider as a "*metricized* $\mathcal{A}$-*module*"; see e.g. A. Mallios - P. P. Ntumba [3: p. 182]. More generally, the preceding are, of course, *particular instances* of a pair,

$$\text{(9.15)} \qquad (\mathcal{E}, \omega),$$

consisting of an $\mathcal{A}$-*module* $\mathcal{E}$, equipped with an $\mathcal{A}$-*bilinear form* ω, in general, as in (9.9).

Now, what one actually means by (9.1.2) (see also Note 9.1), is that *we have attained to*

(9.16)   *translate* (9.12) *into its* "*dynamical analogue*" (cf. thus (9.18.2) below).

Indeed, by assumption, $\mathcal{E}$ stands therein for a *Yang-Mills field* (see (9.5)). Therefore, in order to be

(9.17)   *"physically"*, *in accordance with* (9.12) (cf. also in that context (9.3)), one should *translate the "identification"* at issue, into a *"physical"* one; viz., in other words, one has to *extend*, appropriately, (9.12) *to an analogue of the respective Yang-Mills fields.*

Consequently, what we actually require, in that context, is a

"(gauge) equivalence" between $(\mathcal{E}, \mathcal{D})$ and its *"dual"*, $(\mathcal{E}^*, \mathcal{D}^*)$. We write it, as

(9.18.1) $$(\mathcal{E}, \mathcal{D}) \underset{\bar{\omega}}{\sim} (\mathcal{E}^*, \mathcal{D}^*),$$

or even for short (see also e.g. (9.5)),

(9.18)   (9.18.2) $$\mathcal{D} \underset{\bar{\omega}}{\sim} \mathcal{D}^*.$$

Namely, $\bar{\omega}$ as in (9.12), becomes now a *"dynamical equivalence"*. Yet, *equivalently*, one has the relation,

(9.18.3) $$\mathcal{D}^* = Ad(\bar{\omega}) \circ \mathcal{D},$$

or, even that the $\mathcal{A}$-*connections* $\mathcal{D}$ and $\mathcal{D}^*$ are, as we say, *"$\bar{\omega}$-related"*.

In this regard, see also A. Mallios [VS, Vol. II: p. 24, (5.42)/(5.43), and p. 198, (4.2)], concerning the notation applied in (9.18.3) and the relevant terminology therein, in conjunction with (9.12), as above. Furthermore, still *"technically speaking"*, we claim that

(9.19)   the *"self–($\mathcal{A}$-)pairing $(\mathcal{E}, \mathcal{E}^*)$*, as before (see (9.7)/(9.8), along with the ensuing comments therein), *is compatible with the $\mathcal{A}$-connection $\mathcal{D}$ of $\mathcal{E}$* (cf. (9.5)).

In this context, see also (9.2)/(9.3) in the preceding, pertaining to the *real* (physical) *meaning* of the term *"compatible"*, as above.

Now, it is precisely here that one is confronted with the so-called *"principle of least action"*, pertaining of course to *our framework*, when requiring that the $\mathcal{A}$-isomorphism

$\bar{\omega}$, as in (9.12), is *"parallel"* (with respect) to the $\mathcal{A}$-*connection*,

(9.20)   (9.20.1) $$\mathcal{D}_{\mathcal{H}om_{\mathcal{A}}(\mathcal{E}, \mathcal{E}^*)}.$$

Namely, that *(9.1.2) is in force*. Thus, in other words,

(9.20.2) $\bar{\omega}$ is *insensible to/uninfluenced by*, *the $\mathcal{A}$-connection* (9.20.1), at issue (:*"least action"*).

In this context, cf. also ibid., p. 137, (6.4), and comments following it; yet, together with (9.2) in the preceding of the present account, concerning the previously applied terminology.

On the other hand, by still referring to the notation applied in (9.20.1), and based on the *"identification"* (viz. $\mathcal{A}$-*isomorphism*) $\bar{\omega}$, as above, one further gets at the following relations, that will be also of use in the sequel. That is, we obtain,

$$(9.21) \quad \mathcal{H}om_{\mathcal{A}}(\mathcal{E}, \mathcal{E}^*) = (\mathcal{E} \otimes_{\mathcal{A}} \mathcal{E})^* = \mathcal{E} \otimes_{\mathcal{A}} \mathcal{E}^* = \mathcal{H}om_{\mathcal{A}}(\mathcal{E}, \mathcal{E}) \equiv \mathcal{E}nd_{\mathcal{A}}\,\mathcal{E} \equiv \mathcal{E}nd\,\mathcal{E}.$$

within, of course, $\mathcal{A}$-*isomorphisms*: see also A. Mallios [VS, Vol. I: Chapt. IV; (6.11), (6.16) and (6.1)]. Therefore, one comes to the conclusion that,

(9.22)
> the *"adjunction"* through the $\otimes$-*functor is effectuated by the* $\mathcal{H}om$-*functor*, via the *"dual"* (thus, still another *"adjunction"*(!)); so we also get at a so-to-say *"physical interpretation"* of the *dual*. Thus, more precisely,

(9.22.1)
> > one goes from the $\otimes$-functor to the $\mathcal{H}om$-functor, alias, one gets at the *"$\mathcal{H}om$-$\otimes$ adjunction"*, via the *"dual"* (: canonical *"duality"*, cf. (9.7)/(9.14)).

Now, as already mentioned (see comments on (9.14) in the foregoing) the above identification, as in (9.12), can still be attained *without* any *exterior intervention*, like ω, as above (cf. (9.9)), hence, it can be *more natural*! So this happens, for instance, when in the important particular case, one considers what we named in the foregoing (see (9.6)), the *"Heisenberg analogue"* of $\mathcal{E}$; that is, the $\mathcal{A}$-*module*, in fact, *"vector sheaf"*, $\mathcal{E}nd\,\mathcal{E}$ (see also e.g. (9.24.1) below), so that one obtains,

$$(9.23) \qquad\qquad \mathcal{E}nd\,\mathcal{E} = (\mathcal{E}nd\,\mathcal{E})^*,$$

within an $\mathcal{A}$-*isomorphism* (ibid., p. 306, Corollary 6.3). Of course, as already noticed (cf. (9.13)), one has a similar situation, by considering any *free* $\mathcal{A}$-*module*, when now the *duality is inherent*.

Thus, by recapitulating the foregoing, one is led to the basic conclusion, which also falls within our initial aim (see Section 1.6), that

(9.24)
> by *requiring* (9.1.2), *we do "participate Nature"* (see (9.1)/(9.2)); of course, *equivalently*, in view of (9.21), one can also look, within the same context, at the *"Heisenberg analogue"* of $\mathcal{E}$, viz. at $\mathcal{E}nd\,\mathcal{E}$ (see thus (9.26) below).

On the other hand, within the latter point of view, as above, one further gets at the particularly *useful* relation,

$$(9.25) \qquad\qquad \mathcal{E}nd\,\mathcal{E} = M_n(\mathcal{A}), \quad with \ \ n = rk\mathcal{E}, \ \ "locally"(!),$$

within an $\mathcal{A}$-*isomorphism*; see also, for instance, ibid., p. 138, (6.27). In this regard, we still recall here that, by assumption (cf. (9.5)), $\mathcal{E}$ is a *locally free* $\mathcal{A}$-*module* of (finite) *rank* $n \in \mathbb{N}$, viz. a *"vector sheaf"* (ibid., p. 27, Definition 4.3). Yet, the latter relation further reminds *"Bohr's correspondence principle"* (see (0.2)). Now, as already remarked, based on (9.21), one can further obtain (9.1.2) into the form (cf. also (9.6)),

$$(9.26) \qquad \mathcal{D}_{\mathcal{E}nd\,\mathcal{E}}(\bar{\omega}) = 0.$$

In this context, the above relation can still be construed just, as *another version of* the *"least action principle"*: this is based on our previous remarks in (9.19), along with (9.1)/(9.2). So, to say it once more, it is the outcome of our whole discussion, within the framework of ADG, that

> (9.27) any time we succeed, our theory/equations (here, for instance, $\bar{\omega}$), to be *in accordance with* (the function of) the *physical law* (: $\mathcal{A}$-*connection*, see e.g. (9.4))/*Nature*, as this happens e.g. by (9.1.2)/(9.26), then one gets, in effect, at *another realization of that* fundamental *principle* at issue, cf. also relevant comments in the sequel; see thus (9.28)/(9.29) below.

Yet, the last relation, as in (9.26), is also written, for convenience, and by an obvious *abuse of notation*, in the (familiar, already from the classical theory) form,

$$(9.28) \qquad \nabla\omega = 0.$$

We note on this occasion, that the form of the above equation reminds the familiar *"compatibility"* of an $\mathcal{A}$-*connection with a* given *"symplectic ($\mathcal{A}$-bilinear) form"*; however, still see our comments in (9.3), concerning the proper meaning of the applied terminology herewith. So we have here the basic assumption of what we may call *"Abstract Symplectic Geometry"*, viz. symplectic differential geometry, in terms of ADG. Thus, see e.g. A. Mallios-P. P. Ntumba [3], [2], P. P. Ntumba [1], [2]; yet, see also e.g, K. Habermann-L. Habermann [1: p. 25, Definition 2.2.1] for the classical case. On the other hand, a similar remark to the preceding, referring to (9.28), holds true in the case, for instance, of a *Riemannian/Hermitian* and/or *Lorentzian geometry*, in the sense of ADG, already considered in A. Mallios [VS, Vol. II: p. 165, (8.70), p. 168, Definition 9.1, and p. 174, Theorem 10.1] and [GT, Vol. II: p. 168, (2.51.1), along with p. 169, Scholium 2.2].

Now, within the same vein of ideas, still *connected with (9.28)*, it is worth-noticing here that,

> (9.29) the aforementioned relation *(9.28) permeates* actually, in the appropriate sense, even *more general aspects than* ADG; thus, e.g., a *topos-theoretic presentation*. Hence, one realizes here an indication of the *universal* (: physical) *character of the* fundamental *principle* at issue, as the same is related with (9.28); see also comments following (9.26), along with (9.27), (9.20), and (9.2).

In this context, see also e.g. I. Raptis [5: p. 1504, (1) and comments after it], [6], and E. Zafiris [2: p. 377] for a *topos-theoretic treatment* of *relevant items*, as before.

1.9.1   *Symmetry*

"...*symmetry dictates interactions* ..." (*first used by Einstein himself*; as quoted by C. N. Yang: cf. J. Baez and J. P. Muniain [1: p. 161]).

"...*symmetry*... often *the most profound formulation of a physical law*..." (V. Guillemin-S. Sternberg [1: p. 1]).

Meditating further on the (Physical) *meaning of* (9.28), *and* also by *paraphrasing* thus (9.18), we can still say that,

(9.30)   the $\mathcal{A}$-*isomorphism* $\bar{\omega}$, as in (9.12), defines in fact a "*symmetry*" (: (gauge) equivalence) *between* (the *Yang-Mills fields*) $(\mathcal{E}, \mathcal{D})$ *and* $(\mathcal{E}^*, \mathcal{D}^*)$.

Thus, we are led here to think of a "*dynamical symmetry*" (: equivalence, cf. (9.18.3)), according to Noether. Namely, in that (cf. also (9.53) below):

(9.31)   every "*symmetry*" yields a "*conserved quantity*" (thus, just here, one has something "*covariant*"), while *this association is*, in effect, *bijective*! (E. Noether).

See, for instance, A. Cannas da Silva [1: p. 147, Theorem 24.1]. That is, in other words,

(9.32)   one has a *one-to-one* (and *onto*) *correspondence* between "*conservation laws*" (viz. "ϕ- *related*" $\mathcal{A}$-connections, see e.g. (9.18.3), with $\bar{\omega} \equiv \phi$) and "*infinitesimal symmetries*", arisen from "*dynamical flows*" ϕ (solutions (*locally*) of differential equations).

Cf., for example, V. Guillemin-S. Sternberg [1: p. 311].

Now, the aforementioned fundamental "*geometric*", viz., in fact, "*relational*" (in the sense of ADG, see e.g. (7.3)-(7.5) in the preceding) concept, as in (9.31), can certainly be related, as it actually were (cf. thus (9.32)), with the *physical law* itself (viz. the $\mathcal{A}$-*connection* at issue, cf. (9.4), along with (8.19.4), (8.20)); the latter may be eventually represented here by the *symmetry* in focus (cf., for instance, "*symmetry group*"). Therefore, something in effect, *much more important*; viz. one can thus associate the whole stage with the corresponding *quantum* (!), referring, of course, to the *physical law*, under consideration (see also e.g. (9.40) in the sequel). So one essentially arrives at the conclusion that,

(9.33)   *symmetry can be innately connected with the quantum*!

Therefore, to paraphrase here too D. R. Finkelstein, (see e.g. [1: p. 477]), one is led to say that,

(9.34)   *everything is* (even *locally*, of course, can be reduced to a) *symmetry*!

Yet, the latter reminds us of the aspect that,

(9.35)   *"everything is light"*.

See also e.g. A. Mallios [17: p. 281, (6.21)-(6.24)]; yet, cf. A. Einstein [1: p. 106], in that, to paraphrase him: one might look at the

(9.35′)   *matter, as* "condensed light",

or even (M. Faraday), as

(9.35″)   a *"singularity of the field"*.

Whence, *boson*, therefore again *symmetry*; furthermore, by still looking at (9.32), one might construe

(9.36)   *symmetry* (when *locally* considered), as a *spinoff of* a *"flow"*.   Thus
          again, and *still locally* (!), *"commutativity"* is emerged.

As a result, we get actually here too at another way to remind us of *"Bohr's principle"* (and its *innate meaning*), as in the title of the present study (see also (0.2) in the foregoing).

On the other hand, within the same framework, and still based on classical patterns, one can further

(9.37)   relate *symmetry with isometry*: Namely, in other words, with a *metric*
          (viz. *"gauge"*, see e.g. (8.22)); hence, in fact, again with the concept of
          "adjunction" (cf., for instance, (9.9), (9.12), (9.14), and (9.22.1), along
          with (9.40) below).

In this context, we can refer e.g. to the classical *Cartan-Dieudonné Theorem* in Geometric Algebra, in that

(9.38)   *"every isometry* [: *"adjunction"*, see e.g. (9.12)] *is* ("reduced to") *a* (fi-
          nite) *product of* appropriate *symmetries"*.

See, for instance, E. Artin [1: p. 129, Theorem 3.20], N. Jacobson [1: p. 372], R. Deheuvels [1: p. 152, Théorème IV. 9]; yet, W. A. Adkins-S. H. Weintraub [1: p. 388, (3.24)]. The same result has been recently extended to the realm of *"Abstract Geometric Algebra"*, in the sense of ADG, for suitable *sheaves of A-modules*; cf. P. P. Ntumba [1], along with A. Mallios-P. P. Ntumba [1], [3]. Thus, based further on (9.33), one can still remark now that,

(9.39)   the *quantum* is actually effectuated through a *symmetry* (cf., for example, "*symmetry group (sheaf)*" as in the preceding, along with A. Mallios [GT, Vol. II: p. 85, (1.26)]), which in turn is actually *the innate part of an isometry*; hence, in fact, of an "*adjunction*" as well: see (9.7), (9.11) and (9.12). On the other hand, (9.12) can still be construed, as the function/outcome of an

(9.39.1)   "*observer* ⟷ *state*" *system*!

In this regard, see e.g. G. Lassner-A. Uhlmann [1], along with e.g. D. A. Dubin-M. A. Hennings [1: p. 9, Definition 2.1]; yet, see (10.46) below. Thus, one is led to the following diagram, as a spinoff, so far, of the foregoing (yet, cf., in particular, (6.3) and (9.4)). So we get,

(9.40)   *quantum* ⟷ "*adjunction*" (: "*observer-state*" system) *preserved*
quantity ("*A-invariance*") $\overset{\text{(Noether)}}{\longleftrightarrow}$ *symmetry* (: Klein, "*flow*"; yet, see A. Mallios [GT, Vol. II: p. 33, (4.60)]) ⟷ *physical law*.

Now, in this context, cf. also ibid.: p. 31, (4.51), p. 33, (4.60), as above, together with p. XII, (∗∗), and p. 207, (10.21). So one can further notice that,

(9.41)   *fermions* "*means*" actually *bosons*, viewed of course, relative to a *different* "*geometry*" (viz. multiplication/Grassmann). Within that perspective, one might thus consider *fermions*, as a "*condensed light*"; see the previously hinted at "*mixture of* "*geometry*"" (à la Grassmann). Yet, the same reminds the classical association, à la Faraday, *matter* ⟷ *field*, as quoted by H. Weyl [1: p. 169], "*matter*" being just "*singularities of the field*" (ibid.); moreover, Einstein's perspective that, "*matter* consists of *electrically charged particles*" [1: p. 106] (emphasis in the latter citation is ours). Thus, in other words, by paraphrasing the above, one can further say that,

(9.41.1)   *fermions* "*are*", just, "*condensed*" light (: *bosons*).

All told, we are now finally led, and still *in conjunction with* (9.40), to formulate the following, rather *quite natural perspective*, in that;

(9.42)   all the *fundamental equations in physics* (considered e.g. in vacuo) are nothing more than simply a *particular application of* the "*least action principle*": see thus, for example, *Einstein, Yang-Mills equations*, or even, *more general*, the *Utiyama equation*, in that context; cf. also relevant remarks in the sequel, in particular, (9.51) and (9.58)/(9.59).

Indeed, one can actually notice here that,

(9.43)   the aforementioned fundamental principle, as in (9.42), is in effect *the only way* that *Nature works*! That is, *nothing is in excess*, therein.

Furthermore, to say it otherwise,

(9.44)

> any time we realize/can detect *the principle in focus*, we can still be assured that we have essentially detected/located a *physical law*. Therefore,

(9.44.1)

> *this is the only way*, a *physical law appears to us*; viz., in other words, *in the form of a "least action principle"*.

Now, our aim by the ensuing discussion is to *elaborate* further, somehow in brief, however, *on* the above statement, as in (9.42); of course, always within the context of ADG: Thus, concerning, for instance, the *Einstein's equation* (in vacuo),

$$(9.45) \qquad \mathcal{R}ic(\mathcal{E}) = 0,$$

it has been already realized, by its very definition, that this was actually based on the *principle at issue*, by requiring, in that context, the validity of (9.1.2); in this context, see also our previous comments in (9.18)-(9.20), along with the ensuing general discussion herewith, pertaining to the *inherent mechanism/significance* of getting (9.45) (see thus (9.50) in the sequel): Now, the already classical outcome, in that context, is that;

(9.46)

> the *first variation* of the integral of the *"Action density/ Lagrangian"* is put equal to zero.

Indeed, this is the analogue, in our case (ADG), of the *"least action principle"*; viz., in other words, just the *truth of* (9.1.2), or even of (9.26). See also e.g. (6.7) in the preceding, along with A. Mallios [21: p. 170, (2.60), and p. 180, (4.24.1)]. Yet, cf., for instance, D. Bleecker [1: p. 42], or even M. Nakahara [1: p. 257ff, Section 7.10.1], concerning the classical case. On the other hand, a similar argument is still in force, for the *Yang-Mills equation*,

$$(9.47) \qquad \delta_{\mathcal{E}nd\mathcal{E}}(R) = 0,$$

cf. A. Mallios [GT, Vol. II: p. 36, (4.76)/(4.77)]. Therefore, both the previous two equations (9.45) and (9.47) are usually construed, as the outcome of what we may call, as

(9.48)

> *Einstein-Yang-Mills action principle*: (See ibid. p. 172, (3.11″), together with p. 177ff, (4.20.1)-(4.24)). Thus, both the aforementioned equations are derived from the *"scalar curvature"* of the *field* (viz. *A-connection*, at issue), through a certain particular *"metric"* (that is, still by means of an *"adjunction"*; cf. also e.g. (9.9), (9.22.1), and (9.39.1)/(9.40) in the preceding, along with (9.54) in the sequel).

On the other hand, we have already remarked in (9.42) that,

(9.49)

> *both* the aforesaid *equations* (9.45) *and* (9.47) *are a consequence of* the so-called *"Utiyama's equation"*; hence, in effect, of another version of the *"action principle"* (see also the ensuing discussion).

Now, by looking more carefully at the *essence of the mechanism* afforded the above equations in focus, one is actually led to formulate at this place (see also below), what one may call a

(9.50)

general (mechanism of a) *least action principle*: Thus, this can actually be expressed, through a relation,

$$(9.50.1) \qquad \delta(\alpha) = 0,$$

where one defines,

$$(9.50.2) \qquad \delta := \mathcal{D}_{\mathcal{E}}^k,$$

such that one sets,

$$(9.50.3) \qquad \mathcal{D}_{\mathcal{E}}^k := \Omega^k(\mathcal{E}) \longrightarrow \Omega^{k+1}(\mathcal{E});$$

viz. an *$\mathcal{E}$-valued covariant* (exterior) *derivation, of order $k \in \mathbb{N}$, and $\mathcal{E}$,* a given *vector sheaf.* That is, we set

$$(9.50.4) \qquad \Omega^k(\mathcal{E}) := \Omega^k \otimes_{\mathcal{A}} \mathcal{E} \cong \mathcal{E} \otimes_{\mathcal{A}} \Omega^k, \quad \text{for any } k \in \mathbb{N}.$$

Therefore, (9.50.1) characterizes the *kernel of $\delta \equiv \mathcal{D}_{\mathcal{E}}^k$*, as in (9.50.2); in this context, see also A. Mallios [VS, Vol. II: Chapt. VIII, Section 8], for the above applied terminology. Now, a *particular case* of the preceding is also what we called in (9.48), the *"Einstein-Yang-Mills action principle"*. Yet, *the latter*, as already hinted at in (9.49), can still be construed as a *special* instance of what one may consider, as

(9.51)   *Utiyama's principle.*

See also (6.8.1) in the preceding, along with the ensuing discussion, in particular, (9.53) below. So, roughly speaking, one can say, in that context, that;

(9.52)

any *gauge invariant Action/density* (:*"Lagrangian"*), *referring to* ($\mathcal{A}-$)*connections*, should be also *Ad-invariant* (i.e., relative to the *action of $Aut\mathcal{E}$ on $End\mathcal{E}$*), with respect to the *curvature*, this being also *characteristic* of the whole point of view.

Yet, in other words, and in a more succinct form, one can say that,

(9.53)

"gauge invariance" of an appropriate *"Lagrangian"*, *for* ($\mathcal{A}-$) *connections* should be *equivalent to* a similar *one for the* corresponding *curvature* of the connection at issue.

In this context, see also D. Bleecker [1: p. 153, Theorem 10.2.15], as well as, M. E. Mayer [1: p. 8 (2)], [2]. Furthermore, still based on (9.48), it is *equivalent to say*, in fact, *more generally*, that;

(9.54)

one comes to (9.53), through a suitable *"gauge invariance"* of a given *$\mathcal{A}$-bilinear form*, precisely, via an *"$\mathcal{A}$-metric"*. Therefore, in other words, again in terms of an appropriate *"adjunction"* (see also (9.40) in the preceding).

On the other hand, based further on our previous considerations in A. Mallios [18], in particular, on Section 1 therein, entitled "*A-invariance*", and still, in conjunction with our present discussion herewith, especially Section 1.8 in the preceding (cf. thus e.g., (8.1)/(8.3.1), yet also (8.43)), one realizes that,

> the *point of view advocated therefrom* might be construed, in effect, as a *postanticipation* of "*Utiyama's theory*" [1956]: Indeed, the latter refers, essentially, to an
>
> (9.55.1)   "*invariance of interactions*";
>
> viz., in other words, to what one might consider, as a

(9.55)   (9.55.2)   "*dynamical analogue*" *of* the fundamental
                     "$\mathcal{H}om - \otimes$ *adjunction*"

> (see e.g. (2.41)/(2.42) in the preceding). Yet, *equivalently*, pertaining to the case in focus, as this is translated by the still basic,
>
> (9.55.3) A-*connection* $\longleftrightarrow$ *curvature*
>
> association (:*adjunction*); cf. also e.g. (6.3) above, and subsequent comments therein.

Thus, to paraphrase here R. Utiyama himself [1] (see also M. E. Mayer [1]) along with A. Mallios [18], as above), one can further remark that,

(9.56)
> "*interactions should* always *be* "*A-invariant*", as it actually were, anyhow; indeed, when, in particular, referring to *physical laws* (cf., ibid., p. 1929, (1.1) and subsequent comments therein). Therefore, "*A-invariant*", as it concerns, in fact, *their description* (see also (9.3) in the preceding, and remarks following it). Yet, the latter "*A-invariance*" refers actually to both items of the *fundamental* "*adjunction*", as in (9.55.3), viz. concerning the diagram,
>
> (9.56.1) A-*connection* $\longleftrightarrow$ *curvature*.

Now, by still looking more closely at our previous remarks in A. Mallios [18: Section 1], for convenience, we refer to it as "*A-invariance*", and still, in conjunction with the classical work of R. Utiyama, pertaining to the concept of "*invariant*", in general, one can further notice that;

(9.57)
> the latter is a natural *plausible* (in fact, "*technical*") *vindication* of the former. In particular, Utiyama refers, to the "($A-$)*invariance of interactions*", as this has been already explained in (9.55) above.

See also (9.56), as above; indeed, (9.57) is actually based on the remark/realization that,

(9.58)   "*Nature is functorial*",

see thus, ibid., p. 1930, (1.5), together with (1.3′), (1.6), and p. 1931, (1.7). Thus, to say it, once more, the aforementioned classical

(9.59)   *"Utiyama's Theorem"* is, just, a particular (*"technical"*) vindication/*realization*, or even perspective, *of* (9.58).

See also Section 2.1. Now, all told, since as already said,

(9.60)   *Physis* (: Nature) *functions in a functorial way*, as *we* understand it,

(cf. (9.58), as above, along with ibid., p. 1930, (1.3′)), *we should* always *look after expressing our calculations* (: measurements) *accordingly*; that is, in other words, in a

(9.61)   *"functorial"* (: *"invariant"*) *way*, relative to $\mathcal{A}$.

So it is thus $\mathcal{A}(\equiv$ *"we"*, cf. also (1.4)), viz. our own *"arithmetic"* (or, even *"Calculus"*) that we use at each particular time, and in terms of which we can express our calculations, therefore, to formulate also our *description of Nature*, which is *of importance*, and still plays a *fundamental rôle* in that context (cf. also remarks following (9.67) in the sequel). Therefore, being here in accordance with (9.60)/(9.61), means in effect that,

(9.62)   we are able to *"participate Nature"* (!) in her (*"functorial"*) *function* (still, always as we understand/detect it).

In this context, see also ibid., p. 1932, (1.12)/(1.13), as well as, (1.16) and comments following it, in particular, (1.17) therein. Furthermore, within the above vein of ideas, we can still notice, by paraphrasing, in effect, here E. Galois (see e.g. A. Mallios [17: p. 263, epigraph]), that;

(9.63)   it is actually the *formulas* (: *"general laws"*, hence, *"functorial relations"*) that *are of importance* and *"not the calculations"* (!), through which, eventually, we derive the former.

In this context, we can still remark that the above are *in full accord* with the same idea of the fundamental *"least action principle"*, as considered in the preceding. More on this will be also discussed by the subsequent Section 1.10.

1.9.2   *More Thoughts on a Unified Field Theory*

> *"A fundamental theory needs to be background independent ... capable of being formulated exactly"* (L. Smolin)

This important theme is also presented here, within the whole point of view in the foregoing. The same can still be considered as an *amplification* (in view of

(9.51))/*expansion* of our previous similar considerations in A. Mallios [16: p. 1591, Section 6. (a)]. So we can start with the general perspective that

(9.64)   *"Physis is united"*.

Therefore, as an outcome, it is also quite reasonable to *seek* still *for*

(9.65)   a *unified way of describing Physis*.

On the other hand, based on our present day relevant experience (coming mainly from that one on *relativistic quantum theory*, alias *Quantum Field Theory*, so far), *we are* actually *looking*, in that context, *for a*

(9.66)
functorial (: unified) *relativistic* (: "covariant") *dynamics*, that will be still *localizable* (according to the very definition of this latter term, namely, the "dynamics").

Concerning the above applied terminology, see also ibid., along e.g. with A. Mallios [17: p. 269, (2.14)].

Now, looking primarily at the *conceptual* (see e.g. 't Hooft) *obstacles*, so far, to that important task, as posed in (9.64), one has first to scrutinize the *fundamental ingredients* that *underlie* this endeavor, *as* described *in (9.66)*. Thus, one has to *probe deeply* into the relevant notions there, under discussion: So the task here seems to be in employing, in principle, a so-to-say *"relational"* view (cf., for instance, A. Mallios [20], pertaining to that particular perspective); indeed, this point of view is already indicated in (9.66) by the categorical term, *"functorial"*, whose presence therein is otherwise vindicated by (9.58)/(9.61), as above. Yet, within this same context, the rather *most serious problem* prove to be *located* in

(9.67)
the way we understand and *effectuate*, in other words, *realize*, the *function of "variation"* (: changing of a given situation/state). Yet, otherwise said, *the problem lies in*

(9.67.1)   *the way we "differentiate "*;

that is, once more, in the way we describe/model the changing in focus.

Thus, looking at the relevant situation, one ascertains that,

(9.68)
our experience thus far, still based on historical grounds (Descartes – Newton), is *extremely*, yet one may say rather, *exclusively*, *"spatial"* (!), in nature (thus, *topological vector spaces*, in general, finite-dimensional, or not).

On the other hand, the *function*/procedure, we are seeking for (cf. (9.67.1)), viz. that of *"variation"*, does not have really anything to do with *"space"*, in the sense we usually employ the latter term, i.e. classically (see e.g. (9.68)). So, here we customarily assert that *"the space varies with us"* (cf., for instance, *"tangent space"*), while it is again *"we"*, who vary and still try to carry along, at least, our calculations/tool-kit, viz. our *"arithmetic"* $\mathcal{A}$ (see also (9.62), and comments before it in the preceding), in order to follow, via our corresponding *description of Nature, we observe,* her function; therefore, to achieve thus what we call *"covariance"*, if any (!).

Now, in this regard, one can also remark that, of course,

(9.69)　　　*we are not the "space"*, in any sense, but *part of it*; that is, *part of the Nature* (!) (cf. also (9.70.1) in the sequel). Therefore, the *outcome of the natural laws*; hence, of *relations* again.

In this context, see also A. Mallios [11: Section 1], along with [17]. Furthermore, within this same framework, as above, we still note that,

(9.70)　　　*we vary with the "space"*, as it actually were, anyway (Heraclitus). However, the *"space"*, we are talking about, *mathematically speaking* (cf. also the remarks before (9.69)), is *not*, of course, *the "space" we live*, in its substance (to paraphrase here A. Einstein): Indeed, the latter is *not "spatial"* (!), in our usual sense (see (9.68), i.e. *"arithmetical"*, locally or not), *but* in fact, *"relational"*. Yet, we may still recall, in that context, that

(9.70.1)　　　　　　the (physical) *space is what constitutes it*,

by paraphrasing here, in effect, Leibniz; see also e.g. A. Mallios [16: p. 1557, (1.1), and p. 1558, (1.5)], and [17: p. 267, (2.6)].

All told, it looks thus natural to ask for a *unified way* (cf. also (9.66)) to express/describe *the* manner the *Nature functions*/works; namely, that part of it, we might call the *"dynamical"* one, *this being* otherwise, *in effect, the whole* (!) of it (still, Heraclitus, as above).

Now, we can say that the previous analysis refers to the situation one has, classically speaking, and still in comparison with that one gets, through the view-point of ADG; in particular, this refers to the first item, i.e. *"functorial"*, in (9.66). On the other hand, by referring, in particular, to the *"dynamical"* issue of (9.66), precisely as this concerns the *quantum domain*, it is, at least, useful to recall herewith the decisive rôle, in that context (viz. "dynamics") of the *classical differential geometry* on (smooth) manifolds, i.e., what we denote throughout this treatise by the acronym, CDG. However, in contradistinction to this, we have the classical paradigm of R. P. Feynman, who essentially denied the use of CDG in its traditional/standard form, by initiated instead his own *"Calculus"*, viz. his celebrated homonymous, so-called *"diagrams"*. Yet, within this same vein of ideas, we still have the relevant remarks of C. J. Isham [1: p. 400], by paraphrasing him, in that,

(9.71)   the *"differential geometry"*, we are usually talking about, viz. CDG (cf. also (9.68) above), *does not* actually *exist* (!), at least, in the sense/form we mean, or even employ it; especially, pertaining, as already said, to its potential applications in the *quantum domain* (*"deep"*). Therefore, a f o r t i o r i even as a potential *candidate*/tool for a *Unified Field Theory* (acronym UFT; see also below).

So, as a result of the preceding, one can say that we should really have to

(9.72)   *reconsider* here *our perspective*, pertaining to *the way we look at* CDG. We might thus succeed to become more *"natural"* (!) (viz. *"relational"*, following the point of view, hinted at in the foregoing); therefore, *more effective*, as well, concerning at least potential applications of our *"Calculus"*, even in the *quantum deep*. Consequently, still an

(9.72.1)   *appropriate "Calculus" in formulating a* UFT, as well.

On the other hand, based further on the previous discussion, we can say that; by succeeding to *formulate*, what we might call, as the

(9.73)   *"Utiyama's programme"* (cf. also (9.51)), in terms of ADG, we would have still a *potential proposal for a Unified Field Theory* (UFT). Thus, the corresponding, for instance, within the aforesaid context, *Utiyama's equation* (cf. also (9.49), along with relevant quotations in the subsequent Sections 1.10 and 1.11), would stand e.g. for a *potential "universal equation"* in a UFT, via ADG.

In this context, see also (6.7)/(6.8), together with (9.52)/(9.53) in the preceding. Thus, in other words, the point of view advocated by the above, is that a

(9.74)   UFT might be construed, as an *Einstein-Yang-Mills*, or more generally, as an *Utiyama's principle*, in terms of ADG (cf. also (9.48)). Yet, within the same vein of ideas,

(9.74.1)   *quantize* would thus result in/mean to writing down a *"differential"* equation, exclusively *pertaining* to the *"quantum"* (viz. to the *"field"* itself).

Furthermore, all this in a *non-spatial* manner, in the classical sense of the latter term (CDG); thus again within the spirit, for instance, of ADG. In this context, see also e.g. A. Mallios [11: p. 57, Abstract, along with p. 71, (5.12), and p. 74, (5.22)], as well as [13: p. 79 Abstract].

Now, as an outcome of the above, one still concludes that,

(9.75)   a UFT might also be viewed, as an aspect of an *Einstein-Yang-Mills*, or even, more generally, of an *Utiyama quantized equation*.

Thus, all told, the foregoing suggest that one can further look at the classical

(9.76)   *least action principle*, in the sense, for instance, of (6.4)/(6.5), as *another means toward* an aspect of *a UFT*; of course, this, within the framework, for instance, of ADG.

Now, in this point of view, see also (6.7) and (6.8) in the foregoing, along with the ensuing Sections 1.10 and 1.11 in the sequel.

1.10   Transformation Law of Potentials, in Terms of ADG

We consider below the fundamental *transformation law of potentials*, in the jargon of ADG. That is, in other words, the relation

(10.1) $\delta(\omega) = \widetilde{\vartheta}(g),$

referring, in effect, as we shall see, to another basic aspect of an *"adjunction"*, as this is depicted by the following diagram; its clarification will be provided, right away, by the subsequent discussion. Yet, concerning the *"technical"* part of the above relation, see e.g. A. Mallios [VS, Vol. II: Chapt. VII; p. 115, (2.59), p. 116, Theorem 3.1, and p. 119, Theorem 3.2], along with [GT, Vol. II: Chapt. III, p. 120, Lemma 2.1], and [: Chapt. I; p. 73, (9.24)].

(10.2)

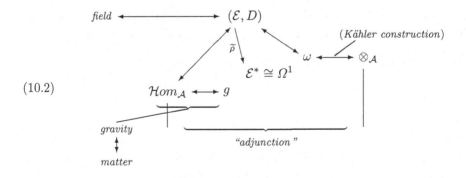

In this context, see also (1.10)/(1.11), and (6.1) along with Note 6.1, in the preceding, as well as, (8.20), (9.5) and (9.12).

Now, as already promised, we further explain, in brief, by what follows, the particular *constituents of* the previous *diagram*: Thus, first, the *horizontal line* in

(10.2) refers to the *standard association*, so far, throughout the entire framework of ADG; namely, one has:

(10.3)   *field* (:⟷ natural law ⟷ elementary particle ⟷ quantum) ⟷
$(\mathcal{E}, \mathcal{D}) \equiv$ "*Yang-Mills field*".

See also (8.20), (9.5) in the foregoing, along e.g. with A. Mallios [14: p. 63, (3.3)/(3.4), and p. 70, (5.9)], and [17: p. 278, (6.6)]. On the other hand, pertaining to the rest of the issues/notions appeared in (10.2), below the top line of it, as explained in (10.3), this refers, in effect, to a *local* (yet, *quantum*-), or even to what one may call a,

(10.4)   "*cohomological*" *aspect of* $(\mathcal{E}, \mathcal{D})$.

In this concern, see also, for instance, C. von Westenholz [1: p. 323]. Thus,

(10.5) $g \equiv (g_{\alpha\beta})$,

in (10.1) stands for a "*spatial*" (!) *realization of* (the "carrier") $\mathcal{E}$ (see also (10.3), viz. a ("*coordinate*") 1-*cocycle*, $\mathcal{E}$ viewed as a *vector sheaf* (: locally free $\mathcal{A}$-module) of *rank* $n \geq 1$ (alias, a "*Yang-Mills field*", in general); see : Vol. I; p. 353, (2.8), and p. 358, (2.43)], along with [: p. 25, (4.14), and comments following it]. Furthermore,

(10.6) $\omega \equiv (\omega^{(\alpha)})$,

in the same relation (10.1) represents the *local matrix form*, that uniquely determines (*locally* (!), always) the $\mathcal{A}$-*connection* $\mathcal{D}$, as in (10.3) (hence, the "*field*" itself, at issue; see also e.g. (9.4)). The same is a suitably defined 0-*cochain of matrix-valued* "*1-forms*" (alias, a "*0-cochain matrix-form*" of $\mathcal{D}$); cf. A. Mallios [VS, Vol. II: p. 112, (2.39), and p. 119, (3.16)]. Indeed, it is now a *fundamental theorem* here, that

(10.7)

> *a* given *pair*,
>
> (10.7.1) $(\omega, g) \equiv ((\omega^{(\alpha)}), (g_{\alpha\beta}))$,
>
> as above, *yields a* (uniquely defined) *field* $(\mathcal{E}, \mathcal{D})$, *if, and only if, one has* the relation,
>
> (10.7.2) $\delta(\omega) = \widetilde{\vartheta}(g)$.
>
> That is, the so-called *transformation law of potentials*.

See ibid., p. 116, Theorem 3.1, or p. 119, Theorem 3.2. On the other hand, the same rel. (10.7.2) can still be construed now, through the *assumption* of the following $\mathcal{A}$-*isomorphism*,

(10.8.1) $$\mathcal{E}^* \cong \Omega^1$$

of the $\mathcal{A}$-*modules* concerned, as the basis of an *"adjunction"*,

(10.8)

(10.8.2) $$\widetilde{\rho} : \mathcal{E} \longrightarrow \mathcal{E}^*,$$

emanated from a given $\mathcal{A}$-*metric*,

(10.8.3) $$\rho : \mathcal{E} \oplus \mathcal{E} \longrightarrow \mathcal{A}$$

on $\mathcal{E}$ (see also (10.9) below).

Concerning the above applied terminology, see also e.g. A. Mallios [VS, Vol. I: p. 318, Definition 8.2] and [GT, Vol. II: p. 4, and p. 145ff]. Thus, by *assuming the* following $\mathcal{A}$-*isomorphisms*,

(10.8′) $$\mathcal{E} \underset{\widetilde{\rho}}{\cong} \mathcal{E}^* \cong \Omega^1$$

of the $\mathcal{A}$-*modules* involved, one actually understands that;

> (10.9)   *(10.1) can be viewed, as an "adjunction"* originating, in effect, from the standard *"$\mathcal{H}om - \otimes$ adjunction"* in module theory (: Homological Algebra), *depicted* herewith still *by the diagram (10.2)*.

In this context, see also (2.42) and (2.38.1) in the preceding, along with e.g. S. Mac Lane [2: p. 269, Note]. Yet, from this point of view, one can further say, *equivalently* that;

> (10.10)   the same *rel. (10.1) can* still *be interpreted as a natural* (: covariant) *transformation of* the *functors* $\mathcal{H}om$ *and* $\otimes$, viz. of the *homonymous* *"adjunction"*, as before.

The above perspective, as presented by (10.10), concerning the meaning in extenso of the *fundamental relation (10.1)* (cf., for example, (10.7)), appears *more natural*: indeed, it points out the *quintessence* in fact of the action *of an "adjunction"*, as this is expressed, technically speaking, by the *"natural equivalence"* of the associated *functors* therein, as already indicated by the diagram (10.2), for the particular case at hand. In this context, see also ibid., along with [CWM: p. 80, Theorem 1], as well as, for instance, (2.35) in the foregoing; yet, cf. relevant remarks by the ensuing discussion right below. Thus, the above afford to the items involved in (10.1) a more *physical content*, pertaining to their *inherent* (:"categorical") *character* (still, cf. for example (10.14) below), this being also the *real essence of Category Theory*, as well (Mac Lane), yet in conjunction with its applications/*impact*, throughout mathematics, also in theoretical physics, as e.g. in the *quantum domain*. In toto, we thus arrive at what one might call a *"relational"* aspect of the whole framework, that actually goes back, still to Leibniz; see thus e.g. A. Mallios [20], and [17: p. 267, (2.6)], or even [16: p. 1557, (1.1), and p. 1559, (1.11)].

On the other hand, the above perspective for (10.1) leads one now to look at the same relation, yet as

(10.11)   another *effectuation of a form of the "least action principle"*: Indeed, by taking for instance into account our previous remarks in (10.7), we have thus actually detected/described, via the aforementioned relation, a *"physical/natural law"*, therefore our claim, according to (6.4)/(5.4) in the foregoing.

Thus, the above affords now, technically speaking, the same relation in focus a *more contiguous character*/nature *to* what we named in the preceding,

(10.12)   *"Utiyama's principle"*

See (9.51)/(9.52). Indeed, as already noticed in the foregoing (ibid.), the latter principle refers, in effect, to a

(10.13)   *"conservation"*/*"invariance"* of action (of a physical law, see e.g. A. Cannas da Silva [1: p. 147]) thus to the *way of variation* of the physical law, under consideration (*Noether*; ibid. See also (9.31) in the preceding).

Now, within the same context, as above, see also (9.44)/(9.44.1) in the foregoing, this being thus further vindicated, by our previous comments in (10.13), still in conjunction now with the technical meaning of (10.1). See thus (10.6)/(10.5), as before, along with the relevant account right away in the sequel.

So by still referring to the above diagram (10.2), we further remark that:

the following *associations*, therein,

(10.14.1) $\omega \longleftrightarrow \otimes_{\mathcal{A}}$, and $\mathcal{H}om_{\mathcal{A}} \longleftrightarrow g$,

(see also (10.5)/(10.6), as above), represent, firstly, on the one hand, the *relation between* the $\mathcal{H}om$-*functor* and g, referring thus, technically speaking, to the *fundamental notions*,

(10.14)

(10.14.2)   *gravity* (viz. *"curvature"*, thus, $\mathcal{H}om_{\mathcal{A}}$-*functor*), and *matter* (viz. $g \equiv (g_{\alpha\beta}) \longleftrightarrow \mathcal{E}$, locally, yet, *"space"*, cf. also (10.5)).

On the other hand, (10.14.1) refers, secondly, to the relation between the

(10.14.3) $\otimes_{\mathcal{A}}$-*functor* and $\omega \equiv (\omega^{(\alpha)})$ (viz. the *"potential"*, cf. (10.6)) $\longleftrightarrow \Omega^1$ (: *"domicile"* of the potentials), *locally* (viz. *"Kähler construction"*; see also e.g. (8.9), (8.15), and (8.19.4) in the preceding).

Thus, the "*adjunction*" (same diagram (10.2)),

$$(10.15) \qquad \mathcal{E} \xrightarrow{\tilde{\rho}} \mathcal{E}^* \xleftrightarrow{\cong} \Omega^1$$

demonstrates, in fact, the *fundamental "adjunction"*,

$$(10.16) \qquad \omega \longleftrightarrow g;$$

that is, in other words, the relations/diagram,

(10.17) *physical law* ⟷ ω ⟷ $\mathcal{D}_\mathcal{E} \equiv \mathcal{D}$ (*locally*) ⟷ (physical) "*geometry*"
$g \ (\equiv \mathcal{E},\ locally)$;

see also (10.2)/(10.3) above, along with e.g. A. Mallios [16: p. 1557, (1.1), and p. 1558, (1.3)/(1.4)]. The preceding is further realized in the form of a "*least action principle*", pertaining in effect to the *mutual tautochronous* (: simultaneous) *variation of the items in* (10.16)/(10.17); thus, the same is finally expressed, through the *fundamental law* (10.7.2), viz. that of "*transformation of potentials*". Yet, in this context, we also notice the *intervention, herewith*, of the point of view, in general, *of that*, which *we called* in the preceding, "*Utiyama's principle*"; see thus e.g. (10.12)/(10.13), or even (9.51)/(9.52), as well as, (9.56) in the foregoing.

> **Note 10.1.** — Applying the previous terminology, consider the diagram,
>
> $$(10.18) \qquad (g, \omega) \longleftrightarrow (\mathcal{E}, \mathcal{D}) \longleftrightarrow \text{ "field" (locally)}.$$
>
> See e.g. (10.2), (9.5), (8.20), together with (10.14); yet, cf. also, for instance, A. Mallios [21: p. 73, (9.22)/(9.24)]. So, within the preceding vein of ideas, we further remark that,
>
> > the pair (g, ω), as above, appears when, *in* "*action*", just, as "*one*" *item*! (*de Broglie*). Namely, as the "*field*" itself,
> > (10.19) whose thus the "*dyadic*" *substance* is identified/expressed *locally* (!), of course, *through* either one of the previous rels. *(10.6)/(10.5)*, as the case might be.
> > Of course, as already explained, throughout the above, the *criterion of getting at the* aforementioned pair, as a "*field*", is always (cf. (10.7)) the fundamental "*transformation law of potentials*". See [VS. Vol. II: p. 116, Theorem 3.1].

Scholium 10.1. — Our aim now by the subsequent discussion is to *further sustain* our claim in the preceding (cf., for instance, (6.4)/(6.5), as well as, (10.11) above), pertaining to *the validity of the following* diagram;

$$(10.20) \qquad adjunction \longleftrightarrow (\text{ a certain particular form of a}) \ least \ action \ principle.$$

More precisely, we actually assert here that:

(10.21)
the presence of an *"adjunction"* (see, for instance, Section 1.2.2 and 1.3 in the preceding) indicates, *physically speaking*, the application/function, in effect, of a certain particular form of the fundamental, as already pointed out, several times in the foregoing, *least action principle*; and, *conversely*(!) (see also (10.43) in the sequel).

As a matter of fact, the above is, in particular, vindicated, by the situation, in general, one is confronted with in *Mechanics*, classical or not, through the way the aforementioned fundamental principle is applied, still from the time already of *J.-L. Lagrange*: So, within the present context (viz. that of ADG), and based also on the preceding discussion, as it concerns the notion of an *"adjunction"*, along with its *physical meaning*, in general (cf., for instance, (1.19)/(1.20) in the foregoing, together with following comments therein), one is led to a quite *perspicuous aspect of (10.20)*, by looking at what we may call herewith a

(10.22)   *"Lagrangian perspective"*.

Of course, the previous point of view refers here to the (classical) *Symplectic* (Differential) *Geometry* (see below). Thus, we come right away by the next Section to a clarification of (10.22), relative to its connection with (10.20). In this context, we still refer e.g. to V. Guillemin - S. Sternberg [1: p. 145ff, along with Chapter II], R. Berndt [1], N. Woodhouse [1: in particular, p. 131, "the geometric approach"], and also J. E. Marsden - T. S. Ratiu [1: Chapter 7], C. von Westenholz [1: Chapter 12], as well as, for instance, to A. Mallios [17: p. 277ff], concerning relevant accounts of the matter, abstract (ADG) and/or classical ones (viz., in terms of smooth manifolds).

1.10.1   *Lagrangian Perspective via "Abstract Geometric Algebra" (AGA)*

In order to elucidate things, already from the outset, we can state, even at this place, that;

(10.23)
everything is an *"adjunction"* between an *observer* (:*"we"* ≡ $\mathcal{A}$, cf. also e.g. (1.4)) and her/his *"observables"* (: *Nature*).

More on the above point of view will be provided by the ensuing discussion. Yet, the same idea is, in effect, our *"creed"*, as it concerns what we may call *(Abstract) Symplectic Geometry* (acronym, ASG, thus, within the frame of ADG, when still differential geometry, in the classical sense, is also into account). This is actually, far more general, than the standard well-known *Weinstein's creed*, the latter being formulated within the context of the classical (smooth) manifolds theory; cf. A. Weinstein [1: p. 5].

On the other hand, as we shall see straightforwardly in the sequel, *(10.20)* is, of course, *a spin-off if (10.23)*: Indeed, one remarks here that,

(10.24)   *the principle at issue*, as formulated by (10.21), *is effectuated, by* (the function/operation of) *observation*; thus, the *technical formulation* of the latter (even, *axiomatically* doing) is made, through an appropriate "*A-bilinear form*"; (we employ thus here "*ADG-jargon*", see below) - therefore, *Lagrange* (!), in nowadays terminology - as explained right away in the sequel.

Thus, we start straightway, by specifying the decisive notion/terminology, as already appeared in (10.22), and still used in the title of the present Section. So, in this context, when talking, within the framework of ADG, about a

(10.25)   "*Lagrangian perspective*", one means a particular "*self-A-pairing*" (see e.g. A. Mallios-P. P. Ntumba [1], [2], [3], [4]) viz.

(10.25.1) $$[(\mathcal{E}, \phi); \mathcal{A}] \equiv (\mathcal{E}, \phi),$$

where $\mathcal{E}$ denotes an *A-module*, and

(10.25.2) $$\phi : \mathcal{E} \oplus \mathcal{E} \cong \mathcal{E} \times_X \mathcal{E} \longrightarrow \mathcal{A},$$

stands for an *A-bilinear form* on $\mathcal{E}$ (ibid.).

Yet, we denote by

(10.26) $$\mathcal{B}_{\mathcal{A}}(\mathcal{E}) \equiv \mathcal{B}_{\mathcal{A}}(\mathcal{E}, \mathcal{E}; \mathcal{A}),$$

the set, in fact another *A-module*, of such forms (namely, *sheaf-maps* on $\mathcal{E}$, by employing here an obvious *abuse of language*, in view of (10.25.1)).

Thus, one arrives at what we may call a "*generalized A-metric*" on the given *A-module* $\mathcal{E}$, as above. That is, in other words, one thus obtains,

(10.27)   an "*A-geometry*" on $\mathcal{E}$ (: the "*Lagrangian perspective*"/*geometry*); namely, *a means to measure* (in terms of $\mathcal{A}$, via ϕ, cf. (10.25)) on (our "*space*") $\mathcal{E}$.

Therefore, according to the very definitions, one comes to the conclusion that,

(10.28)   to have a "*generalized A-metric*" on a given *A-module* $\mathcal{E}$, as before, is *equivalent* (tautologous) *to* having an *A-bilinear form* on it, as in (10.25.2); hence, an element of $\mathcal{B}_{\mathcal{A}}(\mathcal{E})$ (see also (10.26)).

Thus, by further commenting on (10.27), we realize that the so-called herewith "*Lagrangian perspective*" is actually akin, in a sense, to the *ancient Greek way* of doing "geometry" (see also below), by employing, however, here an *algebraic means* to do it, viz. the *A-bilinear form* ϕ, as above. So, even to recall at this place *Emil Artin* in his classic "*Geometric Algebra*" [1], "*geometry*" means "*product*", hence, *algebra*, expressed here, through ϕ, viz. one writes (ibid.),

(10.29.1) $$\phi(x, y) \equiv xy,$$

(10.29)   with x, y elements of the "*space*" we consider. (We still notice that for Artin "*space*" is a (finite dimensional) *vector space* over a (commutative) field $\boldsymbol{k}$, in general; in *our case*, this is an $\mathcal{A}$-*module*, with $\mathcal{A}$, a (unital commutative) *algebra sheaf*, the latter algebra being *over* a classical (: numerical) *field*).

In this context, we further note that,

(10.30)   the "*product*" as given by (10.29.1), is for Artin (ibid.), by the very definition of ϕ, a "*number*" (alias, a "*scalar*"; viz. an element of $\boldsymbol{k}$, cf. (10.29), by looking, of course, at the result/outcome of the operation considered). Now, in *our case*, this is a "(*sheaf-*) *algebra-valued product*" on the "*space*" (: $\mathcal{A}$-module) in focus. As a result, *we have* thus actually *imposed* (: axiomatically), through ϕ, a "*product*" (: "*geometry*", cf. (10.29)) *on the* particular "*space*", we are interested in.

On the other hand, the same aspect, as above, still represents the point of view of a "*metric structure*", as already hinted at in the preceding by (10.28), and also considered in the particular case of the classical treatise of E. Artin (ibid., p. 105ff). Yet, the previous point of view (see also e.g., ibid., p. 16ff) has been employed, in effect, within our more general context, as applied herewith, by considering a "*self-pairing* $(\mathcal{E}, \phi)$", as in (10.25.1). Furthermore, within the same framework, as above, we also remark that;

(10.31)   the "*geometry*", we are usually talking about/study, is in fact the one derived from a certain special "*arithmetic*", alias "*algebra of coefficients*". Precisely, from *a certain* particular *type of* "*functions*", we employ: constant (: e.g. "*numbers/scalars*", yet, classical or "*generalized*" ones), even not such, as e.g. "smooth" (: differentiable), "analytic" and the like. So one can, more generally, study the "*geometry*", *emanated* from any *abstract* (occasionally, appropriate, *sheaf-*) "*algebra of coefficients*", as this happens, in general, in our case (see e.g. ADG). In this regard, see also, for instance, A. Mallios [17: p. 282, (6.26)/(6.27)].

Thus, as we shall also realize by the ensuing account, the preceding refer, in effect, to the fundamental *abstract form of* the "*Lagrangian picture*", started already with J.-L. Lagrange (1808), by his treatise on the "*variation of the elements of the movement of the planets*" (: Celestial Mechanics); see e.g. M. Puta [1: p. 8, Remark 1.2.1]. Yet, also, for instance S. Mac Lane [3: p. 278, Section 6, together with Chapter IX, in general]. The same study led actually to what we call nowadays "*Symplectic (Differential) Geometry*"; yet, the "*geometric/algebraic*" part of it constitutes what essentially refers to the so-called *"Geometric Algebra"*. Now, within the framework of ADG, this is presented by what one may call,

(10.32)   *"Abstract Geometric Algebra"*: that is, the *study of the relations between the objects themselves*, that interest us; and what is more important, this *without* any intervention of *"coordinates"*, based on a particular *"space"*, as this actually happens, even today, in the classical framework (see e.g. *"symplectic manifolds"*).

On the other hand, the *extension* within the present account (viz. in terms of ADG) *of the classical theory*, offers the possibility of applying, through the abstract (differential) machinery of ADG, conclusions of the point of view, advocated by this study, to the standard *"symplectic theory"* (viz. via the theory of (smooth) manifolds). Thus, we can still expect, thereby, *potential applications* of this type *of abstract theory* (this being, mainly, *"spaceless"*, as it actually should be (!), see e.g. (9.71)/(9.72) in the preceding) *to* well-known contributions of *the classical* (symplectic) *theory* in our current understanding of basic notions of *theoretical physics*, even concerning, in particular, the *quantum domain*. In this context, cf. also the Refs, in that respect, within the comments following (10.22) in the preceding; yet, see also e.g. A. Mallios [12], [14], and [GT, Vol. I: Chapter 5].

Indeed, the preceding framework is also in accord, as already hinted at in the foregoing, with the *point of view of Lagrange's work* itself (see e.g. (10.27)/(10.28)). The same aspect can actually still be reduced back to *Leibniz* (cf., for instance, A. Mallios [17: p. 267, (2.6)], or even to the way the *ancient Greek geometers* used to work: viz. by referring directly to the objects, in focus, thus, *without* any *"spatial"* support, so to say, whatsoever (!); modulo, of course, for the case at hand, *"Lagrange's metric"*, viz. the respective herewith $(\mathcal{A}\text{-})bilinear\ form$. See also the comments following (10.28) above.

Thus, the *"Lagrangian perspective"*, we are interested in, herewith, being also *inherent, even in the classical symplectic (differential) geometry* (on smooth manifolds), however now in terms of ADG, *characterized*, by the very definition, through the presence of an important *"analytic"*/algebraic *tool*, as one may call it; namely, of an $\mathcal{A}$-*metric*, alias a particular $\mathcal{A}$-*bilinear form* (see (10.25.2), or even (10.28)). Therefore, one is thus able to do *"$\mathcal{A}$-geometry"* always *in* an *abstract form*, in the sense e.g. described by (10.27)/(10.32), hence, again in a *"non-spatial"* manner, as the latter term was explained in the foregoing. Consequently (cf. also (10.28), as above), the following diagram,

(10.33) $$\mathcal{A} - metric\ \longleftrightarrow\ \mathcal{A}\text{-}bilinear\ form,$$

supplies, through its content/function (see also (10.23)/(10.24)), a *particular* important *realization of* what we usually define in Category Theory, as an

(10.34)   *"adjunction"*: namely, roughly speaking, such a relation *between us* ($\equiv$ *"$\mathcal{A}$"*) *and Nature*, by means of the function/action of the map $\phi(\equiv$*"measure"*; cf. also the terminology in (10.27) above).

The previously used categorical terminology has been also often applied in the preceding, so we refer to the foregoing for technical details and relevant Refs; see, for instance, Section 1.2.1 therein.

Now, by thinking further of the essential meaning of (10.33)/(10.34), as above, one comes to the conclusion that;

> "*measure*" means, in fact, the *realization*/description, via the particular "*arithmetic*" we provide, thus, in our case, the *algebra sheaf* $\mathcal{A}$, as before (cf. (10.34)), *of what we* actually *observe*: So cf., for instance, *Feynman's* famous adage,

(10.35)   (10.35.1)   "*Physics is number*".

> Thus, within the same vein of ideas, we can still say that,

(10.35.2)   every *physical law* is realized, when mathematically speaking, through a certain particular *adjunction*!

In this context, see also (10.23) in the preceding, along with (10.34)/(10.35), as before, hence, again Feynman! In turn, the above testifies, *no more no less*, the fact that we thus ascertain herewith, just, a *particular instance* of the *fundamental*, as often recognized, thus far,

(10.36)   "*least action principle*".

The same principle, as above, dominates/*characterizes*, of course, *the* whole *function of Nature*; cf. thus also e.g. (9.50.1) in the preceding. Yet, within the same framework, one further remarks that,

(10.37)   it is certainly always important, any time we can *express, in physical terms, a* given *mathematical notion/function.*

Indeed, to recall at least, in that respect, a relevant remark here of S. Mac Lane [3: p. 289] (emphasis below is ours),

(10.38)   "... it is *easier to state the manipulations than* ... to formulate *the idea* [beyond them] in words".

On the other hand, within the same vein of ideas, one can still refer here, for instance, *to Fermat's action principle* in *geometric optics*, or rather to what one might also denominate, within the same context, as

> "*Machado's principle*", in that;

(10.39)   (10.39.1)   traveler, *there is no path*, other *than* that *you create* [by walking].

Furthermore, one could still mention herewith a recent and telling advertisement of the air-company (!), *ALITALIA*, in that, *by paraphrasing* actually *Fermat*,

(10.40)   *"light makes its own path"*.

Therefore, in other words, we can still say, by generalizing in effect the preceding, that

(10.41.1)   *Physis/Nature* follows her own path.

(10.41)   Indeed, *the most "economical"* one, concerning the spending of energy, in any sense; this is the *physical law*. Therefore, the denomination in (10.36).

Thus, by further looking at the *real meaning of* (9.50.1) in the foregoing, one can also say that,

(10.42)   *anytime we* achieve to *follow*/participate *Nature*, then *we do not* actually *feel* (:*"gauge out"*) her action/*function*!

Indeed, we thus arrive here at *another version of* the classical *"Principle of Equiva-lence"*; see also, for example, A. Mallios - I. Raptis [1: p. 1894ff], along with R. Tor-retti [1: p. 133ff].

Of course, the above may be viewed, just, as a particular instance of an *adjunc-tion/measuring*, in the sense of (10.34), in general. Hence, one finally comes to the conclusion that,

(10.43)   *least action principle* and *adjunction*, in any form (!), are *equivalent* notions/situations: Hence, *the principle in focus* is mathematically (viz. in categorical language) expressed/*characterized*, in terms of a particular *adjunction*.

Therefore, we arrived just to a *vindication of* (10.20)/(10.21).

On the other hand, *the aforementioned principle is*, of course, still *a physical law*: see, for instance, the remarks following (10.36) above, along with (10.41). Hence, mathematically speaking, it is also of a *functorial nature*/character, as of course we understand it; see e.g. A. Mallios [18: p. 1929, (1.1), and p. 1930, (1.5.1)]. Yet, based on Category Theory, the same holds also true for an *adjunction*, and *conversely*! See e.g. S. Mac Lane [1: p. 81, Theorem 2, and p. 82], along with, anyway, (10.20)/(10.21), as above.

It is now a mere *application of the principle at issue*, that actually *led*, by the very nature of the matters, *to* nowadays *Symplectic* (Differential) *Geometry*, along with the concomitant therein *Hamiltonian methods*. Certainly, the latter are of *fundamental significance* for the *study of motion*, in general; hence, for the *whole Physics*, as well, *macro/micro*. Indeed, to recall here the *Greek antiquity*, we can still remark that,

(10.44)   *"everything flows"* (Heraclitus).

To summarize, one is thus led to the following diagram, which, in view of the preceding, still represents the idea behind the *principle of least action* in (Classical) *Mechanics*, according to *Lagrange-Hamilton*, now in terms of course of ADG. That is, one has;

(10.45) $\qquad$ *physical law* $\longleftrightarrow$ *least action principle*
$\qquad\qquad\quad \longleftrightarrow$ *adjunction/(measuring)*.

On the other hand, the *methods of ADG* might be, on the basis of their own idiosyncrasy (this being thus, by the very definition, *"functorial"*) still of particular significance for the standard theory (manifolds) of the *Calculus of Variations*: indeed, one copes therein with *"constraints"* and *"corners"*, namely peculiarities/*"singularities"*(!), pertaining to the *classical theory of smooth manifolds*; see also e.g. S. Mac Lane [3: p. 281]. Thus, it is exactly here that ADG, by its very structure, can be ahead of such cases; cf., for instance, A. Mallios [12], [14], A. Mallios-E. E. Rosinger [1], [2], I. Raptis [7]. So one arrives here too, to another *potential application* of the techniques of ADG in problems of *Mechanics*, in general.

Moreover, by further looking at the first two items in (10.45), as before, we can also refer, for relevant remarks, within however the framework of the *classical theory*, e.g. to D. Bleecker [1: p. 42ff], V. Guillemin - S. Sternberg [1: p. 145ff, and p. 272ff], as well as, to the above cited work of S. Maclane (in particular, [CWM: pp. 281, 282]).

1.10.2 *More on the Fundamental "Adjunction"*

Now, our aim by the present Section, is to point out further consequences, emanated from the *"adjunction"* $\mathcal{E} \xrightarrow{\tilde{\rho}} \mathcal{E}^*$, that was also considered already in the preceding, see (10.15). In particular, we wish to discuss further the same *"adjunction"*, in the point of view of (10.20)/(10.21); see also, in that context, (10.22), as before:

Thus, starting with the *basic adjunction* (10.16), $\omega \longleftrightarrow g$, along with that one of (9.14), the latter actually strengthens the present one, we arrive at *the following diagram*, that also *supplements* (10.2);

(10.46)

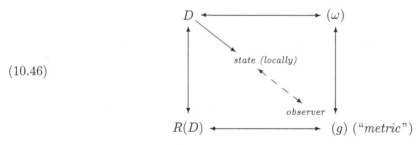

For the notion employed in (10.46), we also refer e.g. to the aforementioned diagram (10.2), along with (10.5)/(10.6), while $R(D)$ stands therein for the *"curvature"*

(alias, *"field strength"*) of the ($\mathcal{A}$-)connection (: potential, *"field"*) involved; see also e.g. (10.17)/(10.18), and A. Mallios [7: p. 191ff]. Yet, the *"diagonal"* in the previous diagram hints at what one may call therein, as

(10.47)   *"Lassner–Uhlmann principle/adjunction"*;

see thus (9.39.1), along with the ensuing comments therein, and still (8.13.1).

Therefore, based on the diagram (10.45), as above, precisely on the part of it, we are interested in, herewith, viz. the diagram,

(10.48)   *"adjunction"* $\longleftrightarrow$ *least action principle*,

we come to the conclusion that,

(10.49) $\mathcal{D}_{\mathcal{E}nd\,\mathcal{E}}(\widetilde{\rho}) = 0$;

see also (9.26) in the foregoing, with $\bar{\omega} \equiv \widetilde{\rho}$. Thus, in other words, one gets at a relation of the form,

(10.50) $\delta(\alpha) = 0$

(cf. also (9.50.1)). Here δ stands for an appropriate *"differential operator"*; hence, one still arrives at a *"differential equation"*, whose *"space"* *of solutions* characterizes the (domain of) validity of the *physical law*, represented by (10.50). In this context, see also (8.23) - (9.25) in the foregoing.

As a result, and based further on (10.35.2) and (10.45), we conclude that,

(10.51)   on the basis of (10.50)/(10.51), one is actually led to an *"adjunction"*. Therefore, still to an eventual *domain of existence of* (a particular form of) *the least action principle*; hence, also we thus arrive at the particular *physical law*, the principle in focus is concerned with (cf. still (10.35.2) and (9.44.1)).

Consequently, one gets at the following diagram, *complementing* thus, as already promised, the previous ones in (10.45/(9.40). That is, we have;

(10.52)   *adjunction* $\longleftrightarrow$ (kernel of a) *"differential operator"* (: *"differential equation"*) $\longleftrightarrow$ *physical law* $\longleftrightarrow$ (particular form of the) *least action principle*.

Now, in this context, by still looking at (10.51)/(10.53), we finally remark that one thus gets herewith at the usual *form/meaning*, which is *attained by classical equations* in physics, when viewed *through the least action principle*; hence, *equivalently*, *by* means of (10.49)/(10.50). In that context, see also (9.1.2)/(9.2) in the foregoing. So cf., for instance, the celebrated *Yang-Mills equation(s)*, or even, more generally, *Utiyama's equation* (!) (see also e.g. (9.49) in the preceding; yet, cf. D. Bleecker [1: p. 153] for the last issue, within the classical framework (: manifolds theory)).

Scholium 10.2. — By looking at the above *"Lagrangian perspective"*, that is, in other words, at a more

(10.53)   *"geometrical/relational"* point of view, than at an *"analytical"/* arithmetical one,

one gains in *diversity*/yet, *profundity* of *vision*; indeed, to paraphrase/generalize herewith C. von Westenholz [1: p. v],

(10.54)   the *"geometric spirit"* working methods, *in terms of* ADG, provide an *alternate description of natural phenomena ... beyond* the description obtained in terms of *analytical methods.*

So we are following, for instance, a *Lagrangian viewpoint* than that of Hamilton, yet, *without any surrounding "space"* (ADG). Our resulting framework becomes thus *still pregnant* with (having already inherent) a possibility *of "quantization".* Therefore, in effect, even of a *situation leading*, most likely, *to a UFT.*

In this context, we further note that (see also Section 2.1.1; (1.1)),

(10.55)   *each* particular *case is characterized* by its own appropriate *"symmetry group"*; hence (Klein) *"geometry"/"cosmos"*(!).

Yet, within the same context, as before, we can still admit that, we do *not rather exaggerate, by saying* that,

(10.56)   *quantization* means, in effect, appropriately *sheafifying*!

See also e.g. A. Mallios [13: p. 85, (2.17)/(2.18)]. Yet, within the same framework, as above, we further remark, to paraphrase/generalize herewith C. von Westenholz [1: p. 321], that:

(10.57)   *"the structure underlying an intrinsic approach to physics is "essentially", sheaf-cohomology"*; therefore, still to paraphrase/recall here A. Grothendieck [1], the same intrinsic approach is, in fact, via *"homological algebra"*, in general.

Cf. also, for instance, A. Mallios [GT, Vol. I: p. 91, Section 7, and Vol. II; p. 70, Section 9], concerning a *"cohomological classification of elementary particles"*. Yet, *Chern-Weil theory of characteristic classes*: [VS, Vol. II: p. 266ff], and/or *"geometric (pre)quantization* (in particular, *"Weil's integrality theorem"*; ibid. p. 238). For the latter topic, cf. also [GT, Vol. I: p. 134, Theorem 3.1]. *Existence of A-connections* is *cohomologically characterized* (cf. *"Atiyah* (cohomology) *class"* [VS, Vol. II: p. 47; (9.14), and p. 70, Theorem 13.1]). Moreover, *"flat line sheaves"*, having *"integrable" A-connections* are similarly characterized, through their so-called *"Selesnick-Chern class"* (ibid. p. 145, (7.5), and p. 148, Theorem 7.1).

1.11   Characteristics of a Physical Law

Our aim by the following Section is to point out further *characteristic aspects*, indicating the presence *of a physical law*. Yet, this in conjunction with similar allusions in the foregoing, referring for instance to the significance, in that respect, of the *least action principle* in its various forms. Indeed, what follows is, but another *different* actually *effectuation of the* same *fundamental principle, revealing* in effect the existence of *a physical law*; see also e.g. (9.44.1) in the foregoing. Thus, what we want to prove by the ensuing discussion is the following basic fact, provided by certain fundamental notions/principles, within the setting of ADG. For the notation employed in the sequel, we refer thus, for instance, to relevant standard already accounts in our previous work in A. Mallios [VS, Vol. I-II], [GT, Vol. I-II], along with [12], [18]. So we come, first, to the following statement:

(11.1)

> Suppose we have a *Yang-Mills field,*
>
> (11.1.1) $(\mathcal{E}, \mathcal{D})$,
>
> and let
>
> (11.1.2) $\phi \in Aut\,\mathcal{E} := (End\,\mathcal{E})^{\cdot}$
>
> be an *$\mathcal{A}$-automorphism of $\mathcal{E}$,* such that
>
> (11.1.3) $\phi_*(\mathcal{D}) = \mathcal{D}$.

That is, $\mathcal{D}$, the given *$\mathcal{A}$-connection* of $\mathcal{E}$, as above, is a *"fixed point"* of an *"Ad-invariant action"* of ϕ on $\mathcal{E}nd\,\mathcal{E}$. Now, our *claim* is that;

(11.2) *under the previous conditions, one has located,* in effect, *a physical law* (presented thus the given "field", as in (11.1.1)).

Thus, to say it, otherwise (see also e.g. (10.3)/(10.17) in the foregoing),

(11.3) *physical laws appear,* as *"fixed points"* of the *"Ad-representation"* of (Yang-Mills) *"fields".*

Concerning more technical details for the preceding material, and part from the ensuing specialized terminology/notation, we refer, for instance, to A. Mallios [21: p. 89, (2.18)/(2.19), and p. 91, (2.34)]. So in that context, (11.1) indicates, in fact, that:

(11.4)

> $\mathcal{D}$, as in (11.1.1), represents in view of (11.1.3), a *"self-related"* $\mathcal{A}$-*connection of $\mathcal{E}$,* relative to ϕ, as in (11.1.2); *equivalently,* an *"Ad-invariant action"* of ϕ on $\Omega^1(\mathcal{E}nd\,\mathcal{E})$: in point of fact, an *"action"* of $Aut\,\mathcal{E}$ on $\Omega^1(\mathcal{E})$, via the *"adjoint representation"* on the latter $\mathcal{A}$-*module;* thus, one sets,
>
> (11.4.1) $\phi_*(\mathcal{D}) \equiv Ad(\phi)(\mathcal{D}) := (\phi \otimes 1) \circ \mathcal{D} \circ \phi^{-1} \equiv \phi \mathcal{D} \phi^{-1}$,
>
> with $1 \equiv 1_\mathcal{E} \in Aut\,\mathcal{E}$, the *identity* of the $\mathcal{A}$-*algebra sheaf,* $\mathcal{E}nd\Omega^1$. (We note here that $\phi \otimes 1 \in Aut(\Omega^1(\mathcal{E})) \equiv Aut_\mathcal{A}(\mathcal{E} \otimes_\mathcal{A} \Omega^1)$, in view of (11.1.2), while $\Omega^1(\mathcal{E}nd\,\mathcal{E})(X)$ is the *"model"* of the *"affine space"* of $\mathcal{A}$-*connections* on $\mathcal{E}$, $Conn_\mathcal{A}(\mathcal{E})$; in this context, cf. also A. Mallios [21: Chapt. VI; Section 7, and p. 24, (5.41)/(5.43)]).

On the other hand, within the same vein of ideas, we further remark that;

(11.5) *"movements"* of $\mathcal{E}$, that is, elements of $\mathcal{A}ut\,\mathcal{E}$, which are *"compatible with $\mathcal{D}$"*, so, otherwise said, providing *"Ad-invariant actions"* on $\Omega^1(\mathcal{E})$, as above (cf. (11.1.1)/(11.1.3)), are in effect *equivalent* notions *with that sort of $\mathcal{A}$-connections on $\mathcal{E}$*, that still might be viewed, as representing the

(11.5.1) *"dynamics" of $\mathcal{E}$.*

See also (9.5)/(9.6) in the foregoing, along with (6.1) and relevant comments therein. Now, in toto, and still taking into account our previous remarks in (10.12)/(10.13), we finally conclude that,

(11.6) *the above* is, but *a spin-off of* what we have named before (ibid.), as

(11.6.1) *"Utiyama's principle".*

Thus, in other words, we still arrive, herewith, at another version of the *"least action principle"*; see also (9.44)/(9.40). Yet, by further looking at (11.1), we also remark that;

(11.7) a *"flow"* (referring to an object $(\mathcal{E},\mathcal{D})$, as before, see e.g. (6.1)), associated with an element,

(11.7.1) $$\phi \in \mathcal{A}ut\,\mathcal{E} \equiv (\mathcal{E}nd\,\mathcal{E})^{\cdot} < \mathcal{E}nd\,\mathcal{E},$$

means actually, (*locally*) a *"symmetry"*; therefore, the *existence of* a *"physical law"* (hence, still of a *"field"*; see e.g. Section 2.1; (2.2) below), which is finally located/realized by the *kernel of a* (differential) *operator* (see below): precisely, for the particular case at hand, the *"locus"* of validity *of the physical law*, at issue, is given by the following diagram (relations);

(11.7.2) $$\begin{aligned}\textit{"flow" of } (\mathcal{E},\mathcal{D}) &\longleftrightarrow \ker(\mathcal{D}_{\mathcal{E}nd\,\mathcal{E}}|_{\mathcal{A}ut\,\mathcal{E}}) \\ &\equiv \{\phi \in \mathcal{A}ut\,\mathcal{E} : \phi_*(\mathcal{D}) \equiv Ad(\phi)(\mathcal{D}) = \mathcal{D}\}.\end{aligned}$$

See also (11.7.1) and (11.1), along with (9.42)/(9.43) in the preceding.

To sum up the foregoing, we further remark that,

(11.8) *"Ad-invariant action"*, as above (see, for example, (11.4)), means (*locally*(!)) *"symmetry"*, even *"à la Lie"*; viz.

(11.8.1) $$[\mathcal{D}, \phi] = 0,$$

see (11.4.1) and (11.7.2). Yet, one gets at an interrelation of this with the *least action principle*, and finally still with *Utiyama's principle* (see thus (10.53) and (11.6)).

(11.8.2)   Hence, *the above characterizes* thus the *"evolution"* (: flow) *of* $(\mathcal{E}, \mathcal{D})$, as before. That is, the very *appearance to/detection by us* (:"$\mathcal{A}$") of the *physical law* itself.

Now, as already, said in the preceding (see, for instance, (9.44)),

(11.9)   the whole of *the function of Nature* can be reduced to some form of the *"principle of least action"*!

Thus, see also e.g. *Noether's principle/symmetries* (cf. (9.40)/(9.44)). On the other hand, within the same context, concerning in particular the last item (: *"symmetry"*, as before), one might still look at the classical *Cartan-Dieudonné Theorem*, as expressed in *"Geometric Algebra"*, that also we already mentioned in the foregoing (cf. (9.38) and comments following it, along with (9.33)/(9.34)). Yet, the same point of view, pertaining in particular (in effect, *more generally*(!)), to the *"abstract"* version of the aforementioned standard result, is still in accord with what we have called in the foregoing, *"Lagrangian perspective"*, in that context; see thus Section 1.10.1 above, in particular, (10.22)/(10.25).

Therefore, as a characteristic corollary of the preceding remarks (for $n = 1$, *"rank"* of the pair $\mathcal{E}, \mathcal{D}$), see e.g. (11.1.1)), one might refer, for instance, quite well, to the classical (*"atomistic"*),

(11.10)   *"Democritian assertion"*

pertaining to the *real substance of* the *physical world*.

Now, within the same vein of ideas, we further remark that the above are still related, however, just epigrammatically, with the ensuing telling (conceptual) diagram,

(11.11)   *curvature* ⟷ *"space"* (: matter) ⟷ (*"functor of points"* (cf. (2.18)/(2.19)); viz. a notion, realized via the *functor*, $\mathcal{H}om_{\mathcal{A}}$, hence,) ⟷ *"events"* (: relations).

Thus, according to the last item of the previous diagram, we can also say that,

(11.12)   it is *not the "events"*, but *the relations, which characterize the events*, that should be primarily of our concern (see also herewith, von Neumann [letter to Birkhoff]); so the latter are, in effect, the real essence of the notion *"functor of points"*, as above. It is thus a *concept*, extremely *categorical in nature*, yet, in point of fact, a *topos-theoretic* one; therefore, still akin to a *"quantum"-theoretic* aspect of the matter (: *no-points*(!)-*"events"*, just always *relations*).

So we have thus arrived again at what we might call, in postanticipation, still a

(11.13)   *"Leibnizian perspective"*,

of what we are actually interested in; thus, for instance, *"geometry"*, in the broadest sense of the term, as this was also hinted at in the foregoing (see e.g. (10.29), yet (8.29)/(8.30), or even, (4.2)-(4.6)); yet, *"analysis"*, in the same point of view (cf., for instance, (1.9)).

Within the same framework, as before, see also A. Mallios [17: p. 268, (2.9), as well as, the ensuing comments therein, and also p. 267, (2.6)], yet [20]. On the other hand, the preceding have certainly a special bearing on our previous discussion, concerning (10.1), along with the relevant remarks in (10.7), and its effectuation, as a form of the *"least action principle"*; see also (10.16) and remarks following it. So, in other words, one is actually concerned, herewith, as this is mainly referred to the *meaning of the principle* in focus, with an

(11.14)   *assemblage of relations, characterizing a physical law* (: elementary particle-quantum, still cf. (9.40)) *and its consequences*: i.e., field, *"events"* and the like, as before.

The above is much reminiscent of a recent *"telling"* terminology of C. Isham, referring to an *elementary particle* in *"t* (:*temporal*)-*topos theory"*, as an *"elephant"* (!), as lately quoted in a relevant account of G. C. Kato [2].

1.12   Complementary Remarks

We come by the present Section to some more, *complementary remarks*, pertaining to the preceding material: Thus, we first note that, the

(12.1)   *deviation from symmetry* leads to/indicates a *change/variation of* situation (:*"state"*); on the other hand, *"symmetry"* means the *"geometric"* formulation/realization of what one can further *"algebraically"* express/ describe (see also below), through *"commutativity"*.

Consequently, one thus concludes that,

(12.2)   *deviation from symmetry* entails a *change/modification* of a given *"geometry"* (: algebra; S. Germain (: *Pensées*))/*product*. Cf. also the famous so-called *"Higgs-phenomenon"*; yet, relevant work of S. A. Selesnick [3], [4].

In this context, see also e.g. A. Mallios [17: p. 282, (6.26)/(6.27)]; yet, I. Arahovitis [1: p. 35, Remark]. Furthermore, within the same set-up, we still recall that,

(12.3)   *"every flow is causally stationary"*.

See A. Mallios [2: p. 133, (5.10)/(5.12)]; yet, I. Raptis [2: p. 87]. Now, the preceding still suggests that the celebrated,

(12.4)   "(spontaneous) *symmetry breaking*" indicates, of course, a certain particular type of "*creation* (:"*genesis*") *of matter*"!

Namely, the *break down* of an existing "*flow*" (: symmetry, see thus (12.3), along with e.g. (11.7.1), as above) assures the appropriate conditions, cf. (12.1)/(12.2), for "*mass creation*". Furthermore, the same is also reminiscent of the famous "*Higgs theme/mechanism*", within the present framework: in this context, see for a nice "*geometrical*" account thereof, concerning the classical counterpart, still S. A. Selesnick [1: p. 400ff], [2], or even a more recent account, by the same author, on "*quantum symmetries breaking*" in [3]. However, cf. also F. Strocchi [1: Part C; Chapt. II].

Yet, as another exploitation of the preceding (cf. (12.1)/(12.2)), one is also led, even classically speaking, to the notion of a

(12.5)   *derivation* (hence, of an ($\mathcal{A}$)-*connection* too); viz. to a concept of a way to "*handling*"/(even, *describing*) a *change* of a given initial set of data; viz. a "*state*".

Thus, see for instance *S. Lie, E. Kähler*, as it concerns, of course, the homonymous classical methods of taking "*derivatives*". In this context, see also e.g. N. Bourbaki [1: Chapt. 3; p. 133, Proposition 18], together with D. Eisenbud [1: p. 389, and p. 407, Theorem 16.4]; the last citation pertaining, in particular, to the rôle of *Topological Algebras* theory, within the same point of view, as adopted herewith.

On the other hand, and in conjunction with our previous remark in (12.3), we further note that, *within the appropriate context*, yet, always an "*abstract*" one, as advocated, hitherto,

(12.6)   a *group action* produces a *flow*, therefore, a *symmetry* too; so, algebraically speaking, it entails "*commutativity*". Consequently, speaking now in physical terms, this implies *conserved* (quantity)/*invariance* (E. Noether).

Now, as a result, we are thus led again to a restatement of/adjustment to our *diagram* in *(9.40)* above. That is, one still gets at the following *complemented version* of the same;

(12.7)   *group action* $\longleftrightarrow$ *flow* $\longleftrightarrow$ *symmetry* $\longleftrightarrow$ *conserved* (quantity) *invariant* (Noether) $\longleftrightarrow$ *commutativity* (: *geometry*) $\longleftrightarrow$ *algebra* (S. Germain).

Of course, concerning the above diagram, one should also take into account the "*local*" *validity* of the pertinent arrows, involved therein; see thus, for instance, (11.6) in the foregoing.

On the other hand, one can also afford the following telling diagram that still pluralistically summarizes the preceding:

(12.8) *invariant* (*local*(!)) *information* (see e.g. A. Mallios [17: p. 266, (2.1.1)]), means, in effect, the *presence*(!) *of* a *physical law*, hence, (*Lagrange*) equivalently, of a *field* $\xleftarrow{\text{Hamilton/Schrödinger}}$ $\mathcal{A}$-*connection* $\longleftrightarrow (\mathcal{P})\mathcal{D}E \longleftrightarrow$ (*locally*) *flow* $\longleftrightarrow$ (*locally*) *symmetry* (cf. (11.6.1)) $\xleftarrow{\text{Noether}}$ "*conserved quantity*" (: alias, "*momentum map*") (see e.g. A. Cannas da Silva [1: p. 127]).

The above also describes the "*dynamical mechanism of physical fields/laws*", via a *purely algebraic* (*re-*)*formulation, independently of any* "*spatial*" *substratum*; hence, following ADG, still capable of *admitting*, along the relevant PDE's, *any sort of* "*singularities*! Yet, in this context, cf. also C. N. Yang [1: p. 96, Figure 6].

On the other hand, by "*normalizing*" the celebrated *Schrödinger's equation* (: taking constants equal 1), one gets at the relation

$$(12.9) \qquad \partial_t = H.$$

Thus, employing now herewith "*ADG-jargon*", one reads the same equation, as a

(12.10) *transmission of* the "*variation of information*" (: "*dynamics*"), viz. *of* our *basic* "*arithmetic*" (even, *Calculus*) on the *reals* $\mathbb{R}$,

$$(12.10.1) \qquad (\mathbb{R}, "dt") \equiv (\mathcal{C}_{\mathbb{R}}^{\infty}, dt, \Omega_{\mathbb{R}}^1) \equiv \mathbb{R},$$

to another niveau, presented herewith through the "*Hamiltonian*".

Yet, in other words, one *thus* arrives, *no more no less*, at a

(12.11) "*quantization*" *of* "$\mathbb{R}$", in the form of (12.10.1).

Consequently, any time we can afford "*relativistic dynamics*", in the sense of ADG, viz. *sheaf* (hence, even *topos-*)*theoretically* (see also e.g., A. Mallios [18: p. 1946, Epilogue]), then, automatically, one is still led, to a "*quantized*" niveau; namely, we thus have already *quantized, via Schrödinger*, our "*basic arithmetic*". Therefore, at the very end, even *in effect* to a sort of a *UFT*. See also Section 2.4; (3.5.1).

1.13 Epilogue

We terminate by the present Section, with some further remarks/*morals*, emanated from the preceding discussion, being also characteristic of the point of view, that has been advocated, so far, within the framework *of ADG*.

So it is instructive, once more, to remark, this being also quite illuminating of our general perspective, thus far (cf., for instance, comments after (9.61) in the preceding), that;

(13.1) *everything is started from and ended in* $\mathcal{A}$ (: "*we*").

See also e.g. (1.4) in the foregoing. Consequently, in other words, it is always, "*we*", of course, who observe and then, *measure*; see also e.g. A. Mallios [18: (1.3)/(1.12)]. Therefore, *the* above *association* (yet, cf. (10.28)/(10.33), along with (9.39.1)/(9.40)),

(13.2)   *observe* ⟷ *measure*, technically speaking (viz., in terms of ADG), *through* that one, effectuated by,

(13.2.1) $\mathcal{D} \longleftrightarrow \omega \equiv (\omega^{(\alpha)}) \longleftrightarrow R(\mathcal{D}) \longleftrightarrow g \equiv (g_{\alpha\beta})$,

(cf. also (10.6)/(10.5)), *clarifies* further *the* fundamental *adjunction* (3.5), or even (10.16). In this context, see also (9.39), together with the ensuing comments therein.

On the other hand, the *quantum gravity* perspective (or even, more generally, that one of a *Unified Field Theory*, cf. also Section 1.9.2), taken always within the present (viz. *axiomatic*) framework, consists actually in looking at a

(13.3)   *sheaf-*(or even, more generally, *topos-*)*theoretic dynamics*, in terms of ADG, employing throughout, *commutative algebraic* (yet, "*functorial*") *localizations*.

The above reminds, of course, the emphasized, throughout the foregoing, *Bohr's point of view*, according to the *homonymous principle* (see, for example, (0.2)). Once more, based on the same aspect, we still note (: it is actually *proved*, see below) that,

(13.4)   *locally prepared measurement environments* correspond to *commutative algebraic contexts*. (Cf., for instance, "Kähler dynamics"; see e.g. (8.19), as well as, (5.4)).

Therefore, in other words,

(13.5)   *global non-commutativity* (: potential, our own descriptions) reduces *locally* to *commutativity* (: our calculations, "*local situations*")!

Consequently, one gets, even herewith, thus by physically describing the situation in (13.5), still as,

(13.6)   another aspect of the classical "*principle of equivalence*" (: Minkowski - Lorentz, Einstein).

Furthermore, it is also instructive to recall, once more here, the *important fact* that,

(13.7)   the above are in force, within a "*non-spatial*" framework (!) (hence, the denomination "*abstract*"), *according to the very essence* of the whole mechanism *of ADG*.

Yet, the previous statement, is, in particular, further corroborated by the *topos-theoretic* considerations throughout the foregoing. Thus, one is finally led to what one might conceive, as a *"relational"* perspective, against the classical one, the latter being always an extremely *"spatial theory"*. In this context, see also e.g. A. Mallios [20], along with relevant comments in the sequel.

On the other hand, the above is certainly ascribed, technically speaking, to the *basic result* in Section 1.3 (see *Theorem 3.1*; yet, e.g. (3.13) therein). Furthermore, the same might still be viewed as a *clarification of the* classical

(13.8)   *"valuation vagueness"* (yet, probabilistic aspects) in *Quantum Theory* (W. Heisenberg), even of the classical *"complementarity principle"* (N. Bohr).

All told, we come thus to the conclusion of adopting a *"relational viewpoint"* of the preceding, as it actually were (!), anyway, yet, in particular, to a conception of a *"relational Calculus"*, referring to Analysis and/or Geometry. In this regard, a characteristic example certainly constitutes, what one might still perceive, as a *"Feynman Calculus"*; thus, something undoubtedly *"non-spatial"* (!), in the standard sense of this term. This, of course, pertaining to his already celebrated theory of the homonymous *"diagrams"* (cf. also ibid., as above).

In this context, it is also worth mentioning here the fact that one usually speaks of a *"non-spatial Calculus"*, by actually referring to the *techniques/mechanism that govern(s) the function of the* particular *"Calculus"* we usually employ; this, is, so far, in principle, the outcome of the *"space"* we use, in the standard sense of this term. However, as one realizes (cf., for instance, the situation we are confronted with, when working within the context of ADG). Thus,

(13.9)   *the* same *mechanism is* in effect finally quite *independent of the "space"*, that has been classically employed/concocted for the job, namely, to define/create that mechanism *Example* again, ADG, Feynman etc. So one actually realizes herewith, what we have often pointed out in the foregoing, as the

(13.9.1)   *"functorial"* (hence, in effect *more natural* (!), see e.g. (9.58)) *way*, that the fundamental techniques we use, function, at each particular time.

Yet, it is just a matter of time of our experience to recognize/discern the *"functorial"* rôle/character of the real essence of what we actually employ in our *"everyday"* function/application of the *"Calculus"* we afford.

Chapter 2

Applications: Fundamental Adjunctions

2.1 On Utiyama's Theme/Principle Through "$\mathcal{A}$-invariance"

2.1.1 *Introduction*

To paraphrase M. F. Atiyah (see e.g. [GT, Vol. II: Chapt. 2, p. 79],

(1.1)   *gauge theories/"geometries"* are characterized by a certain particular *"symmetry group"* (F. Klein), and all *physical/geometric properties* are *"gauge invariant"*.

On the other hand,

(1.2)   *"invariant interpretation of interactions"*

is, actually the point of view, already advocated by R. Utiyama in his classical paper [1]; yet, as it seems, still the same aspect has being independently conceived by C. von Westenholz [1: p. 516]. Now, *Utiyama's Theorem* (see Corollary 3.1 in the sequel) may be considered as a special important example of (1.1): As a matter of fact, both these statements are *very particular cases of* the fact that,

(1.3)   *a physical law is (always) "$\mathcal{A}$-invariant" (of whatever "$\mathcal{A}$" might be).*

Certainly the latter issue is something, which actually ensures the detection of the same physical law. See e.g. A. Mallios [18: p. 1929, (1.1), and subsequent comments therein], in conjunction with A. Mallios [GT, Vol. II: p. 148, (1.13.1), (ii)[†]]. Now, "$\mathcal{A}$", viz., our "*arithmetic*" (see also (1.4) in the preceding), is, of course, *locally employed*, and this, by the very definitions, means in effect the action/employment herewith of a particular "*gauge*" (: method of measurement); see also (1.6) below. Consequently,

(1.4)   (1.1), hence, *Utiyama's Theorem, as well*, is a *particular aspect of* (1.3).

[†]With the kind permission of the reader, I take here the opportunity to restore an oversight elapsed in the cited book of mine, in p. 148, as indicated above: viz. instead of Chapt. VI, please, write therein, Chapt. I.

Within the same context, as above, we still recall here that,

(1.5)   *"physical geometry"* is the *outcome of the physical laws.*

See A. Mallios [16 : p. 1557, (1.1)]. Therefore, by further commenting on (1.1), we also note that, a

(1.6)   *"physical/geometric property"* means actually just a *particular re-alization/description,* hence, *"gauge"* (: measurement) *of* the *physical law,* under consideration.

In this context, cf. also, for instance, (8.22) in the preceding (Wittgenstein's adage). Thus, the above clarify now completely our claim in (1.4), yet, in conjunction with the ensuing discussion.

2.1.2   *Utiyama's Theorem*

We continue, by technically approaching, first, the theorem in the title of this Section. So, roughly speaking, the theorem in focus refers to the *behavior of* (the two items of) *a fundamental* (physical) *"adjunction",* described by an appropriate *"Lagrangian function,* under a suitable *"gauge-action".*

We haste right away to associate/(explain) the aforementioned technical terms with/(by) notions that we already employed in the preceding account of this study: Thus, the aforesaid *"physical adjunction"* refers in effect to a really fundamental one, as described by the *"adjunction",*

(2.1) $\mathcal{A}$-*connection* $\xleftrightarrow{\hspace{2cm}}$ *curvature*

The *physical counterpart* of (2.1) is expressed by the diagram,

(2.2)   *physical law* $\longleftrightarrow$ *field* ($\longleftrightarrow$ $\mathcal{A}$-*connection*) $\xleftrightarrow{\hspace{1.5cm}}$ *curvature/ field strength* (:*"trace"* of the physical law).

The technical part of the previous association between (2.1) and (2.2) has been already adequately explained in the foregoing of the present treatise: see thus (6.3), (2.42) and (2.38), as well as, (10.2)/(10.14), and (8.24), (8.28)-(8.30).

On the other hand, the *"gauge action"* that still was mentioned above, refers, in particular, to that one of a group sheaf of the form,

(2.3) $\mathcal{A}ut_{\mathcal{A}}(\mathcal{E}) \equiv \mathcal{A}ut\mathcal{E}.$

Namely, the group *sheaf of $\mathcal{A}$-automorphisms,* of a given $\mathcal{A}$-*module* $\mathcal{E}$, which will mostly be a *vector sheaf* (: locally free $\mathcal{A}$-module of finite rank). In this context, we may still recall here that, within the terminology of ADG, we have also represented (axiomatically) an *"elementary particle",* by a particular type of a *vector sheaf,* viz. with say, a *"dynamical vector sheaf"* (cf. (6.1)/(8.20) and (8.19.4), as above). Yet, see also A. Mallios [16 : p. 1582, (4.31)], along with [14 : p. 62, (3.2)].

Indeed, we are actually going to consider an appropriate subgroup of (2.3), that is, the group,

(2.4) $$(Aut\,\mathcal{E})_\rho < Aut\,\mathcal{E},$$

where ρ indicates the presence of a suitable $\mathcal{A}$-*bilinear form* on $\mathcal{E}$, as before, occasionally an $(\mathcal{A}$-$)metric$, viz.

(2.5) $$\rho : \mathcal{E} \oplus \mathcal{E} \longrightarrow \mathcal{A},$$

the latter is *symmetric* or *skew-symmetric, non-degenerate* or *strongly non-degenerate*, as the case might be. Thus, (2.4) stands for

(2.6) those elements $\phi \in Aut\,\mathcal{E}(\equiv (Aut\,\mathcal{E})(X)$; cf. A. Mallios [GT, Vol. I: p. 35, (6.21)]), which are "ρ-preserving"; that is, one has,

(2.6.1) $$\phi^*(\rho) := \rho \circ (\phi, \phi) = \rho.$$

Now, according to (1.3), a physical law is $\mathcal{A}$-invariant; yet, *equivalently*, we still say that

(2.7) *any physical law is "functorial"* ,

or even, in other words, *physis* (viz. *its function*) *is functorial*, see also A. Mallios [18 : p.1930, (1.5.1)]. In this context, it is still instructive, at this point, to look at the latter Ref., Section 1, in general, for relevant comments, pertaining to the essential meaning of the previous term, *"functorial"*, and the way *we* usually *employ* it.

Furthermore, by still taking (1.1) into account, we remark that,

(2.8) the *gauge theories* we use, so QFT is for instance one of them, should be *"gauge invariant"*, as well; that is, their physical/ "geometric" properties.

Indeed, as already clarified by the foregoing, this is in effect a *consequence of* (1.3) (!). Therefore, to say it once more this because

(2.9) our theories try to *describe always – not explain* (!) *– the function of Physis*.

On the other hand, the whole *function of* a *physical law* (viz. *of the Physis* herself) is mathematically rendered/described by the *fundamental adjunction* (2.1), thus, *equivalently*, through its *physical transcription via* (2.2). As a consequence, one thus gets at the conclusion that,

(2.10) *any function used to express the adjunction* (2.1), thus, *equivalently, (2.2),* as well, *should be "gauge invariant"* too. In this context, we still note that the function alluded to above, *becomes* in fact, mathematically speaking, a *"natural transformation of functors"* (defining the adjunction in focus; cf. S. Mac Lane [1: p. 78, Definition]).

Of course, the preceding is still based on the *"functoriality"* of (the function of) *Nature*, hence, of the *physical law* itself (see (2.7) and comments following it); yet, on the way *we* actually use to look at/get profit of it. Thus, *"we"*, that is, in other words, *the* function/procedure (: our *theory*) *we employ should be "$\mathcal{A}$-invariant"*, alias, *"gauge invariant"*. See also A. Mallios [18: the whole discussion on p. 1930ff]. Indeed, the above as, in particular, stated by (2.10), constitute essentially the *quintessence of* the *point of view* of the theorem *of Utiyama*. More technical aspects of the same will be presented right away by the following Section (see, for instance, Corollary 3.1 below).

2.1.3   *Utiyama's Theorem (Contn'd: Technical Details)*

Our aim by the present section is to elaborate further our previous conclusion in (2.10). So the *"function"*, we are interested in herewith, according to the terminology employed therein, is the well-known from the classical theory already, fundamental notion of a *"Lagrangian"*: Thus, by applying a more technical language, however, still roughly speaking, we come to the following *"technical reformulation of (2.10)"*; that is, one concludes that,

(3.1)   the *"($\mathcal{A}$-)connection Lagrangian"* is *gauge invariant, if, and only if,* the associated *"curvature Lagrangian"* (see e.g. (2.1)) is Ad-*invariant*.

In this context, we further note that an *"$\mathcal{A}$-valued* (alias, *"generalized"*) Lagrangian"* is in particular just, a *sheaf morphism*, $\mathcal{L}$, as follows, which still will be further specified, as the case might be; thus, we set:

$$(3.2) \qquad\qquad \mathcal{L} : \mathcal{E} \times \mathcal{E}^* \longrightarrow \mathcal{A},$$

where, of course, one sets

$$(3.3) \qquad\qquad \mathcal{E}^* := \mathcal{H}om_{\mathcal{A}}(\mathcal{E}, \mathcal{A}),$$

i.e., the *dual* of the $\mathcal{A}$-*module* $\mathcal{E}$.

On the other hand, the aforementioned in (3.1) *(gauge) invariance* (: "$\mathcal{A}$-invariance") *of* the *Lagrangian* is usually expressed, through a given $\mathcal{A}$-*metric* ρ (cf., for instance, (2.5)), requiring the *invariance of the metric*, with respect to $\mathcal{A}ut\,\mathcal{E}$ (see (2.6)). Thus, one gets at the associated in terms of the above, *"moduli space"* of $\mathcal{E}$, viz. the quotient space,

$$(3.4) \qquad\qquad \mathcal{M}(\mathcal{E})_\rho \equiv \mathcal{C}onn_{\mathcal{A}}(\mathcal{E})/(\mathcal{A}ut\,\mathcal{E})_\rho.$$

See also (2.4), as above, in conjunction with A. Mallios [GT, Vol. II: p. 49, (5.46.2) along with pp. 134-135, (7.2)-(7.7)], concerning the particular notation as employed in (3.4), yet, the ensuing discussion right below.

So the *"generalized Lagrangian"*, as before, pertaining to the $\mathcal{A}$-*connections* on $\mathcal{E}$, viz. the set,

$$(3.5) \qquad\qquad \mathcal{C}onn_{\mathcal{A}}(\mathcal{E})$$

(see also e.g. A. Mallios [VS, Vol. II: Chapt. 7]), yet the aforementioned in (3.1) "($\mathcal{A}$-)*connection Lagrangian*", refers in effect to an $\mathcal{A}$-*valued* function (see e.g. (3.2)) of $\mathcal{A}$-*connections on* $\mathcal{E}$, more precisely elements from (3.4) *and* their (*covariant*) *derivatives*: indeed, one essentially employs here an $\mathcal{A}$-*isomorphism*,

$$(3.6) \qquad\qquad \mathcal{E} \underset{\bar{\rho}}{\cong} \mathcal{E}^{*},$$

that is, via the aforesaid $\mathcal{A}$-*metric* ρ (see (2.5)), while we also assume the following $\mathcal{A}$-*isomorphism of* the $\mathcal{A}$-*modules* involved,

$$(3.7) \qquad\qquad \mathcal{E}^{*} \cong \Omega^{1}.$$

See also e.g. A. Mallios [GT, Vol. II: p. 145, (1.4)], for a relevant application in what we may call "*Abstract General Relativity*".

In this context, it is still instructive to recall/comment on the following adage of M. F. Atiyah (see also (1.1) in the preceding);

$$(3.8) \qquad\quad \text{"...}physical/geometric\ properties...are\ gauge\ invariant\text{"}.$$

The above is, of course, just a consequence of what one might denominate, as

$$(3.9) \qquad\qquad principle\ of\ \text{"}\mathcal{A}\text{-}invariance\text{"},$$

as this has been advocated by A. Mallios [18: p. 1930, (1.3)/(1.5)]. In this regard, one should further note that,

by referring to "*physical properties*", as in (3.8), one actually means,

(3.10)
$$(13.10.1) \qquad \begin{array}{l} properties,\ \text{yet},\ characteristic\ \text{ones},\ of\ the\ function\ of\ Na\text{-} \\ ture,\ \text{hence, of the}\ physical\ laws,\ \text{as well (cf. (1.6)).} \end{array}$$

Thus (ibid.), one can further express in effect, *equivalently*, the same principle, as before, by still saying, that

$$(3.11) \qquad\qquad \text{Physis is "functorial"}$$

(ibid. p. 1930, (1.5.1)); therefore, still what amounts to the same thing,

$$(3.12) \qquad\qquad Physis\ does\ not\ depend\ on\ us\ (\equiv \text{"}\mathcal{A}\text{"}).$$

(ibid., yet, p. 1932, (1.12)).

Now, the previous association, deduced from (3.12), that is,

$$(3.13) \qquad\qquad Physis \longleftrightarrow we \equiv \text{"}\mathcal{A}\text{"},$$

can be very well related with the *fundamental*

$$(3.14) \qquad\qquad Hom - \otimes\ adjunction.$$

See also the preceding *adjunctions* (2.1)/(2.2), and following comments therein. Yet, compare the above with (10.47), (10.48) (: "*Lassner-Uhlmann principle/ adjunction*"), along with (10.45) in the preceding.

On the other hand, the *"connecting function"* of a given *adjunction* (see S. Mac Lane [1: p. 78, Definition]) is, in fact, a *natural transformation of functors*; therefore,

(3.15)   the latter should still preserve *"(𝒜-)invariance"* of the adjunction, with respect to any *"(𝒜-)invariant function"*, pertaining to any one of the two associated functors, through the adjunction.

In this context, see also Section 1.2; (2.35) in the preceding discussion. Yet, take for instance the *fundamental adjunction* (3.14), as above, and any *"Lagrangian invariant"* function, as e.g. in the standard Utiyama's Theorem.

In summa, one thus gets at the following.

Corollary 3.1. (*"Utiyama's Theorem"*). The *"gauge invariance"* of an appropriate *"Lagrangian for (𝒜-)connections"* is *equivalent to* a similar *invariance of the* corresponding *Lagrangian*, pertaining *to the* associated *curvature of the* particular *(𝒜-)connection* at issue.   □

Scholium 3.1. — Our aim by the following lines is to supply *further evidence* in our previous argument, pertaining *to the proof of the* last *Corollary*, as above (see also the discussion in the preceding; (9.52)-(9.54)): Thus, by still referring to the aforesaid *"natural transformation of functors"*, yet, the *"counit isomorphism"* (see also the discussion in the foregoing, (2.34)/(2.35), along with (3.17)-(3.19)), one gets at a *category equivalence/isomorphism*, hence, *preserving* also any *"invariance"* notion of the categories involved.

So, to say it otherwise, the aforementioned *category isomorphism* leads to an *equivalent* in effect *formulation of Corollary 3.1*; that is,

(3.16)   one obtains an *(𝒜-)invariance* (with respect to an appropriate *group of transformations*) of an *(𝒜-)connection Lagrangian, if, and only if,* an *(𝒜-)invariance of the* corresponding herewith *curvature Lagrangian* is still in force.

Of course, as in the classical case, see for instance, R. Utiyama [1], along with e.g. D. Bleecker [1: p. 147ff], the above *conclusion (3.16)* has a special bearing on the classical *Yang-Mills Theory*.

Thus, *solutions of the Yang-Mills equations*, i.e., *"Yang-Mills fields"* (strictly speaking), are by definition (see, for instance, A. Mallios [GT, Vol. II: p. 32, (4.54.1)]), *"gauge equivalent"*. Therefore, *"(𝒜-)invariant"*, under a certain particular *"gauge group of transformations"*; ibid., p. 59, (5.46.2), or even, p. 92, Section 2.1.

Now, by considering any $\mathcal{A}(X)$-*valued function*, on

(3.17) $$\mathcal{C}onn_{\mathcal{A}}(\mathcal{E}) \times (\Omega^*(\mathcal{E}), \mathcal{D}),$$

to have it *"compatible"*, within our framework, *we assume* on the basis of the foregoing (see thus, for instance, A. Mallios [18: p. 1930, (1.3)/(1.9.2)], yet the present account, (1.13)-(1.16)), that *the same function*, as before, *is "$\mathcal{A}$-invariant"*. Concerning the notation in (3.17), see also (3.5), as above, in this section, along with (3.6), herewith; yet, by the second factor in (3.17), we mean, collectively, the "*Grassmannian*" of $\mathcal{E} \cong \mathcal{E}^*$, together with the corresponding, as the case might be, still succinctly, "*covariant differentials*" (cf. also e.g. A. Mallios [VS, Vol. II: p. 228, (8.15)]). Indeed, the above is, in particular, *specialized* to the case of what we call in the preceding, an ($\mathcal{A}$-)*connection Lagrangian* (see also e.g. (3.2)). Hence, the latter should be thus "*$\mathcal{A}$-invariant*". On the other hand, in view of the *fundamental adjunction* (cf. also (2.1) above),

(3.18)   ($\mathcal{A}$-)*connection – curvature*,

the previous "*$\mathcal{A}$-invariance*" should still be *equivalent* with that one of the corresponding *curvature Lagrangian*; yet based e.g. on the "*counit isomorphism*", of the same adjunction, as before.

In this context, we further note herewith (see also A. Mallios [GT, Vol. II] for a detailed account) that,

> the *critical points* of the "*Yang-Mills* (or, even "*Einstein-Hilbert*") *functional*", viz., those $\mathcal{A}$-*connections* $\mathcal{D} \in \mathcal{C}onn_{\mathcal{A}}(\mathcal{E})$ (cf. (3.5)), for which one has,

(3.19)

(3.19.1)
$$\delta \mathcal{YM}_{\mathcal{E}}(\mathcal{D}) \equiv \delta(\mathcal{YM}_{\mathcal{E}}(R(\mathcal{D}))) = 0,$$

> are, in effect, *solutions of the Yang-Mills* (or, even classically speaking, "*Euler-Lagrange*") *equations*.

Yet, *equivalently*, one still says that;

(3.20)
> the same $\mathcal{D}$'s, as before, are "*zeros of the variation*" (cf. (3.19.1)) *of the Lagrangian (action-)density*,

as the latter is given, by its associated "*integral form*", viz. by the relation,

(3.21)
$$\delta \mathcal{YM}_{\mathcal{E}}(\mathcal{D}) := \frac{1}{2}\delta \left(\int_X tr(R \wedge *R) \right) = 0.$$

On the other hand, we also express the last relation, as above, by still saying that:

(3.22)
> the *Euler-Lagrange equations*, as before, *are satisfied* (here one actually realizes another version of the "*least action principle*", or even of the so-called "*Noether's preservation principle*" (cf. also e.g. Section 2.2.3; (1.7) in the sequel), *if, and only if, the* corresponding (action-)*integral* (over time) *of the "invariant Lagrangian" is "stationary"*. (See e.g. A. Mallios [GT, Vol. II: p. 66, (8.6)].

In this context, we can also note here that the above two remarks, as in (3.20) and

(3.22) could still be conceived as *essentially another version of Utiyama's argument.* Cf. also ibid. p. 69, (8.24)/(8.26.3).

> **Note 3.1.** (Terminological). — The *"integral"* appeared in (3.21) can be considered as an appropriate *"continuous linear form"*, defined on the space of functions, in effect *sections of* the sheaves (: $\mathcal{A}$-modules) involved. Of course, this refers to the standard notion of a *"Radon-like measure"*: cf. also ibid., p. 63, Scholium 7.1, in conjunction with (7.29) therein.

For more details, in conjunction with the preceding discussion, as already mentioned, we further refer to the same treatise, as above; in particular, p. 57 therein, (6.44)/(6.47), along with p. 50, (6.2)/(6.3), and p. 48, Lemma 5.1, as well as, p. 68, (8.21).

> **Scholium 3.2.** — Throughout the preceding we considered an

(3.23)
> *invariance* of a Lagrangian, *with respect to a* suitable *flow*; therefore, according to the very definitions, with respect to a *symmetry*, which is entailed by the flow (*Chapman-Kolmogorov*), thus, through a *transformation* (:*"action"*) group.

Consequently (*E. Noether*), one gets at the

(3.24)
> existence of a *one-to-one correspondence* between *invariance of a Lagrangian*, as before, hence, *symmetry*, and *"conserved quantity"* (:*"momentum map"*; see for example, J. E. Marsden-T. S. Ratiu [1: p. 372, Theorem 11.4.1]).

On the other hand, the same *"invariant,* alias *conserved, quantity"* is usually realized by a sort of an appropriate *"integration"* (cf. the previous Note 3.1) along a suitable (*"critical"*) *path* (see (3.19)/(3.21)), on which the so-called *Euler-Lagrange equations* are *in force*, with respect to the aforementioned *invariant Lagrangian*.

As a result, based on the foregoing one realizes of course here too, another instantiation of the celebrated, so-called *"least action principle"*; see e.g. L. Nicolaescu [1: p. 170, Theorem 5.1.3]. Yet, within the same context, as before, we still recall here also in connection with the above principle, the important fact that;

(3.25)
> the aforementioned *Lagrange's equations are independent of coordinates,* as this can be explained by the so-called,
>
> (3.25.1)   *"Hamilton's principle"*.

See e.g. S. Mac Lane [3:pp. 278, 279]. The same is actually expressed, still through the preceding argument in (3.19)-(3.22); thus, in effect, an *equivalent* formulation of the same principle, as in (3.25.1) (ibid.). As a consequence, we can still say that,

(3.26)   *the Lagrange's equations/point of view* enjoy entirely a *"relational"* perspective/character!

Therefore, one gets at what we call in the preceding (see e.g. Section 1.10.1, in particular, (10.32), along with Scholium 10.2),

(3.27)   *"Lagrangian perspective"*.

Cf. also, for instance, A. Mallios [20: p. 64, (2.12)].

In this context, to be more (mathematically) precise, one considers a *"flow"*,

(3.28) $(\phi(t) \equiv \phi_t)_{t \in \mathbb{R}},$

that is, a *(sheaf) morphism*,

(3.29) $\phi : \mathbb{R} \longrightarrow \mathcal{A}ut\,\mathcal{E} < \mathcal{E}nd\,\mathcal{E} \equiv \mathcal{H}om_{\mathcal{A}}(\mathcal{E}, \mathcal{E}),$

alias, a *"curve"* in the *group (sheaf) of $\mathcal{A}$-automorphisms of $\mathcal{E}$* (cf. also e.g. (2.4). Of course, we consider herewith the *reals* $\mathbb{R}$, as a *constant* (group-)*sheaf*; cf., for instance, A. Mallios [VS, Vol. I: p. 17ff]). Thus, one gets at a *symmetry* of $\mathcal{E}$, as above (Chapman-Kolmogorov, see (3.23)). *Then*, following *E. Noether*,

> for any ϕ-*invariant Lagrangian L of $\mathcal{E}$*, viz. such that,
>
> (3.30.1) $\phi^*(L) := L \circ \phi = L$, alias, $\mathcal{L}_\phi(L) = 0,$
>
> (3.30)   there exists a *"conserved quantity"* (:*"momentum map"*). Cf. also (3.24), along with A. Mallios [GT, Vol. II: p. 32; (4.54)], as well as, Section (2.3); (1.7) in the sequel.

Yet, regarding our previous argument in (3.30), see also e.g. (9.39.1) in the preceding, or even *"Lassner-Uhlmann adjunction"*, as in (10.46)/(10.47), and still (10.34)/ (10.54). Furthermore, concerning the notation employed in (3.30.1),

(3.31) $\mathcal{L}_\phi(L),$

viz. what one may call, *"Lie derivative of L, with respect to ϕ"*, see for the classical case e.g. A. Cannas da Silva [1: p. 36, Definition 6.3, along with p. 106, Example].

In the same vein, and *in conjunction* also *with* (3.28)/(3.29), as above, one can thus still refer, herewith, as an *instrumental factor*, to the

> (additive) *abelian group of*
>
> (3.32)   (3.32.1)   *"1-dimensional translations"*.
>
> Namely, *the reals*, $\mathbb{R}$.

See thus e.g. for the terminology applied, H. Weyl [2: p. 193]. Yet, cf. also the relevant comments in the sequel: Section 2.3, (7.34)/(7.36), along with the subsequent Section 2.2.

On the other hand, the *"space"* or, *equivalently*, what we actually mean, in general, about it anytime we are talking for the *"geometry"*, as a *"physical entity"*, seems to be *"infinitesimally"*, yet, *"locally"*:

(3.33)   *linear*, or *affine*, or even *symmetric*; hence, *commutative*!

Thus, the above, constitutes in fact, once more,

(3.34)   a *basic moral* of the present account.

See also e.g. Section 1.12; (12.7)/(12.8) in the foregoing. On the other hand, a *full-fledged function of* $\mathbb{R}$, in particular, through the map/*functor* **exp**, is still given throughout the Appendices III, IV below.

The preceding contribute in the clarification of the situation, which is described by (1.1), and still of what actually was, technically speaking, inherent in the very essence of the same *Utiyama's Theorem*; see for instance, (3.23)/(3.24) in the foregoing. Yet, within the same point of view, cf. also Section 2.4.5 in the sequel.

Now, we come next, by the following Section, to further illuminate another *basic issue of the* same theorem, pertaining to the fundamental *adjunction* (3.14).

2.1.4   *Dynamical Analogue of the Fundamental Hom-Tensor Adjunction*

As already said at the end of the previous Section, our main purpose by the ensuing discussion is to elaborate further on (9.55)/(9.56), as well as on (9.59), as presented in the foregoing of this treatise, being, in effect, in the heart of the aforementioned theorem, as in the preceding Section; the same is also the subject matter of this Section.

Now, the very title of the present Section refers, of course, to the

transcription of the *homology-type term*,

(4.1.1) $\mathcal{H}om\text{-}\otimes$ *adjunction*,

(4.1)   to its *analogue*, *within* the context of *ADG*; viz., in other words, through the fundamental *ADG-theoretic adjunction*,

(4.1.2) $\mathcal{A}$-*connection* ("*field*") $\longleftrightarrow$ *curvature* ("*field strength*")

Indeed, as it has been already hinted at in the preceding (see e.g. (9.55)/(9.59)),

(4.2)   *the* previous *diagrams* (4.1.1) and (4.1.2) are, in effect, *equivalent* formulations, *physically speaking*, of the same fact/situation.

Now, as already said at the beginning of this Section, we proceed to better *clarify* our assertion (4.2), as above: Thus, the idea behind our *claim in* (4.2) is that: on the one hand,

(4.3)   in the *homological-theoretic* (fundamental) *adjunction* (4.1.1), one may associate, by the very definitions, the *functor* $\otimes$ with the notion of an $\mathcal{A}$-*connection*: in fact, a *natural transformation of functors* (see e.g. A. Mallios [17: p. 281, (6.17)]). This can still be rooted, for instance, on what we call in the preceding "*Kähler construction*"; see e.g. Sections 1.5 and 1.8 in the foregoing of this treatise.

On the other hand, we further remark that,

the other *functor* $\mathcal{H}om$ *in* the same *adjunction* (4.1.1), as above, can still be attributed to what we may call:

(4.4)   "*physical geometry*", in the sense, at least, of ADG (see e.g. A. Mallios [16: p. 1557, (1.1)]), hence, the *spin-off of* the *physical law(s)*. Therefore, the appearance again of the (fundamental), notion of an $\mathcal{A}$-*connection*, thus, of the

(4.4.1)   functor $\otimes$, as well; see also e.g. (4.3), as before, along with A. Mallios [17: p.280, (6.15) (6.16.1)], or even, (10.17) and (2.38.1) in the preceding of this text. As a result, one thus gets at the notion of "*curvature*", viz. *the* always existing *outcome of an* $\mathcal{A}$-*connection*!

In this context, see also the same Ref., as before, p. 267, (2.5), along with p. 282, (6.26)/(6.27). Yet, cf. A. Mallios [14: pp. 63, 64, (2.7)/(2.8): in particular, before (2.7) therein, the remarks pertaining to the notion of "*curvature*"; also, see p. 62, (2.1), and p. 66, (3.3), as well as, the comments following it].

The preceding still enlighten the manner the *physical adjunction*, as realized by (4.1.2), is mathematically expressed, *in terms of Homology*, hence, in a *functorial way, through the* (typical/formal) *adjunction* (4.1.1). And, of course, *vice versa*(!), as it actually was our claim, still in (4.2).

On the other hand, in conjunction with the preceding account, and by still looking at the previous *Corollary* 3.1 ("*Utiyama's Theorem*"), we further point out the ensuing comments (cf. Scholium 4.1 below), relating this same issue, as before, with our previous relevant discussion, at least, in Sections 1.9 and 1.10. So we get at the following.

Scholium 4.1. — In other words, one can still construe, what we called in the preceding *"Utiyama's principle"/ Theorem* (cf., for instance, (6.8.1) and (9.51)-(9.53) in the foregoing) as *another formulation*, pertaining to the (classically named) *"minimal coupling/replacement"*: Indeed, the latter refers, in effect, within the *framework of ADG*, to the *fundamental* relation/*"adjunction"* (4.1.1), as above, being thus *inherent* in the *instrumental*, throughout the whole ADG, *couple*,

(4.5) $(\mathcal{E}, D)$;

that is, in our terminology the basic *"Yang-Mills* (gauge) *field"/* (physical law), which relates

(4.6) $matter\,(:\mathcal{E})$, to $field\,(:D)$

(see, for instance, Section 1.10; (10.3), in the preceding). In this context, see also e.g. G. L. Naber [1: pp. 391-392].

Now, the same aspect, as above, is actually further related with the classical *"least action principle"*: Cf. thus, as already mentioned, the previous discussion, in particular: Section 1.9, in the foregoing; (9.1.2)/(9.26), (9.42), and (9.55)/(9.56), along with (9.59). Furthermore, Section 1.10: (10.20)/(10.21) and (10.43)/(10.45), as well as, within the same vein, Section 1.11; (11.8).

Yet, within the same context, as above, see also (2.44)/(2.45) in Section 1.2 of the preceding, as it concerns a *quantal-topos description of* (4.5), as before.

2.2   "Affine Geometry" and "Quantum"

2.2.1   *Introduction*

The present account originated in the adage of H. Weyl as appeared in his classic [2: pp. 85/86], to the extent that;

(1.1)   *"... affine geometry holds in the infinitely small."*

In the same treatise (ibid. p. 86) Weyl further remarks that,

(1.2)
"... in Euclid's geometry the space appears to be ... much more special (namely, non-curved) than the possible surfaces in it, ... [something that] ... the Riemann's space concept has ... the right degree of generality to do away with this discrepancy."

The moral herewith appears thus to be, still in conjunction with the general viewpoint of ADG, that:

(1.3)   *"rigid" geometric structures*, in general, *are not* the *appropriate* ones, *to be confronted with problems in quantum theory* (: the *"infinitely small"*).

Yet, within the same perspective, one can still consider, as an *important spin-off* thereby, what we might call a,

(1.4)   *"Feynman Calculus"*!

In other words, *we* may *employ anything from classical mathematics*, however, with the proviso that this is done, as we can say,

(1.5)   *without* any *fixed "origin"*, at all(!);

therefore, *"affine"*, in quite a *broad sense* of the term. Yet, otherwise said,

(1.6)   we may employ (classical) *notions*, always in the sense of (1.4)/ (1.5), provided they are basically *effective*.

Namely, we are thus more interested in their *"inherent mechanism"*, than in their technical/*classical* *"foundation"*, that usually comes after grasping the essential meaning of the former; and this still, *depending* (even in its simplicity) *on* the level of *our abstraction*, pertaining to the theory employed (cf., for instance, (1.2)). We come next, by the following discussion, to comment further on (1.1); yet, on what we might still call, in general, a

(1.7)   *"Feynman's principle"*.

In this context, see also Section 2.3.5.3.

2.2.2   *ADG vis-à-vis the "Infinitely Small" (: "Infinitesimal")*

Looking at (1.1), and by further expressing in *physical parlance* the mathematical term (in quotation marks) in the title of the present section, one comes to the *conclusion* that,

(2.1)   *"affine geometry"* subsists, when confronting with the *quantum* deep.

That is, essentially, an *equivalent formulation*, physically speaking, *of* Weyl's dictum, as in (1.1). In this context, we might still quote here, Feynman's:

(2.2)   *"... the simple ideas of geometry, extended down to "infinitely small" ... are wrong."*

The emphasis and quotation marks above are ours. See R. P. Feynman [1: p. 167]; yet, his famous, *"fancy schmanzy"* *differential geometry*, when referring to the *Classical Differential Geometry* (acronym, CDG)-*smooth*(: C^∞-)*manifolds*, in general. Furthermore, in connection with (1.1) and (2.1), as above, one may quote herewith, again H. Weyl (ibid., p. 86), to the extent that,

(2.3)   *"... the world must be comprehended through its behavior in the infinitely small [: the "quantum"]."*

The emphasis above, along with the clarification brackets therein, are still ours. Now, the above further supports/actually describes, *"Leibniz's continuity principle"* (ibid.); namely, the *"... lawfulness of nature"* concerns its function, pertaining to *"laws of nearby action"*; therefore (ibid.),

(2.4)   *"... the basic relations of geometry should concern only infinitely closely adjacent points ("near-geometry" as opposed to "far-geometry").*

In toto, the preceding are thus reminiscent of *Riemann's foundation* of *"differential geometry"* (: manifolds), starting from the *"infinitely small"*: Therefore, still the notion of an *"A-connection"*, in general, in terms of ADG, viz. *derivative*, thus, again Leibniz with the *"infinitesimals"*; that is, *"dt"*: in other words, *the way (local) information* (see also the ensuing diagram (2.5)) *is varied*; thus, a *locally*, in principle, *defined function/law*, being still *able to supply global information*, as well. See, for instance, the categorical/algebraic, indeed, *sheaf-theoretic*, or even *topos-theoretic*, formulation of the *derivative/(A-)connection*, within the context of ADG. Therefore, one thus arrives to a *"relational"*/categorical *context*, akin still to the notion of *"quantum"*(!), as well. Within the same context, it is still quite *important* to point out that working in a *sheaf-theoretic framework*, one is able to argue, in terms of the (rightly) *local* (hence, e.g. *"quantum"*, cf. R. Haag [1]) and also *global* perspective, as well!

The following diagram, in conjunction also with that one in (2.7) below, clarifies several notions, alluded to in the preceding.

(2.5)

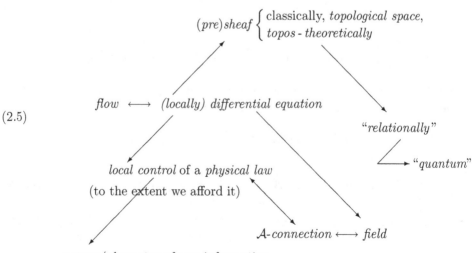

On the other hand, by further arguing within the *framework of ADG*, we already know, always in terms of the same *abstract* (: *spaceless*) *viewpoint* of the latter account, that;

(2.6)   the totality of $\mathcal{A}$-*connections* of a given $\mathcal{A}$-*module* $\mathcal{E}$ (cf. also Section 2.1.4; (4.5)), constitutes an *affine space*. The "*model*" here is the $\mathcal{A}(X)$-module, $Hom_{\mathcal{A}}(\mathcal{E}, \Omega^1(\mathcal{E}))$.

See A. Mallios [VS, Vol. II: p. 32, Theorem 7.1]. So one gets herewith at a first *formal*/mathematical *hint* of Weyl's argument, as in (2.1), when also taking into account an *analogous reasoning* to that one, mainly *conducted by* the previous diagram (2.5) (cf. also the similar one in the sequel). We *further comment on* (2.1) (see also (1.1) in the preceding) right away in the sequel, as well as, through the ensuing Section 2.3. This clarification is actually provided by a *heedful scrutiny of* the preceding diagram (2.5), in conjunction with *the following one*, that *further points out* particular *crucial issues in* (2.5), within always, of course, the *perspective* thus far *of ADG*. So one gets at the following association of *physical/mathematical notions*, referred to therein;

(2.7) $\mathcal{A}$-*connection* $\longleftrightarrow$ *field* $\longleftrightarrow$ *physical law* $\longleftrightarrow$ *differential equation*.

Further interrelations/consequences of the preceding last two diagrams will also be supplied, through the subsequent discussion.

2.2.3 *Flow, and the "Quantum"*

The term "*flow*" refers, of course, to a "*dynamical system*"; therefore, in general, to a "*group action*": take thus, for instance, "*time*" (hence, e.g. $\mathbb{R}$), on a given data/set (of situations). Now, according to the general theory,

(3.1)   a *flow is "causally stationary"*.

Concerning the classical counterpart of (3.1) (: smooth manifolds theory), this is based on the very definitions, pertaining to the *local behavior* of a flow (the same interpreted as the), "*solution curve*" of a "*vector field*", the latter being now the "*infinitesimal generator*" of the flow/field at issue (cf. also (2.5)); alias, one can just refer herewith to the classical "*Chapman-Kolmogorov law*". So see, for instance, R. Abraham-J. E. Marsden [1: p. 61].

On the other hand, by arguing within the framework of the *Abstract Differential Geometry* (ADG), as is actually the case, throughout the present treatise, the previous situation can be transcribed, within the context of a so-called "*Yang-Mills category*"; see A. Mallios [GT, Vol. II: Chapt. 1, p. 26]: Namely, if

(3.2) $(\mathcal{E}, \mathcal{D})$

denotes a *Yang-Mills field*, in general (ibid., cf. also, for instance, Section 2.1.4; (4.5)), one can further consider its *"flow"*; that is, the *"orbit"* (:*"flow"*) of the corresponding to the *field* at issue *A-connection* $\mathcal{D}$ (cf. (2.7)), hence, of the associated with it *physical law* (see also (2.5)). Of course, all this within the context of the aforementioned category; cf. A. Mallios, ibid., p. 92, (2.37), and p. 26, Subsection 4.2.

Thus, (3.1) becomes now the *outcome of a*, so to say,

(3.3)
> *"generalized Gelfand duality"*, in the sense of
>
> (3.3.1)   *interchanging the rôles between functions and variables.*

The previously applied terminology is, of course, reminiscent of the classical *map*, initiated by I. M. Gelfand in *topological algebras*, in general; cf., for instance, A. Mallios [1: p. 73, (7.14)/(7.15) and p. 212, (5.1)/(5.2)]. Concerning the present framework, see also A. Mallios [GT, Vol. II: pp. 31-33], where the entire situation is exposed in detail. Within the same context, one thus obtains, in particular, that:

(3.4)
> the *"flow"*, say $\Phi(\mathcal{D})$, of an *A-connection* $\mathcal{D}$ of a given *Yang-Mills field* $(\mathcal{E}, \mathcal{D})$, appears to be *"locally symmetric"*; that is, one has, by definition,
>
> (3.4.1)
> $$\Phi(\mathcal{D}) := ker(\mathcal{D}_{\mathcal{E}nd\,\mathcal{E}}) \ (\text{"causally stationary"})$$
> $$\equiv \{\phi \in \mathcal{A}ut\,\mathcal{E} : \mathcal{D}\phi = \phi\mathcal{D}\}.$$

See also ibid. p. 27, (4.26) and p. 28, (4.32)/(4.33). On the other hand, we can still express the above, by saying that;

(3.5)
> the *"symmetries"* of a given *A-connection* (cf. (3.4.1)) (hence, of a *physical law*, see (2.5)/(2.7)) *specify its "flow"* (: function/action of the associated, as before, physical law).

See also ibid., in particular, p. 33, (4.61). As an outcome, by still taking into account (2.3), along with (3.4), one comes to the conclusion that:

(3.6)
> the *characteristic feature* of the *"behavior of the world in the infinitely small"*, hence, in the *"quantum"* (deep), as well, is *symmetry* (viz. *"commutativity"*)! This, of course, still reminds (0.2) (N. Bohr), as in the beginning of the present treatise.

Furthermore, the preceding is also *reminiscent of* the perspective that,

(3.7)
> *"everything is light"*; cf. A. Mallios [GT, Vol. II: p. 207, (10.21)]. Yet, *"... all is quantum"* (D. R. Finkelstein); same Ref., p. 207, (10.20). Within this same vein of ideas, see also e.g. A. Mallios [17: pp. 281-282, (6.21)-(6.24)].

On the other hand, one might further refer herewith in conjunction with (3.7), to G. Fichte, as quoted by H. Weyl [2: p. 130], in that;

(3.8)   *"The light is ... within me ... I myself am light"*.

In toto, the preceding leads now to the following recapitulating diagram:

(3.9)   *physical law* $\longleftrightarrow$ $\mathcal{A}$-*connection* $\longleftrightarrow$ *field* $\longleftrightarrow$ *differential equation*
$\longleftrightarrow$ *flow* $\longleftrightarrow$ *symmetry,*

provided, of course, that everything as above, is *locally* perceived, as it actually were, whatsoever! Consequently, the *correlation of* (3.9) *with the quantum* (:*"infinitely small"*); see also (2.5)/(2.7), as above.

Yet, by still following the previous reasoning, and paraphrasing/indeed, generalizing also C. von Westenholz, [1: p. v], one can further note (see also (10.54) in the foregoing), that:

(3.10)   the *"geometric spirit"*, in effect, the *"functorial"* working methods *in terms of ADG*, provide *an alternate* [indeed, more *"natural"*(!), see A. Mallios ̍[18: p. 1930, (1.5.1)]] *description of natural*/physical *phenomena* ... *beyond* the description obtained in terms of *analytical methods.*

Indeed, working within the aforementioned context, we are actually following, for instance, a so to speak,

(3.11)   *Lagrange point of view, instead of Hamilton*; yet, *most importantly*, we perform this *without* the need of *any supporting "space"*(!), in the standard sense of this term, as is classically the case (e.g. CDG). The latter viewpoint is *still pregnant* of an inherent potential applicability in usual problems/for example, *"singularities" of "quantization"*; therefore, as a matter of fact, still of a *situation leading* no less than *to a UFT*, as well.

As a consequence, one thus comes to

(3.12)   *consider each* particular *case* by *its* own *appropriate "symmetry group"* (Klein), viz. its own *"geometry"*/*"cosmos"*(!).

Of course, *the previous viewpoint is* also *akin to* the very notion of an *"affine geometry"* (see, for instance, N. Bourbaki [1: Chap. II, p. 126, Section 9]), pertaining now, in particular, to its conjunction with the *"quantum"*: true,

(3.13)   the *"spacetime"*–*free* perspective of *"geometry"*, that has been otherwise advocated thus far by ADG, is also *very akin to "quantum"*!

Thus, we can further note herewith that:

(3.14)   any time we choose in a given (algebraic) structure a particular *point*,
as a *fixed* one, viz. an *"anchor"*, so to say *of the* same *structure*, thus
depriving it actually, at all(!), *of any "flexibility"*, we have then *already
transformed the structure into a "rigid" one*! Therefore, according to the
very nature of the matter at issue (cf. also e.g. (1.3) in the preceding),
not thus much *relevant to a quantizable one*!

In this context, see also A. Mallios [17: p. 264, (1.3), along with p. 268ff, (2.10)-
(2.13)].

On the other hand, as already remarked in the preceding (see (3.12) and comments following it), a *"fixed point free"-structure/"space"*, and the presence of a
transformation group, alias *"structure group"*, or even *"symmetry group"*, characterizes actually an "affine geometry". Therefore, *the preceding*, along with (3.13)
and (3.14), *fit* quite well *with* the point of view of the present Section; cf. thus (1.1),
hence, with *(2.1)/(2.3), as well*. Yet, within the same context, see also ibid., p. 282;
(6.25), and the following comments therein.

We terminate the present account with some more comments, pertaining to (3.3)
in the preceding, being relevant with our previous remarks in (1.3). We put this
into the form of the following.

Scholium 3.1. — By still arguing within the viewpoint of (3.3.1) (: *"generalized
Gelfand duality"*), it is worthremarking, in that context, that the same notion
(*"duality"*) can be, *more generally, reformulated*, in terms of the so-called *"Yoneda
embedding"*: Now, this actually concerns an interpretation of the important in
Topological Algebras Theory [TA], *"spectrum functor"* $\mathfrak{M}$ (alias, *"Gelfand space"*),
as a (*Sets*)-*valued presheaf* on the *category of topological algebras*, in general, say,
$\mathcal{A}_{topal}$, with (*Sets*) denoting the *category of sets*; namely, one gets, by applying
topos-theoretic arguments (see, for instance, S. Mac Lane-I. Moerdijk [1]), at the
relation,

$$(3.15) \qquad\qquad \mathfrak{M} \in \mathcal{O}b\big((\mathcal{S}ets)^{\mathcal{A}_{topalg}^{op}}\big).$$

Concerning the notation employed herewith, see also (2.12.1)/(3.22) in the preceding
of the present treatise. Yet, cf. also A. Mallios [17: p. 275, (5.10)/(5.13)], for
technical details referring to (3.15), as above.

Now, the general perspective of the preceding is to point out an as-
pect/mechanism that supplies a method of

(3.16)   *transmuting "rigid"*/abstract *objects* (: mathematical structures, cf. also
(1.3)) *into more "flexible"* ones, as e.g. *functions* (indeed, *sections of*
suitable *sheaves* (see below). We are thus led to a procedure, capable of
important *potential applications*, within the *framework of ADG*, even in
quantum physics (!); yet, this by employing methods, still, no more no
less, from *topos theory*!

As a matter of fact, the above represents a *characteristic aspect* of what one might conceive, as a

(3.17) *"relational" way of* doing (classical) *mathematics*; so, through *sheaf theory*, at least; (*yet, topos theory* is also *applicable* herewith). This still provides, apart from simplicity, against classical *"rigid" methods*, also a

(3.17.1)   *"spaceless" viewpoint,*

concerning the standard aspect of the term, *"space"*. Yet, *"dynamics"*, a sine qua non, pertaining *physical applications*, can be *assured*, as well, due to the *presence of ADG, throughout* the whole argument, whenever needed.

Within the same context, we should also note here that the *form of dynamics* supplied by the preceding, is in effect an *important spin-off* here too of the previous categorical indeed *"functorial"* perspective, a

(3.18) *"covariant"-type of dynamics*, yet, *in the sense of* (3.17.1). Therefore, the aforementioned (cf. (3.16)) potential applications in *relativistic* (according to what we just remarked) *quantum physics*; hence, in particular, in *quantum gravity.*

We still refer here to A. Mallios [20] for a more detailed account, in connection with (3.17), as above. Yet, see also A. Mallios [17: p. 276]; in particular, ibid. (5.18), in conjunction with (3.18), as before.

The perspective afforded by the preceding, constitutes also an aspect of what one might phrase, still as,

(3.19)   *sheafifying à la Yoneda*

(see also ibid., as above); the same entails a *further abstraction*, in that context *of* (3.3.1), as well, while it also leads to the particularly *useful idea* of a *"topological algebra scheme"*. See, for instance, A. Mallios [GT, Vol. II: Chapt. 4: p. 177, Theorem 4.1, and also p. 68, Subsection 8.1], along with a relevant more detailed account in [19: p. 69ff, Applications]. See also remarks after (4.4) below.

On the other hand, in conjunction with the latter concept, as above (:*"scheme"*), the idea of an *"affine"* structure is of course always *inherent*, here too: thus, the notion of an *"affine* (topological algebra) *scheme"*, as the *spectrum* (: *"Gelfand space"*), in effect, of a *"geometric topological algebra"* is appropriately applied herewith; see also the last Refs, as before, concerning technical details of the terminology employed here, as well as, for the applications in *quantum gravity*, hinted at in the foregoing (cf. (3.18)). Yet, see also below, concluding remarks following (4.4).

2.2.4   *Final Remarks*

We terminate the present account with *two more* relevant *notes*; thus, we have:

i) The *generality of the internal structure of ADG*, vis-à-vis the *"rigid"* one of the classical theory (CDG, thus *smooth manifolds*), enriches the first with the possibility/advantage of

(4.1)   *inheritance of the basic* initial *structure* to *"sub–structures"*.

That is, in other words, the *category* defined by the notion *of "ADG-structures"* is *closed under* the *pull-back* functor: see e.g. A. Mallios [VS, Vol. II: p. 24, Section 6] ; yet, for an extended complete account thereon, concerning in effect the *"functorial perspective of ADG"*, in general, cf. M. Papatriantafillou [6].

The above is reminiscent, of course, of relevant remarks of H. Weyl [2: p. 86] (cf. also (1.2) in the beginning of this Section), referring to the *generality of Riemann's structure*, affording thus the latter an increased *"flexibility"*, with respect to characteristic features of basic *"sub-structures"* of it, against the classical Euclidean space and analogous subspaces therein; thoughts of Weyl on the celebrated *"Habilitation"* lecture of Riemann.

ii) In connection with the previous discussion about (3.19) and the notion of a *topological algebra scheme* (*t.a. scheme*, for short), one actually realizes, on the basis of the kind of applications, we can attain, that:

(4.2)   by working within the aforementioned framework (t.a. scheme), one has *locally the model, in its entirety* (!): viz. the *geometric topological algebra*, at issue; see A. Mallios [4: p. 308, Definition 2.1], and [19: p. 67, Definition 2.1 and subsequent comments therein, yet p. 68, (2.10)/(2.11)]. This becomes extremely useful, providing e.g. the appropriate framework of performing *"analytical"* calculations, analogous in effect with those in the classical theory (CDG).

Indeed, the situation described by (4.2), still reminds, of course, what actually happens with the standard case in the so-called *"geometry of schemes"* in the framework of Algebraic Geometry; see, for instance, D. Eisenbud - J. Harris [1: p. 7]: Thus, with in this same context, one actually realizes that,

(4.3)   *"... schemes admit much more local variation* [in their inherent structure]*", while "... in* [an affine] *scheme ... a lot of interesting and nontrivial geometry happens ..."*.

Yet, this *in contradistinction with* the situation one has in the case, for example, of *open balls in a* [smooth] *manifold* (ibid.). So analogously, and in conjunction with (4.2), as above, one still remarks (cf. A. Mallios [19: p. 71, (4.14)]) that: by contrast with the standard theory (CDG),

(4.4)   to work within the context of a *t.a. scheme*, one gets at *a situation, that cannot be afforded*, even *when* one *locally* considers *a differential manifold*!

Now, by further referring to the notion of a *topological algebra scheme*, apart from the previous citations, see thus remarks following (3.19), and still (4.2) above, we also quote the recent work by R. Vargas Le-Bert [1]: it concerns a quite interesting application of the same notion at issue, to problems still related with the standard/*"natural adjunction"*:

(4.5)   *"space"* $\longleftrightarrow$ *"observable algebras"*.

See, e.g. Section 2.1.4; (4.6) and Section 2.4; (1.2), (3.2), along with Section 2.3; (6.4)/(6.5) below. Yet, this by means actually of the notion of a *"geometric topological algebra"*: cf., for instance, A. Mallios [4].

2.3   Chasing Feynman

> *"... a singularity in a theory heralds the breakdown of that theory* unless it can be recast in a new form. When the theory is a theory of *space and time* the problem is *all the more severe."* (R. Penrose)

The above adage of R. Penrose is taken from J. Gray [1: p. 210]; emphasis there in is ours.

2.3.0   *Prelude*

The purpose of this Section of the present treatise is an ... ambitious one (!); namely, of *tracking* (?) *Feynman's* celebrated *"diagrams"*, in trying to give evidence to/understand his paths, as this could be accomplished, through the *"spaceless" methods*, advocated so far *by ADG*. Indeed, it was essentially the same evasion of any *"space"*, in the usual/classical sense of this term, which constitutes a *fundamental*/decisive *feature*, in both *Feynman's* general *perspective* and that one of the so-called herewith, *"Abstract Differential Geometry"* (ADG): Thus, to express it fairly, once more, we systematically avoid, throughout, the use of the standard notion of *"space"*, as it is classically realized by the (smooth) *manifold* concept, in order to support/frame the *"differential"*, more generally, *"analytic"* machinery, we eventually have to employ throughout our study.

On the other hand, it is nowadays well-understood that the same issue of *"space"*, which indeed permits the classical framework, therefore, *Calculus*, (Classical) Differential Geometry (CDG), hence too (*smooth*)*manifolds*, creates in effect basically *insurmountable/pestilential difficulties/obstacles*, any time we try to apply the standard (basically *"spatial"* theory in the *quantum* deep, let alone when in conjunction with *general relativity*; yet, to tell it, *differently*, all *the* aforementioned *"trouble"* ((s)...*"with physics"*) appear(s) each time one is encountered with problems of *quantum gravity*, or *quantum field theory*, in general.

Furthermore, *Feynman's perspective* is, in point of fact, rather *Lagrangian in character*; that is, what one might nominate, as a *"relational"* one: see, for instance, A. Mallios [20], or even [17: p. 267, (2.5), and p. 272, (4.2)/(4.3)], [18: p. 1946, Epilogue]. Therefore, not a *"Hamiltonian"* (viz. *"analytic"*) one.

Of course, the previous two perspectives are (classically) *equivalent* (Legendre); however, according to our point of view, being, in principle, as already noted, a *"spaceless"* one (!), hence *ADG*, our aim herewith is to put in the foreground the so to say (natural) *"geometric"* (viz. *"relational"*, as above) *aspect* of the matter. Therefore, a *"physical geometric"* one; cf. P. G. Bergmann [1]. See also e.g. A. Mallios (14: p. 74, end of concluding remarks therein]), *not* the other way round; that is, *"geometrical physics"* (ibid.), in the usual sense of this term: namely, in what we usually consider nowadays, as *"mathematical physics"*, being thus basically *rooted, from the outset*, on a *"spatial framework"* (!); on such a framework is based, in effect, still the *classical analysis*, hence, *differential geometry* (CDG), as well. In this context, see also A. Mallios [16: p. 1557ff, Section 1], along with previous comments of the present treatise: cf., for instance, Section 2.2; (1.5)/(1.6).

2.3.1   *Field Interactions*

A *field* *"interaction"* is taken place, of course, according to a certain particular (*physical*) *law*; or what amounts to the same thing,

(1.1)   any *field* *"interaction"* is the result of a particular *function of the Nature* (: *physical law*), thus, the *spin-off of a physical law*.

Indeed, the above *characterizes*, in effect, *the* whole *function of Nature*! In this context, see also e.g. A. Mallios [16: p. 1557, (1.1)]. On the other hand, a *physical law* is *locally* characterized by a *"symmetry"*, that is, by a *"flow"*. Yet, in other words, by a certain particular *"action of a group"* (of transformations), alias a *"group action"*, for the fields involved. We usually speak, in that context, about the

(1.2)   *"symmetry group"* of the field/interactions concerned.

In this context, we further note that the aforementioned group, as in (1.2), is still the same for the *classical mechanics* (thus, smooth manifolds), as well as, for the *quantum domain* (: *"symmetry axiom"*; or even, an outcome of the adage, *"physis is united"* (!)).

Thus, we *accept* (: *axiom*) *from the outset* that,

(1.3)   *"field interactions"* are the result/*function of the physical laws*.

On the other hand, according to the same general perspective (*axiomatization*) *of ADG*, we posit the following associations (*"diagram"*);

(1.4)   *physical law* ⟷ *elementary particle* ⟷ *field* ⟷ (𝒜-)*connection*.

See also, for instance, A. Mallios [17: Section 6], along with [16]; yet, one has, first Ref. as before, p. 270; (3.3):

(1.5)   *dynamics* ⟷ *differential equations* ⟷ (𝒜-)*connection*.

Now, the last two diagrams, as above, justify, of course, the *intervention in (1.3)* of the notion *of a*

(1.6)   *"flow"*, hence, *locally* of a *"symmetry"*, as well (*"Chapman-Kolmogorov law"*); therefore, of a *local group* (of transformations) *action*!

On the other hand, in view of (1.6), *the* (local) *symmetry*, associated with a *flow*, still *characterizes* (field) *interactions* (cf. (1.3)), thus a particular *function of the Nature*. Yet, the same *symmetry*, as in (1.6), formalizes (E. Noether; see e.g. A. Cannas da Silva [1: p. 127]) a *"conserved quantity"*. That is, in other words, one thus arrives at the following basic diagram;

(1.7)
physical law ⟷ *field* (*interactions*) ⟷ *elementary particle* (: *quantum*)
⟷ (𝒜−)connection ⟷ *differential equations* ⟷ *flow* $\overset{\text{(locally)}}{\longleftrightarrow}$
symmetry $\overset{\text{(Noether)}}{\longleftrightarrow}$ *conserved quantity*.

Yet, the association of *"field interactions"* with the *physical law(s)*, as before, entails, of course, according to our point of view (ADG), the *presence of ourselves* (:*"observer"* ⟷ *"𝒜"*): therefore, the intervention of *"𝒜-invariance"*; namely, the only way, in effect, to *detect physical laws*: see, for instance, A. Mallios [18: p. 1930, (1.3)/(1.5), (1.6)], still in conjunction with (1.1) in the preceding. In this context, see also H.-J. Schmidt [1: p. 1], and G. Ludwig [1], or even A. Einstein [1: p. 8], as it concerns, herewith, the fundamental notion of *"physical" geometry* (!), not the other way round; see also e.g. P. F. Bergmann [1: p. 88].

Thus, the preceding highlight, in effect, the manner one can actually perceive, within the framework of ADG, the intermingling of such fundamental concepts, as for example,

(1.8)   *field interactions* ⟷ *𝒜-invariance* ⟷ *transformation groups* (: *symmetry*),

pertaining, in particular, to the *function of Nature*, alias *"physical laws"*. Yet, the *relevance of the above* (R. Utiyama) with the fundamental, *"functorial"*,

(1.9)   *Hom* − ⊗ *adjunction*,

can still be viewed, within the same vein of ideas, as before, through an appropriate resort to a *"Lagrangian function"*, in conjunction with (1.9); see thus, for instance, the preceding: Section 2.1, Corollary 3.1, or even A. Mallios [22].

More precisely, one thus realizes (ibid.) that:

(1.10)   A function, related with an ($\mathcal{A}$-)*"connection Lagrangian"*, is *"gauge invariant"* (that is, in other words, *"$\mathcal{A}$-invariant"*) *if, and only if*, the same holds true for the corresponding, through the *"Ad-functor"*, *"curvature Lagrangian"*.

In toto, and in a succinct form, one thus concludes/arrives at, what we may consider, as the

(1.11)   *"Utiyama's principle"*: that is, at *the idea that,*

(1.11.1)   *field interactions can be determined through ($\mathcal{A}$-) invariance.*

Indeed, according to (1.1), *every interaction is the spin-off of a physical law*; on the other hand, the latter is (by assumption, see just below) *"$\mathcal{A}$-invariant"*, otherwise said, *"functorial"* ! See, for instance, A. Mallios [18: p. 1930, (1.2)-(1.5)]. Thus, one gets at the *fundamental conclusion* that,

(1.12)   *"field interactions"* are *"$\mathcal{A}$-invariant"* (viz. *"fundamental"*), *for any "$\mathcal{A}$"* (ibid.), *whatsoever*. Therefore,

(1.12.1)   *the previous conclusion* (1.11.1), *as well.*

Now, the above, *in conjunction* also *with the fundamental adjunction (1.9)*, still provides another perspective/justification of *Utiyama's Theorem*; cf. the present treatise, Section 2.1; Corollary 3.1, or even A. Mallios [22: p. 778, Theorem 3.1].

Thus, to recapitulate the preceding,

(1.13)   *"field interaction"*, alias *interactions of elementary particles* (see also (1.7)), means actually (R. Utiyama) *"($\mathcal{A}$-)invariance"*, which still can *finally* be *reduced* to an appropriate *group of transformations* (cf. (1.8), as above).

On the other hand, the aforementioned *"$\mathcal{A}$-invariance"* is, of course, realized/evaluated through our own *measurements*, viz. the *"gauge"*, we employ at each particular instance/case, so that we further refer to a *"gauge invariance"*, as well. Thus, the same is actually effectuated via our own *"arithmetic"* (: *"$\mathcal{A}$"*), which *we concoct*(!) thereat. Cf. also A. Mallios [18: p. 1930, (1.3) and p. 1932, (1.12)], along with [17: p. 277, (6.5)]; yet, see (8.22) in the preceding of the present account.

Consequently, we are thus led, in effect, *calculationally*, so to say, *speaking* (!), to the following *basic*, so far (cf. also (1.8)), *diagram*/viewpoint ;

(1.14)   *field interaction* $\longleftrightarrow$ $\mathcal{A}$-*invariance*.

In this context, see also our previous remarks in (9.31)/(9.40) of the present study, along with the adage of C. N. Yang in the beginning/frontispiece of Section 1.9.1 in the foregoing, pertaining to the fundamental notion of *"symmetry"*, in conjunction with the present discussion (still (1.8), as before). Yet, within the same perspective, see also (9.61), and subsequent comments therein.

Thus, roughly speaking, one thus arrives, so to say, at an,

(1.15)   *"invariant interpretation of interactions"*,

in the sense of R. Utiyama [1], still *within the framework of ADG* (see e.g. Section 2.1; (1.2) in the foregoing). Furthermore, the same can appropriately be associated with a *group* (sheaf) *action*, cf. (1.7)/(1.8), as above. Now, the latter leads actually to a *Kleinian perspective* of the matter; see also, for example, (1.1) in Section 2.1, as before, together, for instance, with (1.7) in the preceding. Indeed, by still based on Section 2.1, (1.7), and also in conjunction with (1.16)/(1.17) therein, one thus gets, in particular, at the *fundamental*, as we are presently going to see, *notion of a,*

(1.16)   *1-parameter group of ("gauge") transformations.*

That is, in other words, we are thus led to an

(1.17)   *"access" to the "geometry"* itself of the field $(\mathcal{E}, \mathcal{D})$,

at issue. So, technically speaking, to a *"curve"* (alias, *"orbit"* of the field) , say,

(1.18) $\alpha : \mathbb{R} \longrightarrow Aut\,\mathcal{E} \equiv (End\,\mathcal{E})^{\bullet} \subseteq End\,\mathcal{E}.$

Now, as we shall see presently below, the latter *"curve"* might have *important* potential *"differential geometric"* consequences, when within the present *ADG-context*; cf., for instance, (12.10) in the preceding, in conjunction also with the subsequent discussion below. (See thus e.g. Section 2.4 in the sequel).

2.3.2 *A Non-Spatial Perspective. Whence, ADG*

Working within the *framework of ADG*, as this is also, of course, the viewpoint of the present treatise, means fundamentally a

(2.1)   *non-spatial* (!)

one. That is, as already often explained in the foregoing,

(2.2)   *we do not assume* a priori *any* (smooth) *manifold, hence,* the so-called classically (: CDG), *"space-time"* too, *to support our "calculations"*!

This possibility instead, against the standard perspective, results, in principle, just from an appropriate *"mechanism"*; this being also *"functional"*, viz. *"functorial"*, in character, as it actually were, any way, even classically speaking, which is employed at each particular case. The same is exactly, in effect, still the *fundamental difference* herewith, between the *classical theory (CDG) and* our point of view, in terms of *ADG*.

On the other hand, *the* previous *perspective*, being *advocated*, so far, *by ADG*, is *still in accord* with what was *already demanded long before* (!); indeed, this even by men like A. Einstein himself (see e.g. A. Mallios [14: p. 59, (1.6)] for a relevant quotation), as well as by H. Weyl (cf., for instance, (2.30) in the preceding of the present study). Thus, one realizes, in effect, the essentially

(2.3)   *"unreal" rôle of the* notion of a (smooth) *manifold* thus far in *"mathematical physics"*.

Therefore, in other words, the concoction, in that context, of a

(2.4)   *"spatially"* rooted *"arithmetic"* thereby (!), as well.

Within the same vein of ideas, see also A. Mallios [19], together with Section 2.2; (4.4) in the preceding; yet, cf. A. Mallios, [22], A. Mallios-E. E. Rosinger [1], [2], A. Mallios-I. Raptis [3], A. Mallios-E. Zafiris [1], [2], [3], I. Raptis [7], and E. Zafiris [1], [2].

In this context, we further note that *"Feynman diagrams"* constitute, indeed, a particularly *characteristic*, and yet *celebrated, instance* of the way one should work, when confronted with the task of *physically associating mathematical notions/tools*; thus, *one is actually* invited/*encouraged, to employ*,

(2.5)   *extremely non-spatial (!) arguments*; just, on the contrary, what might be/correspond also to the essence of the problem, *"relational"* ones (in the sense advocated by the preceding, see e.g. (2.2) above). That is, entirely, *"space-free"*, in the usual sense of the latter term (cf. CDG).

We are thus trying to bring into the foreground just *"relational"*, so also to speak, *"functorial" concepts*, describing in effect the deeper meaning of the *"calculations"*; the latter, to recall herewith, even E. Galois, are in that respect, simply, *nonpracticable*! Of course, the same point of view (viz. the *"relational"* one, as above) is just akin to *the whole mechanism of ADG*.

Indeed, and by paraphrasing here our previous comments in A. Mallios [19: p. 71, (4.14)], one can resume that:

(2.6)

we do also gain *new physical insight*, by replacing the standard notion of a (smooth) manifold, hence, our *"spatial arithmetic"* based on it, as well; this, through an appropriate (*"structure"*) *sheaf of algebras*, that still leads to a *"relational aspect"* of the universe, as it actually were (: *physical laws*, see e.g. (1.1) in the preceding), thus to a *more physical one*! On the other hand, we also note that *the latter viewpoint* provides further a more satisfactory perception of the observed quantities/phenomena, based instead hitherto (cf., for instance, CDG) to only *"calculational"* *data*; see also A. Mallios [20: p. 65, (2.14) and subsequent comments, therein].

We come now by the following 2.3.3 to the *"core"*, in effect, of this Section, the same being also, if any ..., the justification of it, as well. On the other hand, that, which actually triggered off the entire story, can easily be traced back already to the article of the author, mentioned in (2.6), as above.

2.3.3 *Relational Calculus*

By analogy with the term *"Yangs-Mills space"* (see e.g. A. Mallios [GT, Vol. II: p. 36, Definition 4.1), that is, roughly speaking, the presence of an appropriate *set of data*, within which one can formulate/study several notions, along with their consequences, pertaining to a *Yang-Mills field* (ibid., or even Section 2.1, (6.1) in the foregoing), in the same manner, one might speak of an (*abstract*).

(3.1) *"Feynman space"*.

Namely, again, as that (appropriate) *set of data*, within the framework of which, one can *reasonably* (viz. in a *suitable manner*) speak of/study several notions which appear when looking at the so-called *"Feynman diagrams"*.

Consequently, on such a space one then might hope (?)/try to understand/follow what happens with that, which one might perceive, as

(3.2) *"Feynman's Calculus/mechanism"*.

Of course, what is apparent from the preceding discussion and is still supported by the same *character of Feynman's diagrams and* the associated with them *mechanism*, thus far (cf. also (2.5), along with the comments prior to it), *we* further *posit (3.2)*, as a very *particular case of* what we can perceive, as a

(3.3) *"Relational Calculus"*,

in the sense of ADG; thus, an *entirely "spaceless"* viewpoint. Cf. also for instance, A. Mallios [20].

Now, within the same vein of ideas, a (*gauge*) *field* is, by definition, associated/correspond to an (𝒜-)*connection*, in the *terminology of ADG*; therefore, one has, still by assumption the following *basic diagram*;

(3.4)
$$\begin{array}{c} (gauge) \; field \;\; \longleftrightarrow \;\; (\mathcal{A}-)connection \; \mathcal{D} \;\; \longleftrightarrow \;\; \text{``Yang-Mills field''} \;\; \longleftrightarrow \\ (\mathcal{E}, \mathcal{D}(\equiv \mathcal{D}_{\mathcal{E}})). \end{array}$$

In this context, see also e.g. A. Mallios [14: p. 63, (3.3), or p. 70, (5.9)], yet [17: p. 278, (6.6)].

On the other hand, to be fair, *we are not going* to "*differentiate*" and/or "*integrate*", in the *classical* sense (CDG), due to the very character of (3.3), being as already said, by the same definition, *entirely "spaceless"*; therefore, in view of the preceding, a *similar conclusion,* results *concerning the nature of (3.2),* as well (see also the comments about (2.5), as above). Consequently, one is thus led to seek *systematically* for *alternatives,* that should be "*quite spaceless*", in nature; thus, for example, appropriate "*functors*", as it actually were/happened, even classically speaking, this latter fact being, otherwise, a *fundamental moral of ADG.*

Thus, we come now, by the subsequent 2.3.4 and 2.3.5, to our main topic of this Section, as its title indicates; that is, more precisely, to (3.2) in the preceding.

2.3.4 *"Feynman's Calculus", in Terms of ADG*

Of course, our first task herewith is to look for the *fundamental notion* of "*derivation*" within that framework, as indicated in the title of the present Section. Indeed, this also refers to the very character of the same *behavior of the Nature/Physics* herself, pertaining to the "*infinitesimally small*"; hence also, to the *behavior of the Physis,* in general, as a whole. See e.g. Section 2.2; (1.1)/(2.3). Yet, the same viewpoint (viz. starting from the "*infinitesimal*") referring now, in particular, to "*geometry*", is already found still in the celebrated "*Habilitationsschrift*" of Riemann ("*Riemann tensor*" (CDG), cf. also, for instance, ibid.; comments following (2.4)). So, in other words, one thus arrives just at the "*infinitesimal*" of the basic notion, as it is actually that one of being one able to "*follow the change*" of something (varying); that is, in effect, at the *quintessence of* the same concept of "*derivation*", in general! We come thus to examine now, by what follows, the *crucial aspect of "derivation"*, within the context of (3.2), indeed, one of the main issues of the present Section.

Derivation   As already noted just before, we look, through the ensuing discussion, at the classical notion of "*derivation*", *in the sense of (3.2).* Precisely speaking, we are looking at the concept at issue actually "*from scratch*"; thus, from the "*core*" of the notion, that is, "*infinitesimally*"/alias, "*locally*". So one gets at an analogous *diagram (of notions), as* e.g. in (3.4) in the preceding, still in conjunction with those *in* (1.16), and (1.7)/(1.5).

On the other hand, following the same, viz. *mainly conceptual* viewpoint, pertaining to the items, we are interested in, we first note that the latter are, in effect, classically (: CDG) appeared in a quite "*technical*", in principle, *environment*; so it is here our primary goal to *find* and *point out,* yet further employ, the eventually *inherent,* therein, *conceptual character/mechanism* of the same (cf. also (4.1) in the sequel). Namely, speaking in *our technical terms* (: ADG), the *"tensorial"*, hence *"non-spatial" substance* of the issues in focus. Therefore, it is always *extremely important* whenever we can *detect/isolate* other means as, for instance, *"functional"* ones, to express/formulate the basic/instrumental concepts, under consideration. Indeed, this will be, roughly speaking, the *way/philosophy of* all *our* own *technique*, throughout the subsequent discussion. Yet, it is still our *motivating principle*, thus far, that;

(4.1)   anytime we have/realize an especially working/effective (*"technical"*) method, then *something* more *fundamental/"tensorial"*, hence, by itself (!), *"non-spatial"*, is actually hidden.

True, the so-called herewith, *"Feynman's diagrams"* are, in effect, a *characteristic example*. Yet, P.A.M. Dirac (*"(Schwartz) distributions"*), or even *"Utiyama's principle"*, see Section 2.1 in the preceding.
    Furthermore, within the same vein of ideas, it is also worth remarking here that,

(4.2)   we are actually *trapped*, even *today*, in a so to speak, *old-fashioned* (!), still *"pointwise writing"*, or even *conceived, at least*, manner of employing/use *important functions*, though *we do know/*feel *their inherent meaning*. Thus, we loose/bypass then their esoteric *"tensorial"* (even *"functorial"*), therefore, basically *"non-spatial"* character!

As a matter of fact, in the aforementioned *"trap"* our *"spatial"*, indeed, classical tradition thus far, is, of course, certainly *instrumental*!
    Now, an especially important and characteristic situation, *fitting well with* our previous remarks, is really effectuated through the classical notion of a,

(4.3)   *"Hamiltonian"* (function).

In this context, we still note that the *"passage" from* a *Hamiltonian to* a *Schrödinger equation*, constitutes already, classically speaking, a form of *"1st quantization"* (cf. for instance, N. E. Hurt [1: p. 69]). Thus, the same procedure, within now an appropriate *ADG context* (see e.g. (2.6)/(4.9) and (4.10) in the foregoing), provides just a *quantum field theory analogue*; viz. a form of *"3rd quantization"*, simply to employ herewith *I. Raptis's terminology*; cf., for example, the said author [5]).
    Indeed, what one might still consider (cf. (4.3)) as a

(4.4)   *"Hamiltonian function"*, in conjunction with *"Schrödinger's equation"*, as before, it actually signals the transfer from the classical theory (cf., for instance, CDG) to a *"quantum environment"*/situation. Even, to a *"relativistic"* one, *by its very definition*; namely, it *unifies "time and space"*: cf. ADG, along e.g. with E. Prugovečki [1: p. 46, (5.4)]. Therefore, *quantum field theory*, whenever appropriately formulated: thus, for example, *in terms of ADG*, as already noted above.

On the other hand, as already remarked, see e.g. (12.9) in the foregoing and subsequent comments therein,

(4.5)   one gets *through* (4.3), a *way of following/detecting "time evolution"*, achieved, via *Schrödinger's equation*. Namely, one has, (ibid.), the *basic relation*,

(4.5.1) $$\partial_t = H,$$

which still should be related with (12.9) in the preceding.

Now, the importance of (4.5.1), viz. the relation,

(4.6) $$\partial_t = H,$$

referring to the *"basic differential ∂_t"* of the *"reals"* $\mathbb{R}$, see thus (12.10.1) in the foregoing, being effectuated herewith through the *"Hamiltonian"* (function) H, is further illuminated, by looking at what we may call,

(4.7)   *"push-forward functor" in ADG.*

That is, at the *transfer* of a given *"differential structure"*, in the sense of ADG, herewith of that one *of the reals* (see (12.10), as above), via a *continuous map*; in particular, for the case at issue, via a suitable *"curve"* (cf. also (1.18) above),

(4.8) $$\alpha : \mathbb{R} \longrightarrow Aut\,\mathcal{E} := (End\,\mathcal{E})^{\bullet} < End\,\mathcal{E}.$$

Yet, we further note that,

(4.9)   the *"differential structure" on* $\mathbb{R}$, as above, might be here e.g., *the classical one* (CDG, cf. thus (12.10.1) in the preceding).

On the other hand, in view of the above (see also (1.16)/(1.17) in the beginning of the present Section, pertaining to the terminology employed herewith), we further posit that;

(4.10)   the *"differential structure" on* $\mathcal{E}$, as before is, by definition, the *image of that one of* $\mathbb{R}$, *through* a suitably defined curve α, *as in* (4.8).

Consequently, *one thus arrives at the conclusion* that:

(4.11)   to *"differentiate"* (on $\mathcal{E}$), in the sense of ADG (!), affording thereby a *"non-spatial"* manner, as well, *we* can *restrict* ourselves *to a* suitably chosen *curve* α, as in (4.8).

Note 4.1. — The above reminds, of course, the *familiar, from the classical theory* (CDG), basic definition of a *"connection"* (: *"covariant derivative"*), first, through its *action along* a given (smooth) *curve*.

We are going next to decode a *"mysterious"* (!) *curve*, such as that one, alluded to via the last remarks in the preceding, right a way by the following Section. Yet, as we shall see, such an *important curve* is, otherwise, quite familiar, as well as, *extremely effective/instrumental*, already *from* the *classical theory*, again!

2.3.5   The Exponential

Indeed, the (quite familiar) map in the title of this Section is undoubtedly the *"magic"* (idea), that can be *transformed to anything we want*(!), *under* the proviso, of course, that we afford *the appropriate* environment/framework, or even *mechanism, thereat.* On the other hand, it carries, by itself, in view of (4.6) (cf. also below), the *"germ"* of a classical *"differential structure"*, as it actually is, *that one of* $\mathbb{R}$ (in effect, its *"image"*, via the map in focus). See also (4.7), in conjunction with (12.10.1), along with e.g. M. Papatriantafillou [1], [2]); therefore, still of a *potential "differential structure"* (ibid.), *on any other appropriate domain of arrival* (host/*antenna*) (!) of the same function.

Now, a quite interesting fact herewith is that the *reals* $\mathbb{R}$, as above, appear in a, so to speak, *Kleinian manner"*, hence, *more "physical"* (!); that is, via a (group) *action of* $\mathbb{R}$ on the *"geometry"*, one is interested in, at each particular case. Thus, in the form of a

> *1-parameter group of automorphisms* of the (*"geometric"*) structure at issue, as e.g. $\mathcal{E}nd\,\mathcal{E}$ (see (4.8) above); so, in other words, as a,

(5.1)   (5.1.1)   *"curve" in* $Aut\,\mathcal{E} < End\,\mathcal{E}$,

> ibid. Now, this latter, of course, through the *instrumental intervention of* the same map/functor, *"exp"* (!) (see also Section 2.3.6.1 below).

In this context, it is worth noticing herewith, that the aforementioned *"more natural physical"* intervention/appearance of $\mathbb{R}$, as in (5.1.1) (viz., in effect, *via* (4.8)), has to be *construed in relation with* the "diagrams" (1.7)/(1.8), as above of this Section; yet, within the same vein of ideas, cf. also analogous diagrams, as e.g. in the preceding of the present treatise.

2.3.5.1   The Stone–von Neumann Theorem

It becomes very tempting in this place to look at *the theorem, in the title* of this Section, *within* now the framework of *ADG*: This, because the same theorem, by its very content/substance, is traditionally (: CDG and Functional Analysis) connected with the *essential ingredients*, we have already considered in the preceding, within the specified context, herewith, viz. within a *"spaceless"* one, hence, ADG; precisely speaking, *these* items *are*:

> i) *"Schrödinger's equation"*; viz. *the rel.* (4.5.1)
> (5.2)   ii) *the* (group) *action* (4.8), and
> iii) the way the latter is *realized, through* the *"exp"* function (cf. (5.1)).

Furthermore, as we shall see, it is still not thus surprising the fact that the same classical result in focus is also related with *quantum theory*, let alone with the *relativistic* one; therefore, its *potential presence in* what we already called before (cf. (3.2)), *"Feynman's Calculus"*. So we should *first explain* the aforementioned three *crucial issues, as in* (5.2) above, which still constitute actually *its gist*: Thus, by first referring to *"Schrödinger's equation"*, we note that the same, by its very definition, relates the *"basic differential"*, in the ADG jargon (cf., for instance, (8.10) in the foregoing), *"∂t"* (see also our previous relevant comments in (4.5)), with the *"classical Calculus"*; viz. CDG, cf. also (12.10.1) in the preceding of this study. On the other hand, *this* later *perspective* is further conveniently *transferred*, still within ADG, that is, in effect, *through the exponential map*, *"exp"* (see also the comments in the beginning of the present Section), to what one may conceive, as

(5.3)   *"Feynman's framework"*: namely, as that one, in which one performs *whatever makes sense, within the classical context* (CDG), under the proviso, however, that this is *construed* now, *in terms of ADG*; that is, in other words, within a *"non-spatial"* (!) *environment*.

Yet, what amounts to the same thing, one should look at

(5.4)   (the perspective of) *"Feynman"*, *via the aforementioned Theorem* (as in the title of the present Section), but, *within the framework of ADG*.

Indeed, the justification of the above constitutes actually the essential content of that classical result/Theorem at issue, viewed, however, in the aforementioned context (: ADG, see also (5.3), as before).

On the other hand, the *same result* may still explain, in that context, as we further expose presently below, the other fundamental *pillar*, pertaining to the previous Section 2.3.4. So we come right off to the following Section (natural companion, of course, of 2.3.4, as above).

2.3.5.2 *Integration*

This still *classical* operator/*functor*, as in the title of the present Section, is *naturally connected with the same* basic *Theorem*, as above (cf. the previous Section 2.3.5.1), *via the "exp"* map, again! So one may look at the *functor*, at issue, following thus the *classical perspective* of conceiving it, yet, *within the* same *abstract* (hence, *"spaceless"*) *framework* (: ADG), adopted so far; namely, as an

(5.5)   *appropriate* (continuous) *linear form*. This reminds, of course, the classical viewpoint of *"Radon-like measures"*.

Of course, the functions (in effect, *sections of* appropriate*sheaves* involved) can be *"vector-valued"*, in the obvious (*sheaf-theoretic*) general sense of the present context, that again could be *transformed into* *"real"* ones, through standard *"adjunctions"*. In this context, cf. also, for example, A. Mallios [GT, Vol. II: Chapt. 4], pertaining especially to *Einstein's equation* in vacuo; yet, within the same vein of ideas, see this author [19: Section 4. Applications].

We close the present account, with the following concluding comments, referring to the point of view, advocated, in general, by the preceding. So we have the next.

2.3.5.3 *General Moral*

The foregoing lead us now to the subsequent remarks:

(5.6)
> *one might* always *work* within the standard context (CDG), thus, *in terms of the classical notions/techniques*; however, *under the proviso* that *this would be done* in a

(5.6.1)
> *"non-spatial"* way; that is, in a *"functorial"* one, as *e.g. through ADG*!

In this context, a really *instructive example* constitutes, of course, the study of *Einstein* in [1]; indeed, one can look at it, still as a virtual *forerunner of Feynman*, in that context (see also (5.7) below). Furthermore, the case of *R. Utiyama* [1], as exposed in Section 2.1 of the present study, can still be viewed in that same context; see also A. Mallios [22]. On the other hand, within the same vein of ideas (viz. those in accordance with (5.6.1)), one can also mention the earlier work of *P.A.M. Dirac*. Indeed, one might even say, herewith, that;

(5.7)
> *the above* point of view, that is, *as* that one *described*, so to speak, *by* (5.6.1), *is* actually *the way* the *physicists* usually *work*(!), as well (cf. also (5.8)/(5.9) in the sequel).

Thus, one is really tempted to *find out/detect*, at each particular instance, the inherent *"functorial" leading ideas/"Leitfaden"*, that *make things work*, the same being in effect the *quintessence of our working tools*. Otherwise, we would be compelled to look simply at the *"architecture"*/buildings, in place, as it actually were, of the *"architect"*, viz. of the *physical law*, itself; in this context, cf. also, e.g. (6.9) in the sequel.

In toto, and in view of the preceding, one might still conceive (5.6), as what we can name, a

(5.8)   *"Feynman's principle"*.

So, by virtue of (5.7), we can further say that, more or less,

(5.9)
> *physicists work* (I would rather ... dare to say, *they* actually *should* ... (!)), *according to (5.8)* ... !

2.3.6 *Schrödinger–Hamilton Adjunction*

Our aim by the ensuing discussion is, first, to clarify the meaning of the title, and then to point out further potential consequences, in conjunction with the preceding, especially with Section 2.3.5.1 Thus, classically speaking, cf., for instance, "*non-relativistic quantum mechanics*", as e.g. E. Prugovečki [1], the standard "*Hamiltonian operator*", plays a "*dual rôle*" (ibid., p. 354, Section 7.3); namely, that one of relating the "*dynamics*" of the (physical) system at issue, in our terminology (ADG), the basic ("*A-connection*") ∂_t, with the (total) "*energy observable*" of the system. So, in other words, one realizes, here too, an *association* between the "*physical law*", viz. "∂_t", as before, and ourselves, "*we*" (viz. that, which "*we measure*", hence, "*energy observable*"). Therefore, one thus arrives at a *typical occurrence* of that, which was advocated throughout the present study, as a *fundamental* (categorical) *notion* of an "*adjunction*"; cf., for instance, in the preceding. Therefore, we may perceive the above situation, as/(call it) a

(6.1)   "*Schrödinger-Hamilton adjunction*";

hence, the title and also its meaning, of the present Section. Precisely, one can consider/*clarify (6.1), through the* following *diagram*:

(6.2) $\underbrace{(\text{total}) \; energy}_{Hamiltonian} \longleftarrow \underbrace{dynamics,}_{\partial_t}$

so that "∂_t" determines/*represents* the *physical law*: "*A-connection*", viz. $D \equiv \partial_t$.

On the other hand, based further on classical data, however, appropriately conducted, within our framework (viz. ADG), according to the preceding viewpoint, thus far, the previous diagram (6.1) can be supplemented, by the following one:

(6.3)   *Schrödinger* $\cong$ *Heisenberg* $\longrightarrow$ *Hamilton*

$\cong$ (*Euler-*)*Lagrange adjunction.*

In this context, we further note that the last part of the above (conceptual) diagram can still be conceived via the map, one might consider, as a "*Legendre functor*".

Yet, concerning the *first equivalence* in (6.3), see e.g. E. Prugovečki [1: p. 294, Theorem 3.2]. Furthermore, as already pointed out at the beginning of this Section,

(6.4)   "*observability*", in general, is naturally associated with a "*dual rôle*", acquiring in effect a particular, yet, a fundamental/characteristic form of an "*adjunction*" (see (6.1)).

Now, within the same vein of ideas, it is still instructive to add the following remarks. Namely,

the two "*variables*" of the aforementioned

(6.5.1)   "(observational) *adjunction*",

cf. (6.1)/(6.3), as above, is actually:

(6.5)

(6.5.2)   (i) the "*symbol*" representative of the mechanism/procedure *of the* "*observation*"; cf., for instance, in our case (viz. ADG), the "*A-connection*", say $\mathcal{D}$,

together with,

(6.5.3) (ii) the *"result"*/*measurement* of the previous procedure. That is, in the case at issue, the *"field strength"*, alias the *"curvature"* of $\mathcal{D}$, $R(\mathcal{D})$; alias, the realization of the *"symbol"* (:the *"physical law"*; actually the usual way to understand/*"employ"* the latter (see also (Wittgenstein: see Chapter 1; (8.22)) in the foregoing)).

In this regard, cf. also e.g. ibid., as above; in particular, relevant remarks at the beginning of p. 270.

Now, by looking further at the first *"equivalence"* in (6.3), we also note that,

(6.6) the two *"pictures"* (*Heisenberg* $\longleftrightarrow$ *Schrödinger*) in the aforementioned diagram actually express the *"physical law"* (alias, the *"dynamics"*), through always a particular, as the case might be,

(6.6.1) *"equation of motion"*.

Yet, it is actually the same *physical law/dynamics*, as before, viewed now in terms of our language (ADG), that is simply expressed/*"embodied" in the* said *equation*, according to each particular case/*"picture"*. We further note that this *"unification"* is here achieved, via the *"universality"* of the description of *"dynamics"* in our case, through the notion of an *"$\mathcal{A}$-connection"*, following the particular case in focus (cf. also (6.5.2), as above): thus, *"∂_t"* (*Schrödinger*; ibid, p. 286, (31)), and/or *"Lie product"* (*Heisenberg*; see also ibid., p. 294, (3.25)). Of course, we still remark herewith that the *link* to the two pictures, as before (cf. (6.6)), yet, concerning also the *second "equivalence" in* (6.3), is again, at this place too, the *fundamental theorem*, as in Section 2.3.5.1; see also ibid., as above p. 294, proof of Theorem 3.2. Cf. also e.g. J. B. Conway [1:p. 334, Theorem 5.1]. So it is, in effect, the same fundamental result as before, that also finally *justifies the terminology* applied *in (6.1)*. Something, much more actually happens;

(6.7) indeed, it is the same *"adjunction"*, as *in* (6.1), along with the natural *"equivalences"*, referred to *in* (6.3), that seems to constitute, in point of fact, the inherent

(6.7.1) *crux of* the whole *"Feynman's trick"* (!).

We shall try to be *more concrete*, through the subsequent, more or less, rough discussion of the matter; this being also, more precisely, a complemented, so to say, account to that one already presented in the preceding Section 2.3.5.1: So it is actually the same Theorem in Section 2.3.5.1, which

(6.8.1)   *adjuncts/interrelates*,

at each particular case, the two *"states"/*"images" *of the* (physical) *sys-*
(6.8)   *tem* in focus, as these are presented by (6.3). Namely, one has the relation,

(6.8.2)
$$\textit{Schrödinger}\ (\cong \text{Heisenberg}) \rightleftarrows$$
$$\textit{Hamilton}\ (\cong \text{Euler-Lagrange}).$$

In this context, it would be, at least, *instructive* to recall, herewith, some *"general
principles"*, on which the foregoing are actually based. Thus, first e.g. the ancient
dictum of Anaxagoras;

(6.9)   *"everything has been bedecked by mind"*;

this is here paraphrased, so to speak, by *assuming* the

(6.10)   *"omnipresence" of the physical law(s)*.

On the other hand, we further based on the *adage/creed*

(6.11)   *"everything flows"* (Heraclitus).

Therefore, it is natural to *express the* (physical) *law* by a *relation/*"equation", describing thus its function; that is, in other words, by an

(6.12)   *equation of "motion"* (:*flow*).

See e.g. (6.6), as above. Now, the latter (equation) appears thus finally to be, after
all, the *outcome* of the employment *of a particular*,

(6.13)   *"measurement"*; that is, actually in our terminology, of a suitable *"adjunction"*;

of course, as the case might be! See also e.g. L. Wittgenstein, as in (8.22) in the preceding, along with (6.5.1) above.

Now, the foregoing constitute, in effect, the *quintessence* of the whole procedure
in *Feynman's machinery*, the *catalyst* being, herewith, in view of the aforementioned
Theorem (cf. Section 2.3.5.1), the *exponential map* (!), via its various forms. Yet,
within the same vein of thoughts, we also note that,

any *particular "episode"*, within the previous procedure, can now be rooted on *classical devices* (!): namely, the latter can thus be achieved on the basis of our

(6.14.1)   *capability to resort to "non-spatial" arguments, by employing the techniques of ADG.*

(6.14)   Of course, all this when *appropriately*/accordingly *detected*, within the same vein of thoughts, one can further perceive, for instance, the way we understand/meaning of the famous adage, in that;

(6.14.2)   *". . . quantum mechanics is a particular case of classical mechanics"* (!);

in as much as, of course, the latter is transmuted into the first (viz. to quantum mechanics).

In this context, see also e.g. (5.6) in the preceding, along with relevant remarks connected with (4.9)-(4.11) therein.

On the other hand, the foregoing lead us further to think of what one might also perceive herewith, yet, *equivalently to* (6.1), as a sort of a

(6.15)   *"Stone-von Neumann adjunction".*

Thus, see also the relevant Theorem in Section 2.3.5.1, as above, as well as, the subsequent Section 2.4.

2.3.6.1 *The Exponential Map (Contn'd)*

In principle, we are always interested in/looking for

(6.16)   *"differentiation procedures/mechanisms".*

That is, in fact, for methods so that

(6.17)   to being able to follow the *evolution/ "function" of a physical law.*

Now, within the previous context, it appears that;

(6.18)   the *"exponential map"* is, probably, *a/ "the"* best supply.

Of course, the above actually specify the existence of a

(6.19)   *functor*; or even, at the very end, of a *"natural transformation of functors".*

Indeed cf., for instance, the case of what we called, thus far, an *"A-connection"*, in general.

Thus, by further delineating (6.18), one can look at

(6.20)
    the *exponential map*, when *appropriately conceived*, as a

       (6.20.1)   source of a *"differentiation procedure"* (hence, (6.19)).

Of course, an important hint to that perspective, in conjunction also with the content of the present Section, is still the fundamental Theorem in 2.3.5.1 above. So the instrumental connecting link of the preceding with (6.19) constitutes always, herewith, the application in effect of

(6.21)
    the classical viewpoint, however now, *in terms of ADG*, as the latter is described *through* (4.8) *and* (5.1) in the foregoing.

That is, in other words, we are thus based, *first*, on

(6.22)
    the *existence of* a *1-parameter group of automorphisms*, in the form of *an image of* $\mathbb{R}$ (hence, of a *"curve"*), within the *"geometry"*, we are interested in; the latter being represented, herewith, via *"$\mathcal{E}nd\,\mathcal{E}$"*, in general (see e.g. (5.1), as above). Furthermore, the same *"curve"*, as before, can act as a *catalyst* for defining a *"differential structure"* in *"$\mathcal{E}nd\,\mathcal{E}$"*: namely, by *transferring*, therein, *that* (: the *"classical"*) *one of* $\mathbb{R}$, according to the *principle/mechanism of ADG*.

See also relevant comments in (4.9)/(4.10) as above. In this context, one thus realizes here an *"extension, in effect, via ADG"*, of the classical situation one gets, when

(6.23)
    transferring a *differential structure*, in particular, as that one of a *Lie group* (e.g. here, $\mathbb{R}$) into another one, simply by a *"continuous* (group) *morphism"*.

So cf., for example, J. E. Marsden - T. S. Ratiu [1: p. 274, Proposition 9.1.4, and p. 275, Proposition 9.1.5], concerning the classical case; yet, the standard work of V. S. Varadarajan [1] for more technical details. On the other hand, for the corresponding *"ADG account"*, see also the recent treatise of M. Papatriantafillou [1] and detailed, still within in general the framework of ADG, in [8].

    One more word, pertaining to the

(6.24)
               "naturality" of the *exponential* map,

is still here in order: So it is standard from the classical theory (see, for instance, V. S. Varadarajan [1: p. 104, (2.13.7)]), the

(6.25)
    intertwining of the classical *"adjoint representations"*, *"Ad"* and *"ad"*, within the context of *Lie group* theory, *through the exponential* map.

In this context, concerning the classical theory, cf. also Marsden-Ratiu, as above (ibid., p. 275, Proposition 9.1.5, or even p. 276, Proposition 9.1.6 and p. 277,

Corollary 9.1.8, (i)). Yet, P. Tondeur [1: p. 105, Proposition 6.1.6, along with Corollary 6.1.7, therein), for a similar account, akin to the present setting.

Now, the whole setup, as above, reminds actually our previous remarks in the preceding, concerning (6.19): Namely, that

(6.26) the *exponential* map acts as the *"connecting function"* of the fundamental *adjunction* (6.3); thus, a *"tailor-made"* function, in effect, pertaining to the adjunction, at issue, still in relation with the account in Section 2.3.5.1.

That is, in other words, one thus finally gets at the *standard affair*, as it concerns the following *basic diagram*;

(6.27) $\qquad$ *quantum* $\;\rightleftarrows\;$ *observable* $\;\rightleftarrows\;$ *relativity.*

Yet, as a result of the preceding, one is further tempting to come now to the conclusion that;

(6.28) the same point of view, as above, was, in effect, implicitly employed, being thus the *motivating idea* thereat, still, for instance, by *Feynman*, when applying his celebrated, so-called, *"diagrams"*!

In toto, the above still confirms that, which we already talked about in the foregoing, namely, as the

(6.29) $\qquad\qquad\qquad$ *"Feynman principle"*.

Cf. thus (5.8)/(5.9) of this Section. So to repeat/paraphrase here too, once more, Machado's adage, one notes that:

(6.30) $\qquad\qquad$ *"...the route is always made by the traveler"*!

See also (10.39)/(10.40) in the preceding. Yet, for *potential applications of* the above perspective, as advocated, for instance, by (6.29)/(6.30), *within* always, of course, *the framework of ADG*, it is still worth reminding here our concluding remarks in the previous Section 2.2; (4.4).

Finally, we still wanted to emphasize/clarify our comments in the foregoing, pertaining to (6.25)/(6.26): Namely,

(6.31) the fundamental *"operators"* appeared, therein, "ad" and "Ad" (cf. (6.25)), might also be conceived, as essentially representing the *"geometries"*, respectively, of the *Lie group* concerned and of that one related with what we called in the preceding, cf. (9.6), as its *"Heisenberg(-matrix) representation"* in $\mathcal{E}nd\,\mathcal{E}$. In its turn, based on standard arguments, in connection also with ADG, this might further be appropriately interpreted within the whole ("second") *"quantization procedure"*, to which thus the map exp *acts*, in that way, *as a catalyst* (cf. (6.26)). So, in other words, the latter map can still be construed now, as actually constituting *the quintessence of* the same fundamental *Theorem*, as in Section 2.3.5.1 above.

Within this vein of ideas, see also, for instance, K. Habermann and L. Habermann [1: p. 7, Theorem 1.2.6], R. Berndt [1: p. 169, Example B. 16, and p. 128ff], together with A. Mallios [11: p. 151, comments following (3.5)]; yet, see also e.g. R. W. R. Darling [1: p. 223].

Now, within the same vein of ideas, the following general thoughts seems to be in order, as another *moral* of the preceding. So one realizes that;

$$\text{anytime we are confronted with a serious difficulty, we are then compelled/committed to find out}$$

(6.31′)   (6.31′.1)   *our own way, drastically different* (!),

from the (classical) *one*, we are used to employ, so far, if we wanted (it would be natural) to continue exploiting that, which *we already know*!

Scholium 6.1. — In the same manner that we used above the meaning of the term *"geometry"* (cf. (6.31)), we might further understand it, as an *"echo"* of the so-called *"symmetry axiom"* in *quantization* theory (see also e.g. A. Mallios, ibid.); indeed, an *algebraic analogue*, in that context, or even a *direct association* with it, could still be construed the *fundamental moral/Theorem* of the Galois Theory of *"field extensions"* (: *"coefficients"*, or even *"arithmetic"*, in our terminology (ADG)): namely, one can effectuate, herewith, a so to say *"physical* (: *"geometric"* (!)) *transliteration*, in effect, *of Galois Theory*, by appropriately employing, *in terms now of ADG, physical language* to guiding ideas of the aforementioned theory.

In that context, concerning the classical algebraic theory, as before, cf. also, for instance, M. Artin [1: p. 542, Theorem 1.15, and related comments therein].

On the other hand, *the preceding* still *fits in quite well with* our discussion in *Section 1.9.1* in the foregoing of the present treatise; see, for instance, (9.33)/(9.34), therein; thus similarly to the previous account, here too, *"symmetry"* is, indeed, the *primary notion*!

The above lead us thus naturally to the standard concept of a

(6.32)   *"Galois correspondence/connection"*;

namely, to a notion classically attributed to a function, being a spin-off/moral of the *Fundamental Galois Theorem*; cf. e.g. M. Artin, as above. On the other hand, the latter, according to the preceding (Scholium 6.1), can suitably be connected with the so-called *"symmetry axiom"* in *"second quantization"* (see e.g. A. Mallios [11]); therefore, on the basis of the content of Section 2.3.6 in the foregoing, the same is related, in effect, with what we have called in the preceding *"Hamilton-Schrödinger adjunction"*. Of course, the same may naturally be associated with the fundamental Hom-⊗ *adjunction*; hence, still with that one of an (𝒜-)*connection-curvature* (see also Theorem 5.1 in the preceding). So based on the above, one now gets at a natural conception of what we might consider, as

(6.33)   *"Galois dynamics"*.

However, now *"dynamics"*, in the technical sense of ADG ; therefore, a notion embodying, the *"technical"* one (ADG) of an *"$\mathcal{A}$-connection"*, as well! In toto, the preceding leads us to what one can understand, succinctly, as a

(6.34)   *"Galois adjunction"*.

The same is thus simply an outcome of the foregoing; namely, that the latter concept is just *another form of* the fundamental *"Hom-$\otimes$ adjunction"*, as this has been advocated/set off, already, throughout the present account.

Furthermore, another *moral*, that is worth pointing out here, still as an outcome *from the study of ADG* thus far, is that (cf. also Section 2.6 in the sequel) :

(6.35)   (the mechanism of) *"Differential Geometry"* is an *"identity"*; alias, a *"passe-partout"*, provided we supply the appropriate conditions/ environment to employ it.

Yet, in this context, the experience usually suggests (see e.g. *Feynman*, throughout the preceding, still below) that the aforementioned *"environment"* could be provided already from *the ("classical") theory*, we already have/know; however, now *viewed from another perspective*!

On the other hand, it seems that the so-called in the foregoing,

(6.36)   *"Feynman Calculus" is another* such (viz. *"identity"*), as before.

See thus (6.29) above, along with the relevant discussion in 2.3.5.3 of this Section; yet, cf. also Section 2.2; (1.4)/(1.7). Of course, the preceding justify, once more, in general, the point of view in considering/underscoring, throughout, what we called

(6.37)   *"functorial" character of ADG*!

Now, within the same vein of ideas, we can further note that;

(6.38)   an appropriate *"proof analysis"* may take us to *new results*/conclusions (!), against the (: already existing) classical theory.

On the other hand, technically speaking, one still remarks that:

(6.39)   the important aspect we actually get from an *information* (: *function*) comes, in effect, from *the first two terms of its* (Taylor) *expansion* (: *"infinitesimal approximation"*, yet, *"quantization"*).

Of course, the above further reminds/vindicates the

$$\text{(6.40)} \qquad \text{\textit{``affine'' aspect} of the matter. Namely, in other words, one thus arrives, roughly speaking, at the standard expression;}$$

$$\text{(6.40.1)} \qquad\qquad\qquad \text{``}\alpha\boldsymbol{x} + \beta\text{''}.$$

See also e.g. Section 2.2; (1.5) in the preceding, along with H. Weyl, (1.1)/(2.1) therein. Yet, the foregoing lead us too to an idea/conception, of an indeed, deep *"cosmological"* principle/*"phenomenon"*; viz. to that one of the function, in general, of the Physis/Nature herself.

2.3.7   *"Everything is Light"*

"For the rest of my life I will ponder on the question of what light is !"
(A. Einstein)

"The light is within me ... I myself am light."
(G. Fichte)

Our goal by this final part of the present Section is to provide further corroboration, yet, consequences, of the *"telling"* (!), indeed, *title of this part*. As mentioned below, we already have in the past, in some other places, sporadically referred to the point of view advocated by the title of this Section.

So, to start with, and based e.g. on the terminating remarks of the preceding Section (cf. (6.40)), we first note that,

$$\text{(7.1)} \qquad \text{the \textit{structure of the Universe} seems to be,}$$

$$\text{(7.1.1)} \qquad\qquad\qquad \textit{innately ``affine''}.$$

A vindication, even a hint (!) of the above, might be conceived, still epigrammatically, through the following *"cohomological"* (!), in character, *remark*: Namely, within the present context (ADG), one gets, for instance, at the next result;

$$\text{(7.2)} \qquad \text{any \textit{light bundle} (viz. a particular, depending on the ``color'', ``light ray'', still, ``beam of photons''), } \Phi^1_{\mathcal{A}}(X)^{\nabla}_R\text{, is an}$$

$$\text{(7.2.1)} \qquad\qquad\qquad H^1(X, \mathbb{C}^{\bullet})\text{-\textit{affine space}}.$$

Concerning the terminology employed herewith, we refer to A. Mallios [GT, Vol. I: Chapt. III; Subsection 5.3, in particular, p. 168, (5.116)]. On the other hand, within the same vein of ideas, it is also instructive to recall here, even Einstein himself, by saying that;

$$\text{(7.3)} \qquad \textit{``light rays ... might be ... involved in the origin of the concepts and laws of geometry.''}$$

Emphasis above is ours; see e.g. ibid., p. 118, (1.22/(1.23) in conjunction with

p. 117, (1.19), therein. Furthermore, as a particular echo of an utterance of D. R. Finkelstein, in that *"... all is quantum"*, one might still say that,

(7.4)   *everything "is"* (: can be reduced to) *light*.

In this context, see e.g. A. Mallios [GT, Vol. II: Chapt. IV, p. 207, (10.20)/(10.21)]. Of course, the aforementioned Finkelstein's adage vindicates (7.4); yet, the same was actually independently conceived by the present author, based mainly on previous thoughts in A. Mallios [17: p. 281ff, (6.21)-(6.24)], as well as, in the foregoing of this study, (9.35)/(9.35'): the latter referred to remarks on the (physical) *"tensorial"* relations/*interpretations* of *bosons* and *fermions*.

On the other hand, a significant outcome of the *universal aspect*, as expressed by (7.4), *"is"* (: *can be*) that,

(7.5)   *everything might be reduced to line sheaves*!

Yet, as an outcome of the preceding, see thus (7.3)–(7.5), one might also perceive that, *in effect,*

(7.5')   *Nature is "1-dimensional"*, in character!

Thus, still another corroboration of (7.4), namely, of the title of the present subsection. Now, within the same vein of ideas, see for instance the proof of *Einstein's equation* (in vacuo), as presented in A. Mallios [GT, Vol. II: p. 178ff; (4.12), (4.18), and (4.20)/(4.21)]. Even *"Bohr's correspondence principle"*, as advocated by the present treatise (cf. also ibid., p. 206, (10.18)), might be construed as contributing to the same viewpoint along with (7.5); that is a similar aspect to that one in A. Mallios [GT, Vol. I: p. 118, (1.23)], see also below. So the above, in conjunction with (7.1), leads to the aspect that,

(7.6)   *our "geometry" is* (: might be) *"innately/infinitesimally" commutative*!

Yet, a justification of the previous remark in (7.6) comes even from the so-called *"straightening out theorem"*: cf., for instance, R. Abraham et al. [1: 194, Theorem 4.1.4, along with the remarks before the statement of the theorem]. Therefore, still *Einstein's viewpoint of geometry* (!): see e.g. P. A. M. Dirac in [1].

Furthermore, one is also led to another *tenable vindication of (7.4)*, by reminding here the classical,

(7.7)   *"splitting principle"*

(A. Grothendieck); see e.g. M. Karoubi [1: p. 193, Theorem 2.15], or even H. B. Lawson-M.-L. Michelsohn [1: p. 225, Proposition 11.1]. Yet, one gets at a further *backing of* the same aspect, pertaining to *(7.4)*, based on the notion of a

(7.8)   *"pure"* (alias, *"trivial"*) *A-connection*.

See A. Mallios [VS, Vol. II: p. 214, Section 6]; for the classical case (: (smooth) manifolds), cf. also e.g. B. A. Dubrovin-A. T. Fomenko-S. P. Novikov [1: p. 444, Theorem

41.2.2], along with A. S. Schwarz [1: p. 244; (15.1.5)]. The terminology employed in the sequel still follows that one in the author's work, as cited above after (7.8).

Thus, our principal aim by the following lines is to reinforce further our perspective, as presented by (7.4), based now on *differential-geometric notions, within* the context of *ADG* and, in particular, through that one, as indicated by (7.8): So, by employing herewith A. Grothendieck's terminology, as in [1], we first remark that, by definition,

(7.9)
> a *"flat vector sheaf"*, say $\mathcal{E}$, of rank $n \in \mathbb{N}$, is a *fiber space* (over X) of *fiber structure type* $\mathbb{C}_X^n$ and *structure sheaf* $\mathcal{GL}(n, \mathbb{C}) \equiv M_n(\mathbb{C})^\bullet$. Thus, according to the very definitions, see also e.g. ibid. p. 42, Example c, one gets at the relation,
>
> (7.9.1) $\mathcal{E}|_U = \mathbb{C}_X^n|_U$,
>
> within a *sheaf-isomorphism*, with $U \subseteq X$, an appropriate (open) neighborhood, for every particular point $x \in X$.

In this context, see also, for instance, A. Mallios [VS, Vol I: p. 361, Subsection 2.(a); (2.61). Yet, cf. p. 370, Definition 5.1]. For a similar viewpoint, within the classical case, cf. also S. Kobayashi [1: p. 4, Section 2]. Now, as a consequence of (7.9.1), one gets:

(7.10) $\vartheta(g_{\alpha\beta}) = 0$,

where $(g_{\alpha\beta}) \in Z^1(\mathcal{U}, \mathcal{GL}(n, \mathbb{C}_X \equiv \mathbb{C}))$ is a *"coordinate 1-cocycle"*, of $\mathcal{E}$, as in (7.16) below, cf. also A. Mallios [VS, Vol. I: p. 371, Scholium 5.1], that determines it, *"cohomologically"* (ibid., p. 358, Theorem 2.1, and (2.41.2)). Of course, we also assume, herewith, concerning in particular (7.10), as above, that the base space X is equipped with/carries a *"differential triad"*,

(7.11) $(\mathcal{A}, \vartheta, \Omega^1)$,

such that

(7.12) $\vartheta : \mathcal{A} \longrightarrow \Omega^1$,

stands for the standard *"flat $\mathcal{A}$-connection"* of $\mathcal{A}$. See A. Mallios [VS, Vol. II: p. 13, Subsection 4(a); (4.2), along with p. 3, Lemma 1.1]. So, as an outcome of (7.10), one thus concludes that,

(7.13)
> every *flat vector sheaf*, as in (7.9), *is* in effect *"ϑ-flat"* (: by definition, (7.10) is valid), as well.

Cf. ibid., p. 137, Definition 6.2, and p. 138; (6.12). Therefore, one further concludes, in particular, that:

(7.14)
> a *flat vector sheaf* is *locally "trivial"*; viz. it possesses (locally) a *"trivial $\mathcal{A}$-connection"* (cf. also (7.8) for the terminology employed herewith).

We still recall that we always suppose here, as above, the existence of a *differential triad*, $(\mathcal{A}, \vartheta, \Omega^1 \equiv \Omega)$, as in (7.11). As a result, we first remark that, by the very definitions, a *vector sheaf*, say $\mathcal{E}$, due to its *defining relation*,

$$(7.15) \qquad \mathcal{E}|_{U_\alpha} \cong \mathcal{A}^n|_{U_\alpha}, \quad n = rk\mathcal{E},$$

within an $\mathcal{A}|_{U_\alpha}$-*isomorphism*, for some *open covering* of X (precisely, "*local frame*" of $\mathcal{E}$, viz. (7.15) is true),

$$(7.16) \qquad \mathcal{U} = (U_\alpha)_{\alpha \in I},$$

acquires the so-called "*local Levi-Civita 0-cochain*",

$$(7.17) \qquad \mathcal{D}_\alpha \in C^0(\mathcal{U}, \mathcal{H}om_{\mathbb{C}}(\mathcal{E}, \Omega(\mathcal{E}))), \quad \alpha \in I.$$

See A. Mallios [VS, Vol. II: Chapt. VI, p. 42, (8.32.1]; furthermore, the same 0-cochain is *gauge equivalent* with the *standard flat (local) $\mathcal{A}$-connection of $\mathcal{A}^n$*,

$$(7.18) \qquad \vartheta_\alpha^n \equiv \vartheta^n|_{U_\alpha} = (\vartheta|_{U_\alpha})^n, \quad \alpha \in I.$$

We denote this by the relation (ibid., p. 41, (8.28)/(8.29), along with p. 42, (8.32.3);

$$(7.19) \qquad (\mathcal{D}_\alpha) \underset{\alpha}{\sim} (\vartheta_\alpha^n).$$

Now, for convenience, we still recall that (ibid. p. 214),

(7.20)

> a "*trivial*" (alias, "*pure*") $\mathcal{A}$-*connection*, is a *local Levi-Civita $\mathcal{A}$-connection*, whose *0-cochain matrix*, associated with a *local frame $\mathcal{U}$* of $\mathcal{E}$ (see e.g. (7.15)/(7.16)),
>
> $$(7.20.1) \qquad \omega \equiv (\omega^{(\alpha)}) \in C^0(\mathcal{U}, M_n(\Omega)),$$
>
> satisfies the relation:
>
> $$(7.20.2) \qquad \omega = \widetilde{\vartheta}(g^{-1}),$$
>
> for some (*0-cochain*) $g \equiv (g^{(\alpha)}) \in C^0(\mathcal{U}, \mathcal{GL}(n, \mathcal{A}))$.

For the notation applied, see ibid. p. 214, (6.5). Yet, the same relation (7.20.2) is *equivalent with* the following one ("*Frobenius integrability condition*"),

$$(7.21) \qquad \omega g + \vartheta(g) = 0,$$

ibid., p. 207; (5.24).

In this context, we also note that in the *classical case* (see, for instance, the aforementioned work of S. Kobayashi, p. 5),

(7.22)
> the *Frobenius integrability condition* holds actually *true, if, and only if,* $R(D) = 0$.

On the other hand, the relevant situation in our case (ADG), is exposed in A. Mallios [VS, Vol. II: p. 214ff, along with p. 205, Lemma 5.1, and p. 207, (5.23)/(5.24)]. In particular, by considering just a *line sheaf*, which is also our main interest herewith, see e.g. (7.2)/(7.3) in the foregoing, one gets, under suitable supplementary conditions for our *"arithmetic"* $\mathcal{A}$, at an *analogous result to* (7.22): Thus, see e.g. ibid., p. 208, Theorem 5.2; yet, the case of a *vector sheaf*, in general, is still covered therein, under again appropriate *extra conditions on* $\mathcal{A}$, even *topological algebra*-theoretic ones, for the *algebra sheaf* $\mathcal{A}$, as before; cf. thus ibid., p. 353, Section 9, in particular, p. 355, Theorem 9.1.

Furthermore, by analogy with the classical case (: smooth manifolds, see e.g. still Kobayashi, as before; p. 6, (2.3)), one also obtains, in the case of a *"flat vector sheaf"*, in the previous sense (cf. (7.9)), an expression of the classical notion of the *"holonomy group"* of the particular (flat) $\mathcal{A}$-*connection*, at issue; it is thus the *image* of a suitable (*matrix-*)*representation* of the *"Poincaré group"*, through the corresponding *"symmetry group"* of the case in focus. On the other hand, we can further refer to A. Mallios [VS, Vol. II: Chapts. X (Section 5), and XI], for relevant *topological algebra*- theoretic *ramifications*, pertaining to the *topological algebra sheaf* $\mathcal{A}$, involved herewith: the latter have particular repercussions in the classical case, in connection with suitable *"operator-theoretic"*, in character, specializations/applications; see thus ibid., p. 359, Section 10. Moreover, see also e.g. A. Mallios [11], [8: p. 2667, Section 11] and [19], or even [GT, Vol. I]; yet, S. A. Selesnick [1: p. 1283, along with p. 1290].

Yet, within the same vein of ideas, it is also worth mentioning here that, even

(7.23) the renowned *"geometric (pre)quantization"* theory is still of a *"line sheaf"*, *in character*, in the previous sense; see thus, for instance, *"quantizing line sheaf/bundle"*, or yet, the classical *"Weil's integrality theorem"*: cf. the same Refs, as above; see also A. Mallios [3: in particular, p. 199, (9.1)/(9.2)]. In this context, we may still note at this place that the latter Ref. was actually among the first concrete/detailed expositions of the whole perspective of ADG at the time.

In toto, the preceding provide, in effect, a further justification of the viewpoint, as advocated by (7.3)/(7.4) in the beginning of this Section.

On the other hand, the crucial *"Frobenius integrability condition"*, as above (cf. (7.21)), is actually related, in the particular case, in focus, with a sort of the classical *"Poincaré Lemma"* (cf., for instance A. Mallios [VS, Vol. I: p. 243, Lemma 8.2]).

That is, one gets, herewith, at an analogous situation, referring to *"trivial $\mathcal{A}$-connections"*: Thus, under appropriate conditions that cover, of course, at least the classical case of smooth manifolds, see e.g. A. Mallios [VS, Vol. II: p. 355, Theorem 9.1], one comes to what we might call, a *"Frobenius-Poincaré Lemma"*; namely, one thus concludes that:

a (matrix) *1-form*

(7.24.1) $$\omega \equiv (\omega^{(\alpha)}) \in C^0(\mathcal{U}, M_n(\Omega))$$

(see (7.20.1)) *is closed if, and only if, it is "logarithmically exact"*; that

(7.24) is, in other words, one has that,

(7.24.2) $$d\omega = 0 \text{ is equivalent to } \omega = \widetilde{\vartheta}(g^{-1}),$$

for some $g \equiv (g^{(\alpha)}) \in C^0(\mathcal{U}, \mathcal{GL}(n, \mathcal{A}))$. See (7.20.2).

Now, the *condition* (7.24.2) is, of course, *sufficient*, by the very basic definitions; cf., for instance, the last Ref. above, p. 187, (1.8), along with p. 215, Scholium 6.1, in particular, p. 216, (6.17)/(6.18), therein. On the other hand, concerning the *necessity* of the same condition, as before, cf. p. 356, same Ref., proof of Theorem 9.1, as still cited above: thus, by taking the first relation ($d\omega = 0$) in (7.24.2) for granted (for the case in hand, rel. (9.17) of the aforementioned proof), and following the relevant argument therein, one concludes the second rel. in (7.24.2) (viz. p. 357, (9.22) of the same Ref.); of course, under the *particular presuppositions* of the aforesaid theorem, *therein*.

In this context, cf. also ibid., p. 357, (9.26), along with the subsequent comments therein, in particular, p. 358, Note and Scholium 9.1. Yet, for the special case of a *line sheaf*, which actually is still of our main concern, herewith, see also ibid., p. 216, Theorem 6.1, and p. 219, Corollary 6.1. Thus, one infers thereof that;

(7.25) *"triviality"* of an *$\mathcal{A}$-connection* for a *line sheaf* is, in effect, reduced to the *"flatness"* of the same sheaf.

Of course, in the standard case of a smooth manifold, where one affords the classical *Poincaré Lemma*, in particular, pertaining to the case at issue, the relation

(7.26) $$ker\,\vartheta = \mathbb{C}(\equiv \mathbb{C}_X),$$

cf. also e.g., ibid. p. 254ff, the above situation, as in (7.25), is still characteristic of what we named in the preceding, *"ϑ-flatness"* of $\mathcal{L}$, together with the corresponding relation to (7.9.1) (for $n = 1$). Namely, of being $\mathcal{L}$ a *"flat line sheaf"*, hence, *locally*, a *"$\mathbb{C}$-line sheaf"*; for the general case, still see A. Mallios [VS, Vol. I: p. 371, Theorem 5.1].

Now, the preceding lead us to the perspective of just looking at what one might construe, as a

(7.27) *"1-dimensional derivation"*: viz. a *procedure of following the variation of the "field" along its path*. Hence, in other words (ADG), by considering a *line sheaf* equipped *with an $\mathcal{A}$-connection*, say,

$$(\mathcal{L}, D);$$

thus, what we have called, so far, a *Maxwell field*. As a consequence, the above reduce the *variation*, at issue, into a 1-*parameter action*/function; therefore, one comes to the aspect:

(7.28)   *no PDEs* (!);   *just* "ϑ_t"!

Of course, the (first part of the) latter is reminiscent of a famous *Grothendieck's adage*, pertaining to the virtual *existence of PDEs*; see also (7.5'), as above. Yet, cf. relevant thoughts in Section 2.4; (4.20) in the sequel, and comments following it.

On the other hand, one is still led to a similar viewpoint, as that one advocated by the foregoing, when looking at the classical formula, referring to the "*trace of a matrix*"; in fact, to its "*variation*", as above (cf. (7.27)). Indeed, we first remark that the same notion/"*quantity*" is "*gauge invariant*" (!), in our own sense (ADG); namely, one has the relation,

(7.29) $$tr(\alpha x \alpha^{-1}) = tr(x),$$

with $x \in M_n(\mathcal{A})(X) = M_n(\mathcal{A}(X))$ and $\alpha \in \mathcal{GL}(n, \mathcal{A})(X) = GL(n, \mathcal{A}(X))$. See, for instance, L. W. Tu [1: p. 171, Lemma 5.18, (ii)]; yet, the same notion is involved with the "*variation*", in the sense of (7.27), as above of the "*determinant*" function/section, cf. e.g. A. Mallios [VS, Vol. I: p. 294, Section 4], in conjunction with the "*exponential*": ibid., p. 374, Section 6, in particular, Definition 6.1 and relevant notions, therein.

Indeed, the previous situation yields that,

(7.30)   the function/section "*trace*" realizes, in effect, a "*variation*" of the particular sort, as *described by* (7.27).

Namely, one thus obtains the relations:

(7.31)
$$tr(x) = \overline{\det(\exp(t \cdot x))}^{\,\centerdot}\,(0) = \left.\frac{d}{dt}\right|_{t=0} (\det(\exp(t \cdot x)))$$

$$= \left.\frac{d}{dt}\right|_{t=0} (\exp(t \cdot tr(x))) = tr(x) \cdot (\exp(t \cdot tr(x)))|_{t=0}$$

with $x \in M_n(\mathcal{A}(X)) = M_n(\mathcal{A})(X)$, as above; yet, we still assume the "*identification*",

(7.32) $$\left.\frac{d}{dt}\right|_{t=0} \alpha(t) \equiv \vartheta(\alpha(0)),$$

for a given (just, *continuous*) *curve* $\alpha : I \equiv (0,1) \subseteq \mathbb{R} \equiv \mathbb{R}_X \longrightarrow \mathcal{A}$. See also (7.11) in the foregoing, along, for instance, with A. Mallios [GT, Vol. II: Chapt. 4], and/or [19: p. 70ff], with [10: p. 94ff]; yet, concerning the classical case, see e.g. the aforementioned book of Tu, p. 174, Proposition 15.21, proof, and p. 173, Proposition 15.20. It is worthremarking herewith (cf. (7.31)) the *instrumental intervention of* the function exp, indeed, an appropriate *sheaf morphism*, as already noted in the preceding.

Now, the point of view advocated by (7.27), being also in accord with the subject matter/title of the present Section, is that one is actually interested in such sort of procedures that;

the *"variation"* of a *"quantity"*, whatsoever, can be reduced to *that one* of the *restriction* of the same quantity *on a suitable curve*. Hence, one can finally consider, *equivalently*, an

(7.33)

(7.33.1)   *$\mathcal{A}$-connection of a line sheaf.*

That is, in other words, an appropriate *"Maxwell field"*, in the terminology of ADG.

    Note 7.1. — The content of the previous remarks in (7.27), which one might still name, in effect, as a

(7.34)   *"principle of 1-dimensional derivation"*,

has, indeed, its *"dual"*: The latter terminology is actually reminiscent of the situation one is confronted with, when looking at the *classical result of Stokes*. This certainly deserves a further elaboration. Yet, the *"duality"*, hinted at above refers, of course, to the procedure,

(7.35)   *"derivation $\longleftrightarrow$ integration"*.

Thus, one can also consider the *"dual"* notion *to (7.34)*, as the

(7.36)   *"principle of "1-dimensional" integration"*,

pertaining to the point of view of the aforementioned classical result and its consequences.

    In this context, we further remark that both the above principles (7.34)/(7.36) are special cases of the situation, which one might call, in general, as a

(7.37)   *"pull-back procedure"*.

We still note here that the later fits in, quite well, with the *functorial* (alias, classically speaking, "tensorial") *character of ADG!* See also e.g., concerning the *classical case*, Th. Frankel [1: p. 102 and 155ff], along with R. Abraham-J. E. Marsden-T. Ratiu [1: Chapt. 7, and p. 381, Section 6.5]. Yet, M. Papatriantafillou [1], for the analogous *framework within ADG*, pertaining, in particular, to (7.37).

On the other hand, the above particularized to the case of the *gravitational* (gauge) *field*, we follow here the terminology of A. Mallios [GT, Vol. II: Chapter 4], lead us to the following remarks:

The *Ricci operator*,

(7.38.1) $Ric(\mathcal{E}) := tr(R(\cdot, s)z),$

with $R(\cdot, s)z$ the corresponding *"curvature operator"/* endomorphism

(7.38)   (ibid., p. 149, and p. 150, (1.25)), still expresses the *"variation"* of the

(7.38.2)   *(gauge) field* $\longleftrightarrow$ *curvature* (: *field strength*),

with the latter being appropriately *"transmuted"*, through the *exponential morphism*. ·

Thus, on the basis of (7.31)/(7.38.1) one obtains;

(7.39) $Ric(\mathcal{E}) \equiv tr(R(\cdot, s)z) = \overset{\displaystyle\bullet}{\overbrace{\det(\exp(t \cdot R(\cdot, s)z))}} \ (0)$

$$= \overset{\displaystyle\bullet}{\overbrace{\exp(t \cdot R(\mathcal{E}))}} \ (0).$$

The previous relations are still, of course, technically speaking, a spin-off of the classical relevant relation, written herewith in our terms; viz. one has:

(7.40) $\det \circ \exp = \exp \circ \, tr,$

see also (7.29) above, along with e.g. L. W. Tu [1: p. 173, Proposition 15.20]. In this context, see also, for example. A. E. Fekete [1: p. 327, *"Det-Trace Formula"*].

> **Note 7.2.** — It is also worth remarking here the *mathematical terminology*/notation employed in (7.38.1), *"trace"* (of a *"matrix"*, still cf. (7.29) above) and, *literally*, the same term, as applied, for instance, in the preceding: see, for example, Section 2.1, (2.2).

Yet, it is still quite worth relating the above with the terminology of Section 2.3.5 in the foregoing; in particular 2.3.5.1, pertaining to the classical *Stone-von Neumann Theorem*, as exemplified in our case. On the other hand, the appearance of *"t"* ($\in \mathbb{R}$) in (7.36), as above, may actually be viewed as the *"group action"* of $\mathbb{R}$ in $\mathcal{E}$, via $Aut\,\mathcal{E} \equiv (\mathcal{E}nd\,\mathcal{E})^{\bullet} \subseteq \mathcal{E}nd\,\mathcal{E}$; cf. (1.18) in the preceding, or even (4.8), along with (5.3)/(5.4), and also Section 2.3.6.1, referring to the *"exponential morphism"*: see e.g. ibid., (6.20).

In toto, as an outcome of the previous discussion, one thus arrives at a *situation*, as that one *described by (7.27)* in the preceding. Indeed, this appears to be, hence, the title of the present Section, as the *final most appropriate one*, within which *one is* (*naturally*, viz. by the very fundamental definitions/notions) *led to do calculations*; much more when the latter refer to *quantum* notions and *most particularly*, when in a *relativistic framework*, as it actually were!): see, for example, (6.11) in the foregoing, in conjunction also with what one might call, as the *"general principle of A-invariance"*, cf., for instance, A. Mallios [18]. That is, in other words, any time one is confronted with *quantum relativistic* situations; alias, with *quantum field theory*, proper! Thus, to say it, once more, the *environment as*, for instance, *specified* by

(7.41)   the *title of the present Section*, proves to be the *most appropri-ate*/efficient one, within which one can more aptly/*naturally work with*!

See also, for example, (7.28) as above.

On the other hand, even in corroboration of the aspect that,

(7.42)   *"everything is light"*

as just advocated, throughout the present section, according to its title, one can further notice that: this point of view is, of course, still very akin to what one might consider, as a

(7.43)   *"commutative"*, so to say, perspective of the sort of the *"geometry"*, we are essentially employing/confronting with at the very end; hence, at least, in the *"quantum"* (deep).

See also e.g. (7.6) in the foregoing. Indeed, one thus concludes that,

(7.44)   *"multiplication"* in *1-dimensional "extensions"*, that is, in other words, *"𝔸-bilinear forms"*, with *"𝔸"* our basic *"arithmetic"* (: field, or other-wise), see e.g. *"𝒜"* in the preceding, are always *symmetric*, thus, other-wise said, *"commutative"*!

Equivalently, in the language of *Geometric Algebra*, see, for example, the classic of E. Artin [1], one can further notice that,

(7.45)   there are no, in effect, *"symplectic geometries"* in *1-dimensional exten-sions*/*"spaces"*.

The previous assertion, as in (7.45), is, of course, an immediate consequence of the very definitions: thus, see e.g. the definition of the so-called *"Lagrangian 𝔸-planes"*, within the context of an *affine aspect of* (the classical) *Darboux's Theorem*; cf. (also, for example (7.1)/(7.2) in the preceding, along with e.g., for a recent account, A. Mallios, A. Conte-Thrasyvoulidou, Z. Daoultzi-Malamou [1]. The above still vindicated the

(7.46)   *"commutative aspect"* of Einstein, pertaining to the *"pure [physical] geometry"*, according to which $\alpha\beta \neq \beta\alpha$... *"does not fit in very well with [pure] geometrical [physical] ideas"*; yet, by still paraphrasing him, following actually herein, P. Dirac, *"[through] further developments of [such] geometrical ideas, further problems of physics should be solved"*.

See P. A. M. Dirac [1: p. 84]; cf. also Einstein's famous adage, referred to the problem/*question of the "physis of light"* (!), as in one of the epigrams of the present Section of this treatise.

Thus, based on the above, one is actually led, once more, to the following diagram;

(7.47)   (7.42) ⟶ (7.43) ⟷ *"Bohr's Theorem"* (see Theorem 3.1 in the preceding) ⟶ (7.45) ⟶ (7.46).

Yet, the same diagram, as before, reminds us of/quite relevant remarks in a recent study of S. A. Selesnick on *"quantum symmetries and their breaking"*, in that:

(7.48)   *"The apparent "conundrum" of continuous things arising as quantizations of discrete things ... counterintuitive nature of quantum theory ... a continuum of possible generic* (and generically different) *observers must be allowed ... corresponding to the continuum of* maximal *commutative subalgebras of observables ..."*.

See S. A. Selesnick [3: Introduction]. Emphasis in (7.48), along with the quotation marks on *"conundrum"* therein, is ours; yet, within the same vein of ideas, see also E. Zafiris [4], [7], together with A. Mallios [18: p. 1943, (4.17) and remarks following it]. See also Theorem 3.1 (: *Converse of Giraud's Theorem*] in the foregoing of the present treatise, along with [MM: p. 577, Theorem 1 (Giraud) and p. 580, Section 3].

Yet, motivated further by (7.48), as above, we can also remark that:

(7.49)   indeed, there are no *"continuous things"* in physis (*"out there"*), in the sense that,

(7.49.1)   *everything* (in physis) *is essentially discrete*!

Thus, in other words, *"continuity"* (therefore, *"topology"*, as well, of course, not in the sense that is quoted e.g. in (2.27)/(2.28) in the foregoing) is our (*mathematical* (!)) *innovation*, or even *description*, of what we actually perceive, hence, detect as well. Namely, we cannot perceive/discern the *"(infinitesimally) discrete"*: See, for instance, *Planck's constant*! So one resorts to *"approximations"*, *limits* and the like, even before *topology*! e.g., the so-called *"ε − δ" method* (Cauchy). Therefore, within the same vein, we can further say that,

(7.50)   *"commutativity"* might still be construed as an *"algebraic way out"* of the problem at issue! (See also (7.46), as before). Then, comes *"noncommutative"*, being, however, *innately commutative* (!), according to the preceding.

So cf., for instance, (7.48) as above; yet, the so-called throughout the present study

"Bohr's correspondence principle", along with e.g. its *topos-theoretic formulation*, in effect, via Theorem 3.1 in the foregoing.

Furthermore, within the same context, it is still instructive to remark, once more, the fact that:

(7.51)
> we cannot actually detect/perceive the *"point/event"*, when referring to the *quantum deep*; just, *relations which determine that "entity"*, in focus! Yet, to recall here also Einstein himself, it is good to remind after him that,
>
> (7.51.1)   *"... we must give up a complete localization of the particles"*.

Emphasis in (7.51.1) is ours. Within the same vein, see also e.g. A. Mallios [17: p. 268, (2.9) and p. 277, (6.2)]; yet, S. A. Selesnick [3: Section 3.1].

In toto, we thus try to *quantize things* that *there do not* actually *exist!* Cf., for instance, *"spacetime"*; see thus e.g. A. Einstein, as quoted, for instance, by Yu. Manin [1: p. 71]. Yet, H. Weyl [3]. As a matter of fact, we should try instead to change the terminology we employ thereat and reveal the eventual *inner structure/mechanisms* which are *inherent in our* calculations/*arguments*; the latter probability being just analogous to the efficacy of our techniques. Yet, during the previous procedure, we usually *bring into play* several *sophisticated*, in effect, *techniques*, passing thus on more efficacious/*instrumental*, even more *catalytic notions*, that are indeed inherent in our "sophisticated" methods, we usually apply, being actually fashinated/even, *mesmerized*(!), *by the* success of our *"calculations"*! We lose thus their inherent/esoteric, or even *natural*(!), yet *"geometrical"* meaning, in the sense e.g., as advocated, by (7.46); yet, the latter happens to be – but, how could it be otherwise – *more deep* and *simplifying*, as well!

Now, within the same vein of thoughts, as before, one has to look also at

(7.52)
> the *real meaning*, and *goal*, of the so-called, thus far, *"Abstract Differential Geometry"* (in short, *ADG*): So it is *wrong to think*, and we would have *lost* essentially *the point*, by looking at ADG, just, as a *generalization of* the standard *CDG* (: Classical Differential Geometry), namely, the *theory of smooth manifolds*. No! It is, in effect, *something, much more* than that! Thus, ADG *aims at revealing*, in mathematical terms, of course, viz. to *describe in* an *abstract form*(!), the *basic mechanism* of the *differen*[ces]*tial*! (Taubes). Yet, in effect, due to its *inherent character*, thus, *sheaf* or even *topos-theoretic*, the variation of the same.

It is thus clear that any *"differential structure"* (abstract or not) will be only a special case of the previous perspective, the latter pointing out just the particular instantiation, in focus. As a matter of fact, we are thus interested, so to say, *not* in the *"pythagorean triads"*, but in the *"Pythagorean Theorem"*, viz. the *function of the physical law* itself(!), so the *ultimate goal*, after all, *of ADG*, as presented by (7.52).

The preceding still clarify the meaning of *"geometry"*, in the sense of ADG: that is, a *"physical"* one; cf. Einstein, as also quoted by P. G. Bergmann [1: p. 84], along e.g. with (7.46), as above. Thus, to say it, once more,

(7.53)   ADG, through its *abstract viewpoint*, aims at *uncovering* the *inherent catalytic/algebraic*, in character, viz. *"relational"* (cf. also A. Mallios [20]) *mechanism* of our own, so-called *"geometric" calculations*: the former is thus *hidden* under *overloaded*, in effect, *"luxurious"*, so to say (see e.g. Feynman ...) unnecessary, though wonderful(!), *"topological/analytic* machineries"!

See also A. Einstein [1: p. 8, yet p. 142, near the bottom of the page, and p. 143]. Cf. also A. Mallios [22], in conjunction e.g. with L. O' Raifeartaigh [1: p. 208ff], concerning the so-called in the preceding *"Utiyama's principle"*; a really *typical example* of the point of view advocated *by* (7.53). Still, within the same vein, cf. also, for instance, M. Friedman [1: p. xi], and Y. Choquet-Bruhat et al. [1: p. v, concerning *"physical mathematics"*]. Yet, one can still recall M. F. Atiyah, by paraphrasing him, as quoted e.g. by G. L. Naber [2: p. 400], in that;

(7.54)   *"mathematicians will want to take heed ... once again to their historical roots in physics* [(!)]".

How, could it be otherwise (?); since, for instance, the time of Aeschylus (!), already. Indeed, we learn in that context:

(7.55)   «ἀριθμόν, ἔξοχον σοφισμάτων»   (Aeschylus)

See e.g. A. Weil [1: p. v.]. To conclude, still motivated by the preceding discussion, we might further hold, that:

(7.56)   even *gravity* might be, just, *a particular form of light*; as it actually is, after all, *any other fundamental force*/law in Nature. *Light*, in any form (!), appears thus to be something like the *welding constituent/solder* of the whole *Universe*!

In this context, see also A. Mallios [GT, Vol. I: p. 118, (1.22)/(1.23)], and [17: p. 281, (6.21), and p. 282, (6.23)/(6.24)].

A motivating idea of the preceding account, and still something that actually goes back to Leibniz, see for instance A. Mallios [16: p. 1560, (1.17)], we remark that: we should always be aspiring to

(7.57)   apply a so to speak *"geometrical calculus"*, operating directly on the *"geometrical objects"*; yet, this without any intervention of coordinates.

A fundamental *Example* thereon is, of course, what we called in the preceding *"Feynman's Calculus"*; cf. (3.2). See also the next Section 2.4; (6.1).

2.4 Stone – von Neumann Adjunction

2.4.1 *Introduction*

Within the framework of the present Section the *exp*(onential) function/*functor* constitutes in fact the *penetrating/catalytic morphism*, in any sense, throughout the whole procedure!

On the other hand, the *"adjunction"*, alluded to in the title, refers in effect to the classical fundamental notion of *Homology Theory*, already considered in the foregoing; namely to what we called,

(1.1) Hom-⊗ *adjunction.*

See, for instance, S. Mac Lane [2: p. 269, *"adjoint associativity"*], or even [1: p. 78, Definition]. Yet it seems that the same concept could actually be construed, *physically speaking*, as the *"fundamental model"* for all adjunctions! Indeed, the same association, as in (1.1), is intimately connected with the *basic, throughout all physics, correspondence* (in fact *"adjunction"*), see also below;

(1.2)   observable ⟷ observer.

On the other hand, within the same vein and as a result of the classical homonymous theorem/result with the adjunction in the title of this Section, one is still led to the aspect that;

(1.3)   *every real number* when imbedded in (the complexes) ℂ, and viewed as a (1-dimensional) *"self-adjoint operator"*, might be further conceived as defining a *"1-dimensional derivation"*!

In this context, see also (3.4)/(3.5) in the sequel, pertaining to the terminology employed in (1.3), being still in conjunction with the classical result, as alluded to in the title. Thus, the above can be also understood again through the action of the *exp*(onential) *functor* as this will be made further clear by the subsequent discussion.

Yet, within the same context, as before, we also note once again in this place, that the previous diagram (1.2) can be related with what we already named in the foregoing,

(1.4)   *"Lassner–Uhlmann adjunction"*.

That is, in other words, with the basic diagram,

(1.5)   *"state"* [of a *"physical system"*] (locally) ⟷ *observer.*

See thus (10.46) and (9.39.1) in the preceding, together with relevant comments therein. Thus, our objective in what follows is first to clarify the aforementioned *association*/correspondence *of* (1.5), as above, hence in fact *another form of* (1.2) *with* the items in (1.1), a task which we undertake straight away by the next Section.

2.4.2 *Physical Jargon*

Indeed, *physically speaking*, the *bi-functor* Hom represents what we actually observe, being usually the *substitute*, classically, *for* the *"representation space"* of what we are interested in (: *"observables"*); on the other hand, the existence/cause of the latter is again modeled, classically, after the (bi-)*functor* ⊗: cf., for instance, what we have called so far, *"Kähler construction"*, or even *"extension of scalars" functor*; see e.g. Section 1.2; (2.43). Thus, we stick here with the aspect that one might actually consider *the functor*

(2.1)   *Hom, as a "realization" of the ⊗-functor.*

See also e.g., A. Mallios [VS, Vol. I: p. 304, (6.11)]. Of course, we are led to the previous perspective when looking at the "Hom-⊗ *adjunction*", through the *physical/natural viewpoint*, as *specified by* (1.5)/(1.2). Indeed, one can look, *equivalently*, at

(2.2) *(2.1)*, as the *"dynamical analogue"* of the "Hom-⊗ *adjunction*".

See also below. The previous assertion can be further clarified, by relevant thoughts in the preceding; cf. Section 1.2: (2.40)-(2.43); another justification will still be supplied in the sequel. Thus,

(2.3)
> the *"realization" of the ⊗-functor through Hom* (viz. otherwise, the *homological-type* interrelation of the two functors (ibid.)) is, in effect, the very same expression of *"dynamics"*, the latter notion being associated with the *"physical law"* itself; that is, the (𝒜-)*connection*. Yet, *"tensor product"* (viz. e.g. *"Kähler construction"*, thus, *"extension of scalars"*, as already hinted at in the preceding).

In this context, cf. also relevant remarks in Section 2.3 (6.5), as well as, Section 1.2.2; (2.38.1), in the foregoing. So, otherwise said,

(2.4)
> the same notion of Hom-⊗ *adjunction* may be construed as, or even constitute, *"physically speaking*, the *"dynamical analogue"*, of the very same classical *homological-type relation*, (cf. (1.1) and subsequent Refs there; yet, see (9.55) in the preceding, along with (8.7)ff therein. Besides, the expression through the Hom-*functor* of the *"dynamics"* represented by/inherent in the ⊗-*functor*, is already described by (2.3).

Similarly, for a *topos-theoretic* analogous treatment of the previous *"adjunction"* see [MM: p. 347, Chapt. VII, in particular, p. 357; (15)]. Yet, cf. E. Zafiris [7], as well as Section 1.4 in the preceding and (8.7), therein, along with relevant comments after it. See also A. Mallios-E. Zafiris [1], [2], [3].

2.4.3 *Stone–von Neumann Theorem in Action*

Our goal within the present Section is to point out the fundamental (*functorial*) "*adjunction*", effectuated by the *exponential* function/*functor*, already *inherent in* the classical *Stone-von Neumann Theorem*, by the very substance of the latter theorem: Indeed, the same fundamental function/functor is, in effect, the "*transmuting*", or even, the "*quantizing*" *factor* of the whole procedure, entailed by the same theorem, as before; it is actually this aspect that we are going to deal with, right away by the ensuing discussion:

Thus, to begin with, we hasten to note, even at this point, that the *adjunction*, in focus, is already *inherent in the next diagram*, as we are going to explain it right below; that is, we have the following *association*, contained in the aforementioned classical result:

(3.1)   *self-adjoint operator* $\longleftrightarrow$ (*"strongly continuous"*) *1-parameter* (*"unitary"*) *group.*

Concerning the technical language involved in the previous diagram, we refer e.g. to J. B. Conway [1: p. 334, Section 5], R. Abraham et al. [1: p. 541, Theorem D.1], or even E. Prugovečki [1: p. 335, Theorem 6.1]. Now, the notion of a "*self-adjoint operator*" is classically associated (ibid., as e.g. indicated above) with that one of an "*observable*". We have thus distinguished already the (small) category, "*source*", of the adjunction we are looking for; on the other hand the "*host*" category of the same adjunction, can actually be determined by *the second item of* the previous diagram (3.1): Namely, the notion of a "*1-parameter group*" of appropriate objects; yet, otherwise said, an (appropriate) "*action*" of (the *real group*) $\mathbb{R}$ on the source in effect category as before, or even a "*flow*" in the same category, and *we have* actually *done*! Indeed, for the latter argument as before, we can resort e.g. to diagram (1.7) of the previous Section 2.3; see also (5.1) herewith, along with the comments before it.

On the other hand, we have already commented on the *basic association*,

(3.2)   *observable* $\longleftrightarrow$ *functor* $\mathcal{H}$om,

in Chapter 1; (6.3)/(8.2). Furthermore, according to the aforesaid diagram (6.3), we can still understand the *first* fundamental (the *second* being (3.2), as above) *association*, as follows:

(3.3)   *field* $\longleftrightarrow$ *physical law* $\longleftrightarrow$ ($\mathcal{A}$)-*connection* (: *differential equation*) $\longleftrightarrow$ $\otimes$(*-functor*: "*Kähler construction*").

Now, the *conjunction of* (3.2)/(3.3) constitutes thus the *fundamental relation* (1.1) in the preceding; viz. the so-called, Hom-$\otimes$ *adjunction*.

Yet, within the same vein of ideas, we are going to comment further on the rôle of the *functor* exp, in that framework: Indeed, as we explain right away the latter function might essentially be conceived, as the *natural transformation of functors/isomorphism* (see also [CWM: p. 78, Definition]) in the previous *fundamental adjunction* (1.1), within the present context; first, it is still instructive to summarize the preceding account, through the following diagram (3.4):

(3.4)

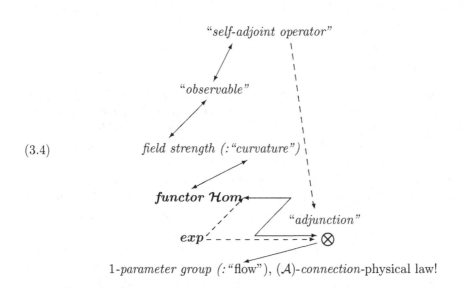

In this respect, by further looking at the *catalytic rôle* of the *function/functor* **exp**, as before, it is still instructive to recall here its *intervention*, through the corresponding *"flow"* (cf. (3.5.1) below); see also Chapter 1; (12.9)–(12.11) above, and following remarks therein. Yet, see also e.g. R. Abraham-J. E. Marsden-T. Ratiu [1: p. 472/473ff], as well as, S. Y. Auyang [1: p. 18, Q2 and p. 198].

In this context, cf. also A. Mallios [17: p. 280, (6.14)/(6.15), and comments following it], concerning an early perspective about the *very character of the* basic *adjunction* (1.1), in the point of view, as indicated by the diagram (3.4).

On the other hand, still as a further clarification of the perspective advocated by the previous diagram (3.4), we can also subsume two crucial aspects of it, as follows: that is, one arrives at the next *succinct form* of the whole story as exhibited by (3.4) through the following *diagram/scholium*:

Stone-von Neumann (Theorem)

operator theory (thus, *quantum mechanics*):

(3.5)   every *self-adjoint operator* is/can be put into the *"trans-muted image"*, via exp, of the *basic ("flat") derivative* (in
(3.5.1) $\mathbb{R}$, (cf. (12.10.1) in the foregoing)/*physical law*, in terms of a *"flow"*; alias, through a (suitable) 1-*parameter group* (of the same sort of operators, as above).

Furthermore, the – sort of a *sheaf* (or even, *topos*) – *theoretic* character of the whole framework, adopted by ADG, affords the present account a *relativistic* and in effect a plain *unified field theory* – perspective. Cf. also A. Mallios [16: p. 1591, Subsection 6.(a)], or even [14: p. 71, (5.12)], along with Subsection 9.2 in the foregoing of this study.

2.4.4 De Broglie–Einstein–Feynman Adjunction

"Feynman ... was ... able to write down the answers [to the problem, at the time of QED] *straight away without using any mathematics"*!
(P. Davies in his *"Introduction"* to R. P. Feynman, *"The Character of Physical Law"*. Penguin, 1992: p. 8).

"... Feynman not only felt that the (differential) geometrical interpretation of gravity "gets in the way of its quantization", but also that it masks its fundamental gauge character".
(B. Hatfield in his forward on *"Quantum Gravity"*, in*"Feynman Lectures on Gravitation"*, notes by F. B. Morinigo, W. G. Wagner and B. Hatfield (Eds.). Penguin, 1999).

Emphasis in the previous two adages is ours. Now our purpose by the ensuing discussion is to comment briefly on another *basic adjunction*, as in the title of this Section, which is actually implicitly employed in the standard *"mathematical physics"* literature. Whence, always of course within a *smooth manifold* framework (!); see thus e.g. W. D. Curtis - F. R. Miller [1: p. 356ff]. So here again, within that context, the catalytic *"transmuting"* operator/*functor* is still the *exponential map*; indeed, the latter works essentially in a similar manner, as in that one, which is indicated by the previous diagram (3.5).

Thus, based in particular on the *fundamental idea* of L. de Broglie (1924), concerning the so-called at the time *"wave mechanics"*, it was actually later R. Feynman (1964), who systematically employed the *exp functor*, as above, in formulating his celebrated ever since idea of the so-called, after him, homonymous *"diagrams"*. Yet, it is still worthremarking herewith, that the same great scientist was *applying*, in effect, classical *Calculus* in his own, *"non-spatial"* way (!), namely, that one we already named in the preceding *"Feynman Calculus"*; see e.g. (5.3) of the previous Section 2.3, along with (5.4)/(5.8) and relevant comments thereon; in particular, Section 2.3.4 therein. Yet, cf. also Section 2.2, (1.4)/(1.7).

Now, within the same vein, as before, we also notice that, the

(4.1)   *trans(lation)/mutation*, via exp, of the *"field"*, in our sense (: ADG, see e.g. (8.20) in the preceding of the present study, or even Section 2.1, (4.5)/(4.6)), is represented by the so-called,

(4.1.1)   *"probability amplitude phase factor"*.

Cf., for instance, the aforementioned Ref., ibid. p. 356.

Yet, see also rel. (4.3) in the sequel.

On the other hand, the same *"phase factor"* as above, is in fact the corresponding herewith, alias, *"Lagrangian-density"* (cf., for instance, Section 2.1, (3.20)/(3.21)), along always, for the case at issue, a suitable *path*; of course, the latter situation still reminds our relevant remarks in the foregoing, pertaining to the "1-*dimensionality*" of the *Nature's character*! (See Section 2.3, (7.5′), as well as, Sections 2.3.2, 2.3.3 - in particular (3.3) - and 2.3.4, therein).

Thus, in conjunction with the *characteristic diagram* (3.4), as above, and by also taking into perspective our previous relevant comments in Section 2.1, (3.2)/(3.7), along with (3.21)/(3.22) therein, one gets at the corresponding

(4.2)   *"Lagrangian (action-)density"*

in the following form:

$$(4.3) \qquad \alpha(t) := \exp\left(i\int \mathcal{L}(\alpha, \dot{\alpha})dt \right);$$

here we put $\alpha : \mathbb{R} \longrightarrow \mathcal{E}$, for a given *"path"* (: curve) in $\mathcal{E}$, the latter being an appropriate $\mathcal{A}$-*module*, as the case might be, according to our terminology, as applied in the preceding: cf. thus for instance the *relevant jargon in* (3.4)/(3.5), as before, with the corresponding one in Section 2.1, as already mentioned above. Yet, for the *"integral"* appeared *in* the previous *rel.* (4.3), see also the analogous (terminological) Note 3.1 in the same Section 2.1 of this study, as before.

Furthermore, the *"Lagrangian"* occurred *in* the same *rel.* (4.3) is the corresponding concept herewith of the *classical theory* (viz. of that one based on (smooth) *manifolds*), associated therein with the standard relevant notion of *"total energy"*: see thus, for instance, Curtis-Miller as quoted above; ibid., p. 356.

Of course, the preceding constitute an approximate only idea of the whole story in focus. However, the *"spaceless"* setup dominating, by definition, the very character of ADG, in conjunction also with the associated therefrom *functorial perspective* of the latter mechanism (ADG), help thus in getting a better (: more *inherent*) point of view of the whole framework! Yet, within the same context, see also e.g. (13.7)/(13.9) in the foregoing.

Yet, in this context, we can still remark here that, what one might call, as

(4.4)   (4.4.1)   "Selesnick's conundrum" (cf. S. A. Selesnick [2]), between

$$\text{"continuum"} \longleftrightarrow \text{"discrete"}.$$

could still be conceived, just, as *another instance* of what we named above, a "*de Broglie–Einstein–Feynman adjunction*".

Thus, in toto and still based on the preceding, we can say that one actually gets at an *interrelation* between:

(4.5)   (i) the *Stone-von Neumann adjunction*,

and also at that one of

(4.6)   (ii) the *de Broglie-Einstein-Feynman one*,

in such a manner that,

(4.7)   the *first*, (i), is in fact a *realization* (: *mathematical interpretation/formulation*) of the second one, (ii).

We express the above, by just setting the following diagram:

(4.8)   *Stone-von Neumann* $\longleftrightarrow$ *de Broglie-Einstein-Feynman*.

In this context, one can further remark here that,

(4.9)   what we named before, as the "*de Broglie-Einstein-Feynman*" part *of* the *adjunction* (4.8), (see also (4.5)/(4.6)), might still be called, a "*Schrödinger-Weyl*" *adjunction*.

See also e.g. Dubin-Hennings, as quoted below (: comments following (4.19) in the sequel).

On the other hand, by meditating on the *function*, so far, *of* what we named throughout the present treatise, (*Abstract*) *Differential Geometry* (ADG), we realize that:

(4.10)   ADG, as a matter of fact, its *mechanism*, constitutes essentially a, so to say, sort of an "*identification*", in order to describe the *function of Nature*. Yet, in particular, its *substance* (viz. its *esoteric character*), in terms of the "*quantum*"! Namely, in other words, through its "*infinitesimal*" behavior, the latter being also in effect the most fundamental *one* (hence, *natural* too).

Thus, once more, we actually effectuate that,

(4.11)   the previous *interrelation* (4.8), still describes/explains the *mechanism/action of the fundamental pair* $(\mathcal{E}, \mathcal{D})$; that is, of the *field* or/even, *alias*, of the *physical law* (!), itself.

Yet, in that respect, based also on the foregoing, and still on historical grounds, we might name/conceive the same pair,

(4.12) $(\mathcal{E}, \mathcal{D})$,

as above, as the *"de Broglie representation/duality"* of a (natural) *field*; in fact, of an *elementary particle*, as well, see e.g. A. Mallios [20: p. 63, (2.3)].

On the other hand, another *moral* that seems to be *in force*, *herewith*, is that:

(4.13)   *"analysis"*, as it particularly concerns the concept of *"continuity"*, does not actually fit in well, pertaining to questions of *Quantum Field Theory* (QFT): the same mathematical discipline proves, in effect, to be *insufficient* in formulation *quantum-theoretic situations*; see e.g. the notion of *"event"*: cf., for instance, A. Mallios [17: p. 268, (2.9)]. Yet, see e.g. W. Heisenberg [1: p. 48], or even A. Einstein, as e.g. in Section 2.3, (7.51.1) above, together with (4.17) in the sequel.

Indeed, in that context, we can assert that,

(4.14)   *Physis* is *"discrete"*, in character!

See, for instance, *"Pauli exclusion principle"* (see e.g. J. M. Ziman [1: pp. 32, 33]), or even *"cohomological classification"* of elementary particles, à la ADG; cf. A. Mallios [GT, Vol. I: p. 226, Theorem 5.1, and p. 230, (6.16)], along with [GT, Vol. II: p. 70ff, in particular, p. 77, (9.50)]. As a consequence, we still realize that,

(4.15)   *"analysis"*, in the manner *we usually employ it*, proves to be *not the appropriate tool* in confronting with problems of *quantum field theory*; hence, the theory of *smooth manifolds* (viz. *classical differential geometry*), based on this type of "analysis", as well!

Thus, see also e.g. C. Isham [1: p. 400], R. Feynman (cf. Section 2.3; (7.53)); yet, the famous adage of A. Einstein in [1: p. 106], or even H. Weyl [1: p. 15, ftn. 5; p. 308].

Furthermore, to paraphrase, and even generalize C. von Westenholz [1: p. v], one can further remark that:

(4.16)   the *"geometric spirit"*, viz. the *categorically*, in fact, *working methods*, as this is really achieved, by means of ADG, provide an *"alternate description of natural phenomena beyond* the description obtained in terms of *analytical methods"*.

So we can follow for instance a *Lagrangian viewpoint*, as we did it already in the preceding, *than* that one of *Hamilton*; moreover, this *without any surrounding "space"* (!): indeed, a *"steady impediment"* of *"analysis"* (classical or not, thus e.g. a smooth manifold-based one), by looking directly at the objects/concepts that actually define the "space" that interest us; something that indeed goes back even to Leibniz. Thus, to paraphrase him,

(4.17)   *"... we lack another analysis, properly geometric or linear, to express location directly, as algebra expresses magnitude."*

Emphasis above is ours; see e.g. G. E. Bredon [1: p. 430]. Yet, cf. also for instance, A. Mallios [17: p. 267, (2.6)], pertaining to a relevant perspective of Leibniz for the *"space"*. No more no less, a *"fatal error"* (!); cf. thus A. Einstein [1: p. 3], as it concerns our classical perspective of the same "space". Therefore, our relevant remarks so far, for that sort of *"analysis"*, usually employed! See thus also Section 2.3, (7.53), along with A. Mallios [20: p. 62, (2.1)].

On the other hand, we further remark that, what we might, look at, as a,

(4.18)   *"Platonic viewpoint"*,

has been several times pointed out, for instance, by W. Heisenberg in [2: see e.g. p. 83, therein]; indeed, the same could be construed quite well as being in effect in accord with the *"relational"* aspect of the *"geometry"*/even, *"calculus"*, which has been employed throughout ADG: cf., for instance, A. Mallios [20]; yet, [22].

Finally, the same *adjunction* as advocated by the present Section, in conjunction also with our considerations in 2.4.3, constitutes in fact a *further corroboration of* the idea presented in 2.3.7: That is, of what we considered in the preceding, as the

(4.19)   *"1-dimensional"* (in effect, *"affine"*, cf. Section 2.3; (6.40)) *character of Nature in the* quantum deep (viz. in the *"infinitesimal"*; thus, the *"quantum"*).

See also (3.5) above, along with Section 2.3; (7.5′). Yet, cf., for instance, D. A. Dubin-M. A. Hennings [1: p. 38], pertaining to an analogous *meaning* within the previous perspective *of* the same *Stone-von Neumann* Theorem; moreover, cf. also K. Habermann-L. Habermann [1: p. 7, Theorem 1.2.6], for a similar point of view, in that context. Furthermore, within the same vein of ideas, cf. also Section 2.2; diagram (2.5), and 2.2.3, therein along with 2.3; (1.7).

On the other hand, a similar point of view as in (4.19) above, viz. pertaining to the

(4.20)   *"1-dimensional character of Nature in the "infinitesimal"*,

it seems to be still *in accord with* the substance of recent *"(super)string" theories*: see thus e.g. M. Nakahara [1: p. 31ff, along with p. 461, Chapter 14]. In the same vein, cf. also P. C. W. Davies-J. Brown [1].

2.4.5 *"Invariance"*

We consider below the general concept of *"invariance"*, as this is suited *within the framework of* ADG: Thus, by referring in general to the notion in the title of this section,

we mean, by definition, the word/concept,

(5.1)   *"invariance"*, *with respect*, always (!), *to* a certain particu-
    (5.1.1)   lar *"group action"*; alias, *"transformation group"*, or even
                "group of transformations".

Now, the latter appear usually, as *"automorphisms"* of some specific *"structure"*, so by taking for instance a mathematical structure (Klein), this represents in fact the *"model"* for a certain *physical* situation/*"structure"*, we are interested in: More precisely, the aforementioned transformations/automorphisms refer actually to the (mathematical, again(!)) *images* of decisive *issues*/elements of the corresponding *"physical group"*; still construed on the basis of the previous association (*"observable"* ⟷ *"observer"*, take e.g. Lassner-Uhlmann adjunction), the same group being *characteristic* of the physical structure at issue: consider, for instance, the *"symmetry group"*, thus, also *"symmetry axiom"*: see e.g. A. Mallios [11: p. 151].

On the other hand, and still in conjunction with the *framework of* ADG, applied throughout the present essay, we *steadily remark* that ;

(5.2)   *a physical law is* (always) *"$\mathcal{A}$-invariant"*, for whatever *"$\mathcal{A}$"*, viz. our
         "arithmetic", might be!

Thus, see also, for instance, A. Mallios [18: p. 1929, (1.1), and comments following it]. In this context, we further remark that,

(5.3)   it is actually *"we"* (if ever (!)), *who*, at each particular time, based (al-
        ways (!)) on our theories, are able to *write* down/formulate the *physical*
        laws (: function of Nature), *independently of* the sort of the specific *"$\mathcal{A}$"*,
        viz. of *our* own *"arithmetic"*, which we employ to describe the laws.

See ibid., p. 1930; (1.2)/(1.3). So to put it indeed in *the right way*, we further notice that;

(5.4)   *it is not* really *Physis*, which is, so to say *"functorial"* (anyhow, *Physis*
        is anything (!)), *but* actually *we* ($\equiv$*"$\mathcal{A}$"*), who *accomplish*, on the basis,
        always (!) of our theories, to be such; that is, *to work*/express ourselves
        "functorially" (!), in the sense of (5.7) below.

However, see also in that context A. Mallios [18: Section 1, p. 1929ff]; yet, Section 1.9; (9.58) in the preceding. Thus, *"invariance"*, as above, means actually, *in terms of* ADG,

(5.5.1)   *"$\mathcal{A}$-invariance"*,

ibid.; that is, otherwise (see also (5.1)),

(5.5.2)   *invariance, with respect to $\mathcal{A}ut\,\mathcal{E} := (\mathcal{E}nd\,\mathcal{E})^{\bullet}$.*

(5.5)

Cf. also e.g. A. Mallios [GT, Vol I: p. 34, (6.12)] for the terminology applied in (5.5.2). The latter (group sheaf, as in (5.5.2)) is essentially *inherently defined, independently of "$\mathcal{A}$"*, by the very definitions, that we may use to define the *physical object*/system, in focus,

(5.5.3) $\mathcal{E} \equiv (\mathcal{E}, D)$.

See also for instance, (6.1), (8.20), and (9.5) in the foregoing, along with (4.12)/(4.13) of the present Section.

As a result, we achieve thus to have

the form of *the equations*, describing the (physical) laws, *independent* of the so-called,

(5.6)   (5.6.1)   *"inertial frame of reference"*!

That is, *equivalently*, independent of the particular *"gauge" (in terms always of "$\mathcal{A}$"*, as before), which is employed thereat.

Indeed, the above as an *outcome*, based on the very definitions *of the* same *mechanism*/perspective *of* ADG. Yet in other words, according thus to its *functorial* behavior/*function*, viz. the way of action (cf. also (5.3), as above); see also A. Mallios [18: p. 1930; (1.5)].

Now, the preceding give us, in effect, the context of what one might understand, through the term,

(5.7)   *"relativistic invariance"*.

The above is thus accomplished on the basis of the *sheaf-theoretic*, and even *space-free* (!), *dynamics* of ADG; see also the same Ref., as before, p. 1936, Section 3. Yet, we further remark that the preceding can still be extended to a more general, entirely *"relational" framework*, of a *topos-theoretic* formalization: See e.g. E. Zafiris [5], along with A. Mallios-E. Zafiris [1], [2], and [3]; yet, A. Mallios [18: p. 1940, Section 4], as well as, [20].

On the other hand, by looking at *"Heisenberg's uncertainty principle"*, one has the relation;

(5.8) $\Delta p \cdot \Delta q \geq \dfrac{\hbar}{2} := \dfrac{h}{4\pi}.$

See e.g. G. G. Emch [1: p. 294; (87), and remarks following it]; yet see also, for instance, Th. Frankel [1: p. 531]. Thus, we might say that,

(5.9)   *Heisenberg explains*/vindicates *Planck's constant.*

As a consequence, one thus realizes that:

by working *within ADG*, and based on the

(5.10.1)   *"principle of general covariance"*,

(5.10)   (: Physis does not depend on us/ "there are no *"singularities"* in Physis", thus, *"indeterminacy"* (: Heisenberg)), we come to the *conclusion* that;

(5.10.2)   (5.9) *and* (5.10.1) are *"physically coherent"*!

See also A. Mallios [18: p. 1933; (1.22)]. Therefore, in other words,

(5.11)   by working *in terms of ADG, we* can *absorb Planck's constant.*

In that context, see also e.g. A. Salam in P. C. W. Davies et al. [1: p. 172/173]; furthermore, in conjunction with relevant remarks as in the latter quotation, cf. also C. von Westenholz [1: p. 323, (ii′)], in that:

(5.11′)   *"quantization is provided by the physical law itself."*

Emphasis above is ours. Yet, see also the same Ref., p. 322, (i′), for relevant remarks in connection with our usual assumption, as in (5.27)/(5.28) in the sequel of the present treatise.

Now, within the same context, we still remark that;

"dynamics", *in terms of* ADG, means in effect by the very definitions of the whole procedure, an (axiomatic) issue/theory,

(5.12)   (5.12.1)   *"...* in which the *fundamental ingredients are* [just] *fields* ...".

See also (5.16) below, as it concerns their

(5.12.2)   *"differential-geometric"* behavior (à la ADG).

The emphasis in (5.12.1) is ours and the relevant phrase is due to S. Weinberg, see (5.16) below, characterizing *Quantum Field Theory* (QFT). Yet, within the same context, cf. also (e.g. M. Nakahara [1: p. 30]), by saying that:

(5.13)   "QFT is occasionally called *"particle physics"* since it *deals with the dynamics of particles"*.

Emphasis above is ours, together with the quotation marks. So by the very definitions, and in order to clarify, once more, our previous comments in (5.12), we further remark that:

(5.14)   *ADG* provides in fact a *"functorial"* attempt to a *QFT*. Therefore, by the same also mechanism of ADG, viz. by its *sheaf* (or even *topos*) – *theoretic character*, we further afford too *strictly speaking*, an *"innately"*/inherently relativistic one, as well.

So, to be still in accord with (5.7), *as it* actually *were* (!), we further note that: we still note that;

(5.15)   any *group action*, that might appear (thus, physically speaking), *"gauge transform"*, should retain(!) that *inherent structure/* situation, *already existing*, by definition, as in (5.14). That is, in other words, the same (action, as before) be *"gauge invariant"*!

See also A. Mallios [18: p. 1930, (1.3)], along with p. 1932, (1.17)]. Moreover, within the same vein of ideas, as in (5.13), we can still recall herewith, S. Weinberg [1: pp. 78/79] (cf. also (5.12.1) above), in that:

(5.16)   *Quantum Field Theory* means actually a theory "... in which *the fundamental ingredients are fields ...*".

The above still remind us of Einstein's words, in the sense that:

(5.17)   "... [the] *innovation ...* of the field concept ... [with] *field ... an independent, not further reducible fundamental concept.*"

See A. Einstein [1: p. 140]. Yet, to turn again to S. Weinberg, as before (cf. (5.16)), we still recall, ibid., p. 79, that:

(5.18)   "... *there is a ... field ... for each* truly *elementary particle* " [that is, the *"quantum"*].

Emphasis in (5.16)-(5.18) above is ours. So the preceding still reminds our *"diagram"*, often used so far; viz. one gets:

(5.19)   *elementary particle* ⟷ *field* ⟷ *quantum*.

See e.g. A. Mallios [14: pp. 62, 63; (3.2)/(3.3)], and (4.11) above, along with A. Mallios [17: p. 278, (6.6), and p. 280; (6.16.1)]. Yet, see also e.g. Th. Frankel [1: p. 536, Section 20.2d, in particular, p. 537]. Furthermore, we still have, with S. Weinberg, that;

(5.20)   "... *particles are little bundles of energy* in the field."

Emphasis is always ours; see also (5.18), as above. The same adage of Weinberg, as before, reminds for instance a relevant

(5.21)   *"cohomological classification of elementary particles"*; both, bosons or fermions.

Thus, see e.g. A. Mallios [GT, Vol. I: Chapter 4], along with [GT, Vol. II: p. 70ff]. On the other hand, concerning the notion of an *"elementary particle"*, as this is, for instance, advocated, in terms of ADG, it is worth remarking herewith a relevant utterance of W. Heisenberg in that (emphasis below is ours):

(5.22)   "... *particles* ... are *representations of symmetry groups,* ... and *resemble the symmetrical bodies of the Platonic view.*"

Therefore, Physics: (classical) group representations! Yet, once more, we realize in fact that:

(5.23)   "*symmetry*" (: commutativity) is inherent in Nature!

See W. Heisenberg [2: p. 83, yet p. 88 ("*particle spectrum*")]. In this context, see also A. Mallios [11: p. 151, "*symmetry axiom*"]. So one can supplement here (5.19), via the following diagram:

(5.24)   *elementary particle* ⟷ field/quantum ⟷ "*symmetry group*"
           ⟷ (Platonic) "*symmetrical bodies*".

Furthermore, in conjunction with (5.20)/(5.21), we might still refer herewith to relevant remarks of Einstein pertaining to our ability of "*localizing*"/detecting what we understand as an "*elementary particle*"; see e.g. A. Mallios [17: p. 277, (6.2)]. This is also in accord with a similar phraseology of W. Heisenberg in [2: p. 15, 71ff, 83]. However, cf. also the same great physicist, concerning the entanglement herewith of (group) *representation theory*, in this regard, as in W. Heisenberg [1: p. 48], in that:

(5.25)   "... *the wave picture* ... *has its limitations, which may be derived from the particle representation*".

In this context, one thus is led again to a

(5.26)   "*relational*", so to say, *perspective*, when confronting with QFT (: "*elementary particles*", see e.g. (5.16), as above, along with (5.19), related with ADG).

So we may still refer herewith to our previous remarks, as in A. Mallios [20], or even [17: p. 276, (5.18), and p. 277, (6.1) along the comments following it. Yet, p. 280, (6.14)-(6.16) and subsequent remarks therein].

On the other hand, within this same context, it is further instructive, as it actually were (!), to remind here, once more, that (emphasis below is ours);

(5.27)   "... *it is* [always] *dynamics that count* [(!)]."

See thus e.g. W. Heisenberg [2: p. 88]. Yet, cf. the above, in conjunction with (5.19) as before together, for instance, with A. Mallios [17: p. 269, (2.14), along with p. 573, (4.4), and the relevant comments preceding/and following it].

Yet, within this same vein of ideas, it is good to remind here the excitement (!) of Abdus Salam in P. C. W. Davies et al. [1 : p. 171], concerning *"string theory"*: namely, we notice thus with him that:

(5.28) *"... this promises-* for the first time-*a quantum theory of gravity"*. Moreover, *"... a string with finite size of the order of* 10^{-33} *centimeters ...* [and also] *... in spite of this* finite length *the theory is still local ... the incredible part of it."*

Now, by the very definitions of the same concept of the *fundamental pair*,

(5.29) $$(\mathcal{E}, \mathcal{D}),$$

within the perspective/framework of *ADG*, see thus e.g. (5.5.3) as above we further remark that:

the same *liberation* of the whole mechanism *of ADG from* the concept of *"space"*[-time], in the classical sense of this term (: CDG, e.g. (smooth) *manifolds*, yet the famous *chez* C. Isham, *"golden shackles"*), allows, in that context, the possibility for the whole theory (of ADG) of an

(5.30)

(5.30.1) *upmost* (in fact *"relational"*), *localization, through* the concept of *sheaf*, even *still more* via that one of topos.

Yet, not less important, all this

(5.30.2) *without losing anything,* at all (if appropriately interpreting, namely still *"relationally"*, the classical theory), from the inherent (actually algebraic) *"analytic" machinery* of the latter (classical) theory!

In this context, see also e.g. A. Mallios [22].

Now, concerning the preceding *remarks in (5.30.1)*, one is actually referring to

(5.31) *"localization"*, in terms of *sheaf–* or even *topos – theory*, as this is accomplished *within ADG*.

Moreover, see also our previous comments in (4.17) in the preceding, pertaining to Leibniz, and then after it to Einstein; yet, cf. Section 2.3, (7.53) in the foregoing. Thus, still the problem here is *always* (!) *"spatial"*, *in character*, in the classical sense of the latter term. Namely, CDG and even, in general, any other *"manifold-based"* analogous perspective.

Furthermore, another corroboration of the same *"non spatial"* viewpoint, as advocated, just before comes still e.g., quite recently from relevant remarks of A. Connes in Sh. Magid (Ed.) [1: p. 236], in that;

(5.32) *"... the idea of trying to *quantize* the *gravitational field* in a *fixed* background *spacetime manifold ... becomes irrealistic* [(!)]".*

Emphasis above is ours. Moreover, this still in accord with similar [really caustic (!)] remarks of A. Einstein, since *already from 1916*, on the subject of *quantum gravity* (!), as quoted e.g. by J. Stachel in [1: p. 280] (see Refs); cf. also A. Mallios [14: p. 59; (1.6), together with relevant comments before/after it].

Yet, by further referring to A. Mallios [11: p. 149, (2.2)], one finds therein an expression of the (actual) *"state"* of a given *"physical system"*, in terms always of ADG, which resembles an analogous formulation, pertaining to the *"virtual state"* of the whole matter, when still taking into account the *"standard model"* action, yet, *"perturbation state space"* (ibid.), of the situation, as presented e.g. in A. Connes: Sh. Magid (Ed.) [1: p. 196, (4.1)].

Now, a *general moral* that springs out herewith by actually paraphrasing H. Weyl [1], is that:

(5.33)   anytime we succeed to express a *physical law "functorially"* (alias, *"relationally"*, as above), this means that we have achieved in fact to grasp it better!

Yet, *"functorially"*, or *"relationally"*, herewith, means essentially, according to our viewpoint (ADG), that one, which is advocated e.g. by A. Mallios [18: p. 1930, (1.3)], or even in general [20]; namely, through *"A-invariance"*!

So, in that context, and by still following H. Weyl (ibid.), we conclude that: by working within the context of (5.33), we achieve thus to

(5.34)   formulate *physical laws, "mathematically"* (ibid.); i.e., *independent of coordinates, dimensions and the like.*

See also Th. Frankel [1: p. 617; frontispiece to Appendix A]. Within the same vein, cf. also e.g. H. Weyl, as quoted in (2.30) of this treatise. Yet, relevant quotations in A. Mallios [14: 58, (1.3)/(1.4)], due to R. Feynman and C. Isham, respectively, as just indicated. Furthermore, by

(5.35)   formulating a (*differential*) *equation, within the context of ADG, is equivalent*, in fact with expressing the *physical law, described by the equation*, at issue, *in a functorial* (hence in effect a *quantized*, see also e.g. (5.39)/(5.40) below) *way*!

As a consequence, one thus concludes that:

(5.36)   a *differential equation*, as before, falls within the *"principle of general covariance"*, or even, the *"principle of general relativity"* (!); indeed, *the latter* (two) *principle*(s), is a *spin-off of* what we might call,

(5.36.1)   *"principle of A-(co)variance"*.

In this context, see also A. Mallios [17: pp. 270, 271; (3.7)/(3.8), and subsequent remarks, therein]; yet [18: p. 1931; (1.7) and the following comments, together with (1.9)]. Therefore, according to the same viewpoint of ADG, namely, in that;

(5.37) | *Nature is "functorial"* (ibid. p. 1930, (1.5.1)), the *existence of "singularities"*, in the usual sense of this term (CDG), would actually be *absurd*, on the basis of the previous principles (see e.g. (5.36.1)).

Thus, one can look, in particular, at the

(5.38) | *Klein-Gordon equation*, as being, in fact, the *"relativistic form of the Laplace equation"*. See also e.g. A. Connes: in Sh. Magid [1: p. 213].

So one actually gets, still in conjunction with our previous comments, following here (5.30), at realizing, in fact,

(5.39) | *a: quantized form of* the same *Einstein's equation*, this being also *background independent*, and *A-invariant* (: "functorial"), as well. Whence, still cf. (5.33) as above.

In this context it is still relevant to notice again, together with e.g. C. von Westenholz [1], that (cf. also (5.11′) in the preceding):

(5.40) | *"quantization is provided by the physical law itself* [(!)]*."*

On the other hand, within this same vein, being also in accord with the *"principle of general covariance"* (see e.g. (5.36.1)), we should further notice, herewith, that:

(5.41) | the (actual) *quantized form of Einstein's equation (in vacuo)* is not going, of course, to be destroyed, by just adding, as a second member, therein, any *"standard model"* (action)/*"perturbation state space"* of the initial equation. Indeed, the final form of the same equation *has* still *to be*/should retain its *A-invariance* (: *"functorial"* form), anyway, as well!

See, for instance, A. Connes [1: p. 196, (4.1)]; yet, A. Mallios [11: p. 149, (2.2)]. One can also recall, herewith, what we named in the preceding, *"Utiyama's Theorem"*/ *principle*: See Section 2.1; Corollary 3.1, and Scholium 4.1, therein. Yet, A. Mallios [22: Section 3].

Finally, in connection with our previous comments in (5.36)/(5.39), *we* can further *remark that*:

(5.42) | the *first variational formula* of the *"Einstein-Hilbert functional"*/action (see e.g. A. Mallios [GT, Vol. II: p. 177, (4.2)], for the terminology employed, herewith); that is, the relation (ibid. p. 179, (4.18)),

(5.42.1) $$\delta S(\mathcal{E}) := \widehat{\delta S(\mathcal{E}(t))}(0)$$

is actually construed, via what *one might call*,

(5.42.2) | *"1-dimensional Euler-Lagrange equations"*.

See thus, for instance, L. I. Nicolaescu [1: p. 169ff]. The same is also in accord with our own viewpoint, in terms of ADG, see the above Ref., in conjunction with (5.42.1): ibid., p. 180; yet, cf. also A. Mallios [19: p. 70(4.4)-(4.11), along with the subsequent comments, therein]. Furthermore, within the same vein, we can still refer to our previous discussion herewith in Section 2.3: see e.g. (1.18) and remarks before it, or even 2.3.7, therein. Yet, (4.19)/(4.20), and subsequent comments there in 2.4.4.

On the other hand, the term

(5.43)   *"newtonian action"*

of Nicolaescu, as before, can still be quite akin to what we called,

(5.44)   *"newtonian spark"*,

in A. Mallios [16: p. 1574, (4.3) and p. 1575, Note 4.1, along with, (4.5)/(4.6), therein]. Yet, in the aforementioned work of Nicolaescu, cf. ibid. p. 173, one reads;

(5.45)   *"... the Euler-Lagrange equations are independent of coordinates"*.

In this context, one actually realizes, herewith, the

(5.46)   *independence, in fact, in terms of "1-dimensional" coordinates.*

Ibid.; thus, see also (5.34) in the preceding, together with (5.42), as above. At this point, still in conjunction e.g. with L. I. Nicolaescu [1: p. 172, Remark 5.1.6, (a)], one gets at a convenient, indeed, use of the notion of a

(5.47)   *"topological algebra scheme"*.

In this context, see also, for instance, A. Mallios [GT, Vol. II: p. 62ff], together with [17: p. 271, (3.11)-(3.13), and p. 274, (5.4)-(5.6)], and [19: p. 70, along with p. 71, Scholium 4.1]. Yet, a quite recent, relevant account/particular application, pertaining to the previous notion as in (5.47), can also be found in R. Vargas Le-Bert [1: p. 11; *"Schwartz schemes"*].

Thus, by coming back to the *"(A-)invariance" of* the *equations*, referred to in *(5.41)*, hence, by the very substance of the matter, still *in (5.45)*, *as well*, one further concludes their looking for *"relativistic invariance"* too, *in the sense of (5.7)*. So one is finally led to the following diagram:

                                                              (Legendre/Fourier)

(5.48) $(Euler\text{-}Lagrange)/Klein\text{-}Gordon \longleftrightarrow Einstein$ $\longleftrightarrow$

           $Hamilton\ equations \longleftrightarrow (A\text{-})invariance.$

Now, to paraphrase herewith Nicolaescu (ibid., p. 170), we actually construe the last term in the previous diagram (5.48), by looking at what we may call (thus, in terms always of ADG),

(5.49)   *"holonomic gauge"*.

In toto, based on the preceding, we essentially realize that:

(5.50) the whole situation, so far, still reminds, in effect, at least the glorious (!) story with *Gauss-Bonnet*!

That is, in other words,

(5.51) we do not actually need to resort to any *"outside geometry"*, as for instance, *"spacetime"* (viz., in fact, *outer space*), in order *to formulate*/study *gravity*. The latter continues to be, in effect, a *physical law* (viz. a *"field"*), after all!

In this context, we also note that:

(5.52) *"algebra"* always *entails* *"geometry"*; but, so to say, *"fuzzy"* (!), in retrospect. Only, *in the "infinitesimal"* (domain) *the two* aspects *are* essentially reduced to *the same thing* (cf., for instance, S. Germain: *"Pensées"*). Yet, this in terms e.g. of the so-called, *"microlocal analysis"*.

See also e.g. M. Kashiwara et al. [1: p. 226]: Namely,

(5.53) *"algebra"* cannot be, in general, uniquely determined by *"geometry"*; unless we let an *inner automorphism* act on the underlying geometry. That is, in other words, we have

(5.53.1) first, to *"quantize"* the geometry; ibid., Example 4.2.5 (*"quantized Legendre transformation"*).

Whence, the *"true"* S. Germain is valid in the *"infinitesimal"* (!), as above (cf. (5.52)).

On the other hand, within the same vein of thoughts, we further remark that:

(5.54) *"differentiation"* is actually a subtle *"algebraic"* (viz., in fact, *"relational"*/yet, *natural*) notion/procedure, in character (!);

(5.54.1) that is, *not* an *"analytic"* one, as we usually, quite erroneously (!), are of the impression!

Cf. thus for instance R. Feynman, in his famous, *"fancy, smanzy"* ..., utterance, by referring to CDG. Yet, H. Weyl, in his classic [2: p. 193], referring to the

(5.55) "... *awkward* [otherwise] assumptions of *differentiability* [(!)] *involved in the notion of a Lie group* [(!)]."

Emphasis in (5.55) is ours.

Now, within the same vein of ideas, we can further note that it is indeed a *general moral* of the whole perspective *of ADG* so far that:

(5.56)   the so-called *"space"* *is* always *determined by* a (certain particular, appropriate) *algebra*. The latter corresponds of course to our own (:*"we"*) organization of the *information we can gather*. Yet, this is always based on our theories that we provide at each particular period of time, about the *physical laws*. In effect, the *"space"*, viz. what actually we perceive as such, is just *a spin-off of the same laws*!

On the other hand, yet, what we usually call,

(5.57)   *"observable"* (algebras) are, in fact, those objects (still concocted by us) that can further be determined, by the *"chosen"* (manufactured) *algebra*, as above.

Thus, we arrive successively to what we may consider, as the *roots of the "Calculus"*: So we first remark that;

any *physical law functions by itself*. Thus, it is still *"we"* who perceive that function, by (instantly), realizing its

(5.58.1)   *"differ"*(ence)*s/diversifications of it*.

(5.58)   Hence, *"flow"* (Heraclitus: *"all flows"*, cf. also (10.44) in the preceding). Therefore, it is exactly here that one needs the appropriate *apparatus*/method, procedure to *detect/describe this flow*. As a spin-off, our great *innovation* (:*"Calculus"*, Fermat, Leibniz, Newton), viz. the *"differential"*; yet, in current jargon e.g. the ($\mathcal{A}$)-*connection*.

Thus, *it is* really very *important*,

(5.59)   *anytime* we can afford the aforementioned detecting/describing *instrument*/tool, as *provided directly* (!) *from the* initial *algebra* already, as *in (5.56)*. (See e.g. Leibniz, yet ADG, along with the following comments below).

Now, the problem thus arises, as to whether, one can

(5.60)   *invent* a short of a *"differ*(ential)*"* (procedure) *from the physical law* itself! True, via the *"host"* of it (so through us again, viz. still, *"we"*).

Indeed, this type of situations can actually be reduced back even e.g. to D. Hilbert (thus, his famous *"5th problem"*) referring to a relevant perspective in the case of *Lie groups*; one can still mention herewith, the telling remarks of H. Weyl in his classic [2: p. 193]: namely, to repeat it again, (cf. also (5.55), as before), one should try to *rescind the*

(5.61)   *"... awkward assumptions of differentiability* [(!)] involved in the notion of a Lie group".

Emphasis above is ours.

A further recent account on the aforementioned celebrated problem is provided, *"in full generality"*, regarding *Lie group actions on manifolds*, by the work of E. E. Rosinger [2: p. 5, and Chapt. 11]. Herewith, the intervention of the so-called *"generalized functions"*, à la Rosinger, is instrumental, hence, the link of the whole story with the point of view of ADG.

On the other hand, relevant recent accounts can still be found, from the same perspective of ADG, via the theory of *Topological Algebras*; see e.g. *"geometric topological algebras"*, or even *"topological algebra schemes"*, cf. A. Mallios [VS, Vol. II: p. 311], and [19: p. 69, Section 4]. Here, concerning in particular the first of the latter two Refs., what we named in the preceding *"Kähler construction"* is essentially involved. Therefore, what one might further consider, within the same context, as a

(5.62)   *"Kähler-Utiyama principle"*.

Cf. also in this context, A. Mallios [22]. Of course, the aforementioned, just before, point of view, still in terms of ADG, is always *assured/characterized*, through *"A-invariance"*!

Furthermore, within the same framework as before one can also consider the quite recent work of M. Papatriantafillou [4], [5], in the spirit of (5.59)/(5.60), with a systematic *looming up of the* fundamental, really instrumental, *"functorial"* character *of ADG*. Yet, the work of R. Vargas Le-Bert [1], the latter mainly pertaining to the important *complex case*. Moreover, the similar accounts by A. Mallios-E. Zafiris [3], by means of *"differential sheaf theory"*, in the spirit of (5.60), as well as, by E. Zafiris [4], still within an analogous perspective.

2.4.6   *Conclusions*

Within the same vein of ideas, and in connection with (7.57) of Section 2.3.7 in the preceding, while still based on Section 2.4.5, we can further remark, see also (3.5.1), as above, that:

the celebrated *"Feynman's trick"* (!) might still be conceived, just, as an

(6.1)   ingenious, in fact, *appropriate*/systematic *effectuation*
          (6.1.1)   of *"Stone-von Neumann adjunction"*, for each parti-
                      cular case at issue.

Yet, we can also notice herewith the deliberate, in fact, avoidance, in that context, of any intervention of the classical background (smooth-)manifold substratum throughout our whole rationale; this is based, of course, on the particular methodology of ADG, which thus turns out to be indeed an instrumental factor towards (6.1.1). Within the same point of view, as before, see also Section 2.3.6.

2.5   Quantized Einstein's Equation

2.5.1   *Introduction*

Referring to the title of this Section, we should actually indicate, first, that: our goal, by the following discussion, is in effect to stress further the point of view, an outcome, in fact, of the *general perspective of ADG*, so far, that:

(1.1)   the *function of a physical law*, as it concerns the relevant *differential equation(s)*, written *in the framework of ADG, is*, therefore *the* same *equations*, as well (*are*), by the very definitions, already, *quantized*!

See also, for instance, (2.2)/(2.3) in the sequel.

Thus, due to the same *"functorial" mechanism of ADG*, the *equations* obtained are also *functorial*, hence, *"A-invariant"* too. Moreover, the whole theory of ADG is upfront *"background independent"*, in the real sense of the term: namely, *without* any *entanglement of any smooth manifold*, at all, as a means *to* enroll/*support the rationale* we employ, *the latter* aspect *being* essentially *our main concern*, otherwise; this, even for theories which are usually viewed as, *"background independent"*, though developed, in fact, within the *framework of CDG, as* this is actually *applied in Physics* (viz. what we usually call *"Mathematical Physics"*).

On the other hand, it is still essentially pointed out, among other things, that *the* whole *mechanism of ADG* constitutes, indeed, what one might perceive, as *an "identity"*! See thus 2.6.1 in the sequel.

2.5.2   *Einstein's Fundamental Equation in Vacuo*

It is already a common aspect, concerning the problem of *quantum gravity*, that within the relevant context therein, viz. in the framework of the so-called *"space-time"* theory, whence, *classical general relativity*,

(2.1)   *"geometry"* becomes a *"dynamical variable"*.

See, for instance, J. C. Baez [1: p. v, Preface]. Now, our aim by this note is to stress the fact that:

by analyzing the notions, as in (2.1), *"geometry"* and *"dynamical variable"*, within the framework of ADG, one actually realizes that the *fundamental equation* (cf. Section 1.9; (9.45)),

(2.2)   (2.2.1) $$\mathcal{R}(\mathcal{E}) = 0,$$

is, in fact, the

(2.2.2)   *quantized Einstein's equation* (in vacuo)

of the *General Theory of Relativity*.

Indeed, the previous assertion is sustained on the basis of the subsequent *fundamental diagram*; the latter is, in effect, a result of the whole, so far, edifice/theory of ADG. Thus, one has the following relations (see also (6.3)/(8.28), (8.30), and (9.55.3) in the preceding):

(2.3)   *physical law* (in particular, for the case at issue, *gravity*) ⟷
         field ⟷ $(\mathcal{E}, \mathcal{D}) \equiv \mathcal{D}_{\mathcal{E}} \equiv \mathcal{D}$, *elementary particle* ⟷ *quantum*.

Yet, within the same context, see also e.g. Section 2.3; (1.7), as above. Now, before we proceed further, it is still instructive to recall, within ADG always, the following basic *identification*:

(2.4) $\mathcal{E} \longleftrightarrow \mathcal{D}_{\mathcal{E}} \equiv \mathcal{D}$ (*locally*),

based actually on the fundamental, so-called, "*transformation law of potentials*" (cf. (1.10)/(1.11), yet (10.7), as before); in fact, a "*dynamical relation*" (see also below). We may call it also, a

(2.5)   *de Broglie-Einstein-Feynman* "*duality*"/adjunction.

As a consequence, we thus obtain that:

(2.6)   (2.2.1) *is referred*, in effect, *to the field* (: physical law, cf. (2.3)) *itself*;
         whence, *to the quantum* of the same (physical) law, ibid.

Furthermore, this same relation (2.2.1) *is* actually derived, by the very theory of ADG, *without resort to any* background "*space*". That is, in other words, it is thus

(2.7)   *background* (-manifold) *independent*,

pertaining, by its concoction (see A. Mallios [GT, II]), *to the* very *field* (physical law/*gravity*) *itself*, cf. (2.3)! Yet,

(2.8)   *without* having thus to lean upon *any background space*, a standard obstacle (!), thus far, otherwise, for the theory of QG(: *quantum gravity*), in terms of CDG.

See also the following diagram together, for instance, with A. Mallios [16: p. 1557; (1.1), and p. 1558; (1.5)), along with Section 1.6]:

(2.9)   (physical) "*geometry*" ⟶   (spin off of the) *physical law*
             ⟵ (cf. (2.3)) ⟶   *field*.

So one further concludes that the equation,

(2.10)   *(2.2.1) is*, in fact, *the equation* (in vacuo) of the *quantum*/field (: *physical law/gravitation*): whence, by definition, the *quantized* one of the law at issue (*gravity*). See also (2.3)/(2.6), as above.

Furthermore, by still employing herewith the same reasoning, as before, one thus obtains that:

(2.11)   the *"geometry"* hence, still *the field* itself (cf. (2.9)), *is made into* the *"dynamical variable"* of the same equation at issue.

The latter is, of course, a *response to the standard quest* of QG. Yet, to stress/clarify, once more, (2.7), one should notice here that,

(2.12)   all the above are *really background-independent* since *no "space-time" notion is* actually *assumed, from the outset!* This, by contrast with the *Classical* (viz. manifold-based) *Differential Geometry* (CDG), together with its applications to the so-called *"Mathematical Physics"*: Furthermore, within the latter context, as of today, the *"background-independent"* theories, as e.g. *string-theory, loop quantum gravity* and *TQFT* are not actually such, in the sense of ADG, as above!

See thus, for instance, L. Smolin [1: pp. 149, 195/196], S. A. Selesnick [2: pp. 424, 427], M. F. Atiyah [1], and E. Witten [1].

Yet, in this regard, we can further remark that:

(2.13)   *topological quantum field theory* (TQFT) means, in effect,

          (2.13.1)   *quantum field theory*, in terms of *sheaf cohomology*.

          Even, *cohomology of the étale topos*, or even, *"étale cohomology"*.

Within the same vein of ideas, see also e.g. C. von Westenholz [1: p. 323, remarks on the rels. (6.2)/(6.3)]; furthermore, J. S. Milne [1], along, for instance with, S. I. Gelfand-Yu. I. Manin [1], and A. Mallios-Zafiris [3], for relations of the present account with the above.

So the *main impediment* of the aforementioned theories, as in (2.12), is essentially their *footing*, in effect, *on the smooth-manifold* concept (: vector bundles and the like): this *does not* actually allow to *"localize properly"*. Whence, they are not *"properly geometric"*/or linear (Leibniz); cf. also e.g. A. Mallios [23: Scholium 6.1; (6.30)]. This, by contrast with ADG, the latter being, by definition, *sheaf-theoretic* in character, hence, more *sensitive in localizing*! We may still say here that;

(2.14)   *sheaf theory* is, mathematically speaking, *the most* sensitive/*appropriate mechanism*, thus far, in *"organizing* (local) *information"*!

Indeed, see e.g. R. Haag, as in A. Mallios [14: p. 72, (5.16)], or even [18: p. 1935; (2.7)]. Yet, the *"global"* aspect of the story is still expressed through e.g. *sheaf cohomology*! See also A. Mallios [23: (6.23)/(6.24), and (6.29)].

2.5.3   *Einstein's Equation: The "Standard Model"*

It is a universal demand that *"interactions"* (: the subject-matter/theory, study
of physics) should respect *"invariance"*. Indeed, this is essentially a *characteristic
situation* of the function *of Physis*, in general; cf., for instance, A. Mallios [18: p.
1930; (1.2)/(1.5)], and [23]. True, this in fact, always happens (!), actually, according
to the very definitions: here by *"invariance"*, we really understand,

(3.1)   *"A-invariance"*,

in the sense of ADG; in technical terms, this means, in effect,

(3.2)   *"functoriality"* of the physical laws.

See, for instance, ibid., p. 1929, (1.1), and p. 1930, (1.2)-(1.5), along with p. 1931,
(1.7)/(1.8). Otherwise said, ibid.,

(3.3)   *Physis is independent of us* (: *"A"*)!

Consequently, the well known *"standard model"*, involved in *Einstein's equation*,
that represents the *interaction* of the *gravitation law*, the latter being already encoded therein by the term $\mathcal{R}(\mathcal{E})$ (cf. (2.2.1)), should still respect *A-invariance*; yet,
this along with the *"fundamental adjunction"*. Therefore, if we further refer to it,
as usual, by the *"tensor"*, say,

(3.4) $*T$,

(cf. also e.g. [GT, I: p. 66; (10.22)]), then again one has the *complete Einstein's
equation* in the form:

(3.5) $\mathcal{R}(\mathcal{E}) = *T.$

Now, by a similar argument to the preceding, *the same equation* still represents the
quantized one!

   In toto, the second member of Einstein's equation, namely, the so-called,
"energy-momentum tensor", stands for the *"interaction"* with the left-hand side
of the same equation; thus (*Newton's law*), it is *equal to it*, while *the whole system*,
has also to obey *"A-invariance"* (: *general action principle*). Whence, whatever
the tensor in question might be, *the* same *equation*, as above, should be *in force*!
This, again, based on the *"principle of A-invariance"*, and still, après *the functorial mechanism of ADG*! Within the same context, see also A. Mallios [GT, II: p.
173, remarks at the end of Scholium 3.1, on *"Einstein's equation"*: p. 172; (3.11),
therein]. On the other hand, we can further note, see e.g. [FP: p. 12, Theorem 4.6,
along with Final remark], that:

(3.6)   *"... what we* already *know, in terms of CDG, it is* also *valid within ADG,
           as well"*!

However, to supplement even the previous statement, we still remark that, this is
really true, yet, i n e x t e n s o (!): namely, *by absorbing* in effect, through the (*generalized*) *functions* involved (cf. below), eventual *"singularities"*(!), in the classical
sense of the latter term; see thus e.g. A. Mallios-M. E. Rosinger [2].

2.6   The Essence of ADG

Our aim by this final short Section is to point out the *essential character* of the mechanism *of ADG*, as this is revealed, through its various applications, till the moment; still, based also on the general standpoint, adopted thus far, throughout the development of the whole procedure. As a matter of fact, this perspective is actually stated already by the title of the following 2.6.1: See thus, in particular, (1.1) and (1.14) below.

2.6.1   *ADG Viewed, as an "Identity"*

Indeed, the *overall general moral*, so far, emanated from the whole venture *of ADG*, is that:

(1.1)   the *differential mechanism*, as we know it, e.g. in terms of the *Classical Differential Geometry* (CDG), however now, *in the light of ADG*, is in effect, *an "identity"*!

So let us start, for instance, with what we called in the preceding (cf. Section 1.1; (1.20)),

(1.2)   *"Goguen's principle/adjunction"*.

Thus, according to that principle, ibid.,

(1.3)   *"realization" is "universal"*!

On the other hand,

(1.4)   the *physical law* (viz. the *function of Physis*) is *"functorial"* (see e.g. A. Mallios [18: p. 1930; (1.5)]); therefore, *"universal"* in the previous sense (cf. (1.3)).

As a consequence, see e.g. (1.2),

(1.5)   the *"realization"* as above, being in effect the *"trace"* (thus, *"representation"* theory, again) *of a physical law, is* equally well, *"universal"*/functorial too!

Now, as a matter of fact, *Goguen's principle* speaks actually of a *"minimal" realization*: see e.g. S. Mac Lane [1: p. 87]. Thus,

(1.6)   *"minimal realization"* reminds, of course, *"least action principle"* (l.a.p.), therefore, *quantum*! Yet, the same being such a *functor* (ibid.) is *left adjoint to "behavior"* (functor; ibid., see also below).

Consequently, mathematically/categorically speaking, we might *associate to* the *"minimal realization"* (functor) the

(1.7.1)   functor *"observable"*.

(1.7)   The latter (action, viz. *observability*) is (alias, associated with) always, of course, *according to us*: namely, it is always, *"we"*, who *"fathom"* (Wittgenstein); whence, to employ the *ADG jargon*, we are thus essentially led to, *"A"*, again!

Therefore, see also (1.5), we can further

associate to the (minimal) *"realization"* functor, as before, *the*

(1.8.1)   functor $\mathcal{H}om$.

(1.8)

So in other words, in fact, *"representation"* theory, still *in sight*: see also, for instance, $\mathcal{E}nd\,\mathcal{E}$ in the preceding: in particular, cf. for example, Chapter 1; (9.6).

In this context, see also e.g. Section 2.4 in the foregoing, (3.2); yet, Section 2.1, (3.12)-(3.14).

On the other hand, *in view of* (1.6), one can consider *"minimal realization"*, in conjunction with the *"least action principle"*. Namely, otherwise said, as a *particular form of a physical law*; hence (see also e.g. (2.3), as before), one gets at the following diagram (ibid.):

(1.9)   *physical law* (l.a.p.) $\longleftrightarrow$ *quantum* $\longleftrightarrow$ $(\mathcal{A}-)$*connection*.

That is, *"derivative"*, whence, the interference (Kähler) of the

(1.10) $\otimes$-*functor*.

As a consequence (cf. also (1.7)/(1.8), as above), one realizes, with J. A. Goguen, the

(1.11)   *"minimal realization"*, as a *left adjoint* (functor) *to behavior*.

Thus, in point of fact, one can perceive the above as a *particular instance of the* fundamental,

(1.12) $\mathcal{H}om$-$\otimes$ *adjunction*!

As a result, in other words, we further remark that:

(1.13)   (minimal) *realization*, viewed according to the preceding, as a form of a physical law, *is universal*; therefore, an *"identity"*!

As a matter of fact, the above in conjunction with the *principle of A-invariance*, which *prevails the* whole *mechanism of ADG*, as it concerns its potential physical applications, thus far, lead us, so to speak, to the *"creed"* (Theorem), that;

the (differential) mechanism of ADG is, in effect, *an "identity"* ! That is, *it continues to hold*, even when its *"constituents"* become *variable*. So, otherwise said, *"differentiation"* is *"functorial"*; hence, its association with the *"physical law"*: see also e.g. A. Mallios [18: p. 1930; (1.5.1), and p. 1932; (1.17)], and still 2.3.3; (6.11) too, in the preceding.

(1.14)

Now, to point it out, once more, it was actually that perspective, as above, which we considered as *"Goguen's principle"*, and which also, *in conjunction with* the principle of *"A-invariance"*, *led us* finally *to* (1.14). Within the same context, see also e.g. M. Artin [1: p. 456, Section 3], together with S. Mac Lane [1: p. 18; *"natural transformation"*].

Furthermore, in connection with (1.11), we also recall that,

(1.15)   *"left adjoints preserve colimits"*.

See, for instance, S. Mac Lane [1: p. 114, Theorem 1, and p. 115, concluding comments]. This might be profitably connected/compared with the following *fundamental relation/Theorem* (Bohr):

(1.16) $$\mathcal{E} \cong \varinjlim((\mathcal{S}ets)^{\mathcal{A}^{op}}).$$

That is, an *equivalence of* the corresponding *categories*. See e.g. A. Mallios [18: p. 1943; (4.17)], along with Section 1.3: Theorem 3.1; (3.3), in the preceding.

On the other hand,

all that we know, by employing CDG, in particular, its applications, for example, in the so-called *"Mathematical Physics"*, is *still in force*, while working in the framework of ADG. So the former case (CDG) proves now to be, *just a very particular instance* of what we could actually obtain, and this let alone, still in a *simpler*, hence, *more "natural"*(!) way, *provided* one worked, *in terms of ADG*.

(1.17)

In this context, see also (3.6) in the preceding, along with subsequent comments therein. Furthermore, within the same vein of ideas, see also, for instance, [FP: p. 12, Final remarks]; thus, we are also led to a, so to speak, formal/*categorical vindication of* (1.17).

Yet, within the same point of you, and in connection with the *classical theory* (: "Mathematical Physics"), we can still refer to the following diagram, by paraphrasing C. von Westenholtz [1: p. 323; in particular (6.2)/(6.3)]; the same further corroborates in fact the notion of the *fundamental pair* $(\mathcal{E}, \mathcal{D})$, as in our axiomatics (cf., for instance, (2.3) above). That is, we have:

(1.18)

$$(\mathcal{E}, \mathcal{D}) \xleftarrow{\ (locally)\ } (g \equiv (g_{\alpha\beta}),\ \omega \equiv (\omega_\alpha)) \equiv (c, \omega)(ibid.)$$

$$\xleftarrow{\qquad} \int_c \omega \in \mathcal{A} \xleftarrow{\ "adjunction"\ } \int_{\mathcal{E}} \mathcal{D}.$$

Thus, still to paraphrase von Westenholtz, ibid., we can consider (1.18), as a (*generalized*) "*Poincaré(-Lagrange) scalar product*"/"*duality*", yet, *adjunction*. In this regard, pertaining to the latter terminology applied, see also, for instance, Chapter 1, Scholium 10.1, together with Section 1.10.1.

Whence, once more, we realize/may assert that (1.18), i.e. "*sheaf cohomology*", *characterizes*/in fact, *describes the fundamental laws of Physis*; indeed, through the pair, $(\mathcal{E}, \mathcal{D})$, as before. Therefore, "*topology*", translated into algebraic terms (thence, algebraic topology, thus, in particular, *sheaf cohomology*), proves to be *more accessible to a formulation*/description *of physical laws*. That is, something actually that goes back already, even to Leibniz: see e.g. A. Mallios [17], and [20]; namely, still in other words, "*AB-effects*": cf., for example, A. Mallios [23: Subsection 6.1; (6.3)]; yet, the same can be associated, for instance, with the so-called, "*topological quantum numbers*", whence TQFT, as above.

2.6.2 *Final Remarks*

We close the present account with some more "*final*" comments, pertaining to the "*essence of ADG*", by just referring to the title, itself, of this Section 2.6.

Thus, another issue which is actually involved, and also worthwhile of attention, within the mechanism/framework of ADG, so far, is the real *meaning*, in that context, *of the* following *diagram*:

(2.1)   "*arithmetization*"/"*calculations*" $\longleftrightarrow$ (*sheaf*)*cohomology* $\longleftrightarrow$ *physics* (:*physical laws*).

As a matter of fact, *the* previous *diagram is*, indeed, *self-explanatory*, by taking into account the whole perspective, thus far, of ADG, as this is already exposed all along the preceding discussion of this work. Yet, the whole subject, as in (2.1), might be subsumed in the really *equivalent diagram*, as follows. That is, one has;

(2.2)   (*sheaf*)*cohomology* (whence, "*characteristic classes*") $\longleftrightarrow$ "*arithmetic*" of *elementary particle* physics (: *quantum* domain).

In that context, it is still instructive to recall here that;

(2.3)   it is actually the *sheaf* (even, *topos-"étale*") *cohomology* [on a space], which *determines* the *topology* (: *topological properties*)/ *geometry* of a space, and *not the* "*space*"!

So see, for instance, A. Grothendieck [2], and S. I. Gelfand – Yu. I. Manin [1: p. IX]. Yet, H. Weyl [3: p. 86], along with e.g. Chapter 1; (2.30), and subsequent comments therein, in the preceding of the present treatise; also ibid., Note 9.1. Yet, A. Mallios [23: (6.23)/(6.26)].

Indeed, it is also worth noting, herewith, that:

(2.4)   the viewpoint of *"A-invariance"* still constitutes a *fundamental/instrumental* item, ensuring the *"functorial"* character of the matter. See also ibid., (6.31), as before; the same corresponds, in effect, to the behavior of the physical law!

On the other hand, still based on the rel. Chapter 1; (9.12), as above, we can further remark that:

(2.5)   the existence of a *"duality pairing"*, a very particular case, in fact, of an *"adjunction"*, realized specifically by an appropriate $(\mathcal{A}\text{-})bilinear$ *form* (ibid., (9.9)-(9.11)), or even, an *"A-metric"*, gives rise to such an *"arithmetization"*, as above. See also, ibid., (10.2). Yet, the *"dynamical analogue"* of the situation at issue, might be further construed, according e.g. to ibid., (9.18).

Now, it is also worth remarking, herewith, that:

(2.6)   the aforementioned *"calculations"* really concern/might be reduced, at each particular case, to such operations, pertaining actually to *"curves"*(!), in the general sense of the latter term. Namely, mathematical entities, depending only on 1-*parameter*! This is, indeed a far reaching viewpoint, extending in fact, at least, from Riemann (*"hypothesis"*) already to Kronecker (natural numbers), with intermediate stops, still at Einstein, and Fichte (light), as well as, Grothendieck (PDE's). See, for instance, 2.3.7, in general: its relevant content corroborates the preceding thoughts.

Moreover, the previous comments, as in (2.6), lead to the perspective, as it concerns the *"geometric structure"* of the Universe, as given, in terms of *light rays* (Einstein); so *"curves"*, as above. Whence, one might further perceive the aforementioned *"universal structure"*, as consisting of *"curves"*; that is, just, through the *"trajectories of the physical laws"*: yet, of their *"carriers"*! This, of course, still reminds Leibniz, by further referring to the *"relational character"* of the *"space"*: see e.g. A. Mallios [17: p. 267; (2.6)], and [20].

2.7 Peroration

"Time and space are modes by which we think, not conditions in which we live".
(A. Einstein)

The present treatise might be certainly viewed, on the basis of its whole development, still, as a sound vindication, among other things, of the above (as in the frontispiece) famous telling adage of Einstein. On the other hand, that same standpoint predominates, in fact, the entire framework of *ADG* (acronym of the so-called, *Abstract Differential Geometry*), within the context of which we actually work throughout, the above utterance of Einstein being, in effect, one of its *basic fundamental assumptions*! That is, in other words, we adopt, technically speaking a, so to speak, (smooth) *manifold-free context* for ADG, to work with, already *from the outset*!

As a result of the previous perspective, one is really able to choose an entirely,

(1)   *"relational"* (thus, *"categorical"*, so to say) *viewpoint*, to describe *"physical"* entities/objects, employing a mathematical language.

The above is, at least, *more "natural"* (!), from several points of view: so cf., for instance ; a) *Plato*: see e.g. A. Mallios [16: p. 1557, frontispiece, along with (1.1)]. Yet, the present treatise, ftn in Section 1.0. b) The so to speak, *"Einstein-Bergmann viewpoint"*, as it concerns, what we call, throughout the ensuing discussion,

(2)   *"physical geometry"*

This being, in fact, in the spirit of (1), as above; see e.g. Section 1.4, (4.2)/Note 4.1, and (7.3). c) S. Sternberg [1: p. 2-7], talking about *"physical conception of geometry"* by means of field theories. d) Within the same vein cf. also Choquet-Bruhat et al; p. v and ftn. 1 therein, referring to a similar aspect of A. Lichnerowicz.

Indeed, the above goes essentially back to Leibniz, already by referring to the real nature of the notion of *"space"*: see thus e.g. A. Mallios [17: p. 267, (2.6)]; on the other hand, a *collective aspect*, of this whole matter, pertaining to *"relational mathematics"*, in the previous sense, as for instance in (1), has been given in A. Mallios [20], with particular emphasis on *quantum gravity*: see e.g. ibid., p. 63, comments following (2.6), and subsequent discussion therein: pp. 64, 65; (2.12)-(2.14).

Thus, as an outcome of the preceding we realize that:

(3)   *"physical geometry"* means, in effect, *"relational geometry"*; whence, the aspect:

        (3.1)   *"relational mathematics"*,

        as just indicated above.

Indeed, the right way of looking at the *"things"*; that is, at the *"wirklich existierende Dingem"* (Leibniz) ; see A. Mallios [17], as before. Yet, cf. A. Döring-C. Isham [1]. Now, the same vein of ideas still leads naturally to the viewpoint of *"Feynman diagrams"*: see also e.g. A. Mallios [12: (3.8)].

Therefore, everything refers in effect to the so-called,

(4)   *"functor categories"*

See thus, *sheaf*, or even, *topos theory*. Whence, the title* of the present study. As a matter of fact, one is actually confronted with

---

* The original title of Mallios's handwritten notes has been *"Bohr's Correspondence Principle (: the Commutative Substance of the Quantum), Abstract (: Axiomatic) Quantum Gravity, and Functor Categories"*.

(5)   *"interrelations" of the above notions,* as in (4).

See also the subsequent discussion. Yet, it is still worth remarking herewith, that the above perspective of ADG, as advocated by ((3.1), might be considered as a *postanticipation,* in fact, of a similar viewpoint of A. N. Whitehead as this is presented recently in relation to quantum theory from a category-theoretic perspective in M. Epperson-E. Zafiris [1].

Now, the *second fundamental assumption* of the whole treatise, i.e. still of ADG, in general, is the so-called *creed,* that:

(6)   there are *no "singularities"* in Nature!

The first typical formulation of (6), in print, was actually that one of our earlier study in A. Mallios [12], and then in [14]: see e.g. ibid., p. 73; (5.21). Yet, a detailed account, in connection with *quantum gravity,* is presented in A. Mallios-I. Raptis [4], still *to "appear";* already available, however (not in its final form, anyway), in arXiv: gr-qc/0411121, v13/7, Mar. 2005. Independently, as it concerns the work [12], as before, there were already the relevant, *telling remarks,* of A. Einstein in [1: p. 164]; indeed, the latter not in the generality and emphasis, as in (6) above. Cf. also (7), as follows. That is, as a matter of fact, one can say that,

(7)   a *"singularity",* as we usually mean it, (CDG), *is* certainly *ours! Not of the Nature!*

Yet, in other words, anytime we speak of a so-called *"singularity"* this is actually *due to our own mechanism/theory,* we apply, to describe that, which *we* observe/realize, in general. Otherwise, *the physical law acts/functions,* always, *independently of us!* See also e.g. A. Mallios [17: p. 1930, (1.2)/(1.5)]. An independent study of the famous *Schwarzschild* solution/*"singularity",* in the same vein of ideas (ADG), is given also by I. Raptis [4].

Within the same context, it is still useful to recall here, the relevant remarks of A. S. Eddington [1], in that:

(8)   *"... the laws of motion of the singularities must be contained in the field-equations."*

Emphasis above is ours; cf. also, for instance, A. Mallios [17: p. 269, (3.2)].

So now, we next arrive at what we may actually consider, as the *fundamental,* indeed,

(9)   *Utiyama-von Westenholz-Yang-Mills principle/"adjunction".*

First of all, the previously applied terminology, as in (9), refers, in fact, to *Utiyama's Theorem:* That is, to that, which one actually understands, when essentially referring to the spirit/perspective of (9): namely, in that;

(10)   *"...interaction between ... fields* can be determined *by* postulating *invariance ...".*

See thus R. Utiyama [1: p. 1957]; yet, A. Mallios [22], along with Section 2.1; Scholium 4.1 in the present study.

Now, within the same context, we further remark, along with R. Utiyama, as quoted e.g. by L. O' Raifeartaigh [1: p. 210], that:

(11)   *"... the concept of* [($\mathcal{A}$-)] *connections is indispensable in establishing* a theory of *interactions."*!

Emphasis above is ours. The concept at issue is, of course, of *paramount importance*, from the outset already, throughout the whole mechanism/foundations of ADG: see thus e.g., Section 1.6; (6.1)/(6.3).

On the other hand, still within the previous vein of ideas, it is our general standpoint through ADG, thus far, that:

(12)

> *quantum gravity*, and, more generally, a *unified field theory* (UFT), could be achieved, just by appropriately
>
> (12.1)   *transmuting CDG*, as applied in the so-called *"Mathematical Physics"*, *into* the Jargon of *ADG*.

In this regard, see also Section 1.9.2. Indeed, the *quintessence of the above* is, in fact, our *ascertainment* in (7); namely, a result after all of the same,

(13)   *"functorial mechanism"* of *ADG*!

See e.g. Section 2.6; ibid., (1.14). Yet, Section 2.3.6. Thus, starting from our creed (cf. (6)) that,

(14)

> *there are no "singularities" in Nature*, we are essentially led to the *non-existence of impediments*, in the classical sense (CDG): that is, e.g. *singularities-infinities*, and the like. This, of course, always pertaining, in particular, to the *mechanism of ADG*! Whence, finally, arrives at one thus prelude to *a form of a UFT*(!), as in (12) above.

On the other hand, by further referring to what we named above, *"relational mathematics"* (cf. e.g. ((3.1), in conjunction with (1)), it is still worthwhile to stress the *importance, efficacy* and *"naturality"* of the *sheaf-theoretic* viewpoint adopted already, in that respect, by ADG. The *preference of* the aforesaid sheaf-theoretic *perspective* against that one of CDG (: $\mathcal{C}^\infty$-manifolds concept), is also presented, for instance, by relevant telling remarks of J. Dieudonné, in that:

(15)

> "the *definitions of differential* ($\mathcal{C}^\infty$-)*manifolds*, through the notion of an "*algebraized space*", in the sense of ADG (: here "$\mathcal{A}$" is the *sheaf of germs of* $\mathcal{C}^\infty$ *functions*"), has the "*advantage* of being *intrinsic*" (yet, "*innate*")! Within the same context, one gets a "*simplification*, (of the whole), *by the use of sheaves*"!

See thus J. Dieudonné [1: p. 240], along e.g. with A. Mallios [VS, Vol. I: p. 23]. Yet, concerning the *"naturality"* of character (whence, the *"geometric"* one, in our sense) of the aspect at issue, as before, see also A. Mallios [20], as well as, the recent book of M. Epperson-E. Zafiris [1], on *"relational realism"*.

Now, the *natural adaptation* of the above *to physics* can be achieved, *through the* following *diagram*:

(16)   *"state"* (of a physical *system*), *locally* (fathomed-Wittgenstein, cf.), *always* (!) (Haag, cf.) $\longleftrightarrow$ *observer*.

That is, in other words, *as it concerns* (16), the latter is in fact realized, *in terms of the* so-called in the text,

(17)   *"Lassner-Uhlmann adjunction"*.

See Section 1.8; (8.13), and remarks following it, along with (9.39.1), and (10.46). Yet, Section 2.4; (1.4). Furthermore, the same diagram, as before (cf. (16)/(17)), i.e.,

(18)   *observer* $\longleftrightarrow$ *"state"* (system),

is in effect, otherwise said, the *physical realization*, mathematically/categorically speaking, of the *fundamental* (homological),

(19) $\mathcal{H}om$-$\otimes$ *adjunction*.

See Section 2.1.4. Thus, within the same point of view, ibid., we can still say that:

(20)   the *"dynamical"* *analogue of* (19) can be associated, equally well, with the *fundamental pair*, throughout ADG; namely, with,

(20.1) $(\mathcal{E}, \mathcal{D})$.

See, ibid., (4.1)-(4.4), along with *Scholium* 4.1 therein. On the other hand, the same *fundamental pair*, as before, being associated with the notion of a *"field"* (whence, of a *"physical law"*), is also *naturally* associated with the notion of a *differential equation*, describing the *law/field* at issue: cf., for example, Section 2.1; (8.20)-(8.24).

As a matter of fact, the *categorical* (homology) *notion* of an *"adjunction"* permeates essentially through *all of physics*, being thus, *realized*, mathematically speaking, *in terms of* (19).

Now, the previous

(20)   *fundamental intervention* of the notion *of "adjunction"* in describing the way of *function of* a *physical law/"field"*, in general, is indeed *generic/catalytic*!

Yet, the above is further clarified, in our case (: ADG), in conjunction with what we may call, so far, as the "*principle of A-invariance*": that is, in other words, the realization that;

(21)   (21.1)   *Physis* (: *Nature*) *is* "*functorial*"!

Cf., for instance, A. Mallios [18: p. 1930, (1.5.1)]

As a point of fact, the above, i.e. as in (19)-(21), constitute, indeed, the prevailing aspect of describing, generally speaking, "*interactions*", in terms of ADG. See also e.g. A. Mallios [22], [23].

Chapter 2 of the treatise is devoted, in effect, to applications of the above *two fundamental principles* in particular standard cases of the previous perspective; that is, to the aforesaid principles/concepts of

(22)   "*A-invariance*" (always (!)), in conjunction, as the case might be, *with* the particular "*adjunction*" at issue, that may occur; thus the latter characterizes, just the special essentially situation under discussion.

As a matter of fact, all the Sections of Chapter 2 are particular important instantiations of this same notion of *adjunction*: Precisely, introductory comments and preliminaries lead in each particular case, to the specific topic, under consideration. The whole matter takes place always, of course, within the framework of "*A-invariance*" (cf. (22)). See also e.g. Section 2.6.2;

To sum up, the intrinsic "*functorial character*" of the instrumental viewpoint/mechanism of ADG, allows the application herewith of the *fundamental notion*, of "*A-invariance*"; viz. invariance of our own "*arithmetic*" (: thus, of our results too), with respect to the *function of Physis* (that is, *of the physical laws*). *Not the other way round*, this being, as usual, the view point in the classical theory! Thus, one still gets to ascertain that *all can be* (uniquely) *determined by* (appropriately) *postulating*/ associating "*A-invariance*". This, yet in connection with the corresponding *functorial/physical meaning* of the *fundamental description* of the physical law/*function*, through the "*adjunction*" $D \longleftrightarrow R(D)$!

Bibliography

M. Abel–N. N. Ntumba

1. *Universal problem for Kähler differentials in A-modules: Non-commutative and commutative cases*. Indian J. Pure Appl. Math. 45(2014), 497-511.

R. Abraham–J. E. Marsden

1. *Foundations of Mechanics* (2nd ed.). Benjamin/Cummings, Reading, MA, 1978.

R. Abraham–J. Marsden–T. Ratiu

1. *Manifolds, Tensor Analysis, and Applications*. Addison-Wesley, Reading, MA, 1983.

Y. Aharonov–D. Bohm

1. *Significance of electromagnetic potentials in the quantum theory*. Phys. Rev. 115(1959), 485-491.

2. *Further considerations on electromagnetic potentials in the quantum theory*. Phys. Rev. 123(1961), 1511-1524.

I. Arahovitis

1. *Deterministic fractals, chaos and topological algebras*. in *"General Topological Algebras"*, Proc. Intern. Workshop, Tartu, 1999. Mati Abel (Ed.), Estonian Math. Soc. Tartu, 2001: pp. 32-36.

E. Artin

1. *Geometric Algebra*. Wiley-Interscience, New York, 1957/1988.

M. Artin

1. *Algebra*. Prentice-Hall, Upper Saddle River, NJ, 1991.

M. F. Atiyah

1. *Topological quantum field theories*. Publ. Math. Inst. Hautes Etudes Sci. Paris 68 (1989), 175-186.

S. Y. Auyang

1. *How is the Quantum Field Theory Possible ?* Oxford Univ. Press, New York, 1995.

J. C. Baez

1. (Ed.). *Knots and Quantum Gravity.* Oxford Univ. Press, Oxford, 1994.

J. Baez–J. P. Muniain

1. *Gauge Fields, Knots and Gravity.* World Scientific, Singapore, 1994.

P. G. Bergmann

1. *Unitary Field Theory: Geometrization of Physics or Physicalization of Geometry?* in *"The 1979 Berlin Einstein Symposium"*. Lect. Notes in Phys., No. 100. Springer-Verlag, Berlin, 1979, pp. 84-88.

R. Berndt

1. *An Introduction to Symplectic Geometry.* Amer. Math. Soc., Providence, RI, 2001.

M. V. Berry

1. *Quantal phase factors accompanying adiabatic changes* Proc. R. Soc. Lond. A 392(1984), 45-57.

D. Bleecker

1. *Gauge Theory and Variational Principles.* Addison-Wesley, Reading, MA, 1981.

A. Böhm

1. *Quantum Mechanics.* Springer-Verlag, New York, 1979.

N. Bourbaki

1. *Algèbre.* Chapt. 1-3. Hermann, Paris, 1970.

2. *Algèbre.* Chapt. 10. *Algèbre homologique.* Masson, Paris, 1980.

G. E. Bredon

1. *Topology and Geometry.* Springer-Verlag, New York, 1993.

Y. Choquet - Bruhat, C. DeWitt - Morette with **M. Dillard - Bleick**

1. *Analysis, Manifolds and Physics.* (rev. ed.). North-Holland, Amsterdam, 1982.

A. Cannas da Silva

1. *Lectures on Symplectic Geometry.* Lect. Notes in Math. No 1764. Springer, Berlin, 2001.

J. B. Conway

1. *A Course in Functional Analysis.* Springer-Verlag, New York, 1985.

R. W. R. Darling

1. *Differential Forms and Connections.* Cambridge Univ. Press, Cambridge, 1994.

P. C. W. Davies–J. Brown (Eds)

1. *Superstrings: A Theory of Everything?* Cambridge Univ. Press, Cambridge, 1988/1991.

A. De Paris–A. Vinogradov

1. *Fat Manifolds and Linear Connections.* World Scientific, Singapore, 2009.

R. Deheuvels

1. *Formes quadratiques et groupes classiques.* Presses Univ. France, Paris, 1981.

J. Dieudonné

1. *Algebraic Geometry.* Adv. Math. 3(1969), 233-321.

P. A. M. Dirac

1. *The early years of relativity.* in *"Albert Einstein. Historical and Cultural Perspectives".* G. Holton and Yeh. Elkana (Eds). Dover, Mineola, New York, 1982, pp. 79-90.

A. Döring–C. Isham

1. *What is a Thing?: Topos Theory in the Foundations of Physics.* in *"New Structures for Physics".* Lect. Notes in Phys. No 813. Springer, Berlin, 2011. B. Coecke (Ed.), pp. 753-940.

W. Drechsler

1. *Gauge Theory of Strong and Electromagnetic Interactions Formulated on a Fiber Bundle of the Cartan Type.* Cf. M. E. Mayer [1] for the exact quotation: Part II.

D. A. Dubin–M. A. Hennings

1. *Quantum mechanics, algebras and distributions.* Pitman Res. Notes Math. Series 238, Longman, 1990.

A. S. Eddington

1. *Report of the Relativity Theory of Gravitation.* Fleetway Press, London, 1920.

A. Einstein

1. *The Meaning of Relativity* (5th ed.). Princeton Univ. Press, Princeton, NJ, 1956.

D. Eisenbud

1. *Commutative Algebra with a View Toward Algebraic Geometry.* Springer-Verlag, New York, 1995.

D. Ellerman

1. *A Theory of Adjoint Functors with some Thoughts on their Philosophical Significance.* in *"What is Category Theory?"* G. Sica (Ed.), Milan, Polimetrica, 2006, pp. 127-183.

G. G. Emch

1. *Mathematical and Conceptual Foundations of 20th-Century Physics.* North-Holland, Amsterdam, 1984.

M. Epperson–E. Zafiris

1. *Foundations of Relational Realism: A Topological Approach to Quantum Mechanics and the Philosophy of Nature.* Lexington Books, Lanham, MD, 2013.

R. L. Faber

1. *Differential Geometry and Relativity Theory. An Introduction.* Marcel Dekker, New York, 1983.

A. E. Fekete

1. *Real Linear Algebra.* Marcel Dekker, New York, 1985.

R. P. Feynman

1. *The Character of Physical Law.* Penguin Books, London, 1992.

R. P. Feynman–S. Weinberg

1. *Elementary Particles and the Laws of Physics. The 1986 Dirac Memorial Lectures.* Cambridge Univ. Press, Cambridge, 1987.

D. R. Finkelstein

1. *Quantum Relativity. A Synthesis of the Ideas of Einstein and Heisenberg.* Springer-Verlag, Berlin, 1996.

M. Fragoulopoulou

1. *Topological Algebras with Involution,* North-Holland, Amsterdam, 2005.

M. Fragoulopoulou–M. H. Papatriantafillou

1. *Smooth manifolds vs. differential triads.* Revue Roum. Math. Pures Appl. 59(2014), 203-217. [For convenience, this will be cited also, as FP].

Th. Frankel

1. *The Geometry of Physics. An Introduction* (rev. ed.). Cambridge Univ. Press. Cambridge, 1997.

M. Friedman

1. *Foundations of Space - Time Theories. Relativistic Physics and Philosophy of Sciences.* Princeton Univ. Press, Princeton, NJ, 1983.

S. I. Gelfand–Yu. I. Manin

1. *Methods of Homological Algebra* (2nd ed.). Springer-Verlag, Berlin, 2003.

R. Godement

1. *Topologie algébrique et théorie des faisceaux* (3ème Éd.). Hermann, Paris, 1973.

J. Gray

1. *Ideas of Space. Euclidean, Non-Euclidean, and Relativistic.* Oxford, Univ. Press, Oxford, 1979.

A. Grothendieck

1. *Sur quelques points d' algèbre homologique.* Tôhoku Math. J. 9(1957), 119-221.
2. *A General Theory of Fiber Spaces with Structure Sheaf* (2nd ed.). Univ. Kansas, Dept. Math., Lawrence, Kansas, 1958.

V. Guillemin–S. Sternberg

1. *Symplectic Techniques in Physics.* Cambridge Univ. Press, Cambridge, 1984.

R. Haag

1. *Local Quantum Physics. Fields, Particles, Algebras* (2nd ed.). Springer, Berlin, 1996.

K. Habermann–L. Habermann

1. *Introduction to Symplectic Dirac Operators.* Lect. Notes Math., No. 1887, Springer-Verlag, Berlin, 2006.

M. Hakim

1. *Topos annelés et schémas relatifs.* Springer-Verlag, Berlin, 1972.

W. Heisenberg

1. *The Physical Principles of the Quantum Theory.* Dover Publ., 1930.
2. *Encounters with Einstein. And Other Essays on People, Places, and Particles.* Princeton Univ. Press, Princeton, NJ, 1983.

G. 't Hooft

1. *Obstacles on the way towards the quantization of space, time and matter.* Spin-2000/20.

N. E. Hurt

1. *Geometric Quantization in Action. Applications of Harmonic Analysis in Quantum Statistical Mechanics and Quantum Field Theory.* D. Reidel, Dordrecht, 1983.

C. J. Isham

1. *Canonical groups and quantization of geometry and topology.* in *"Conceptual Problems of Quantum Gravity".* A. Ashtekar, J. Stachel (Eds.). Birkhäuser, Basel, 1991, pp. 351-400.

M. Kashiwara–T. Kawai–T. Kimura

1. *Foundations of Algebraic Analysis.* Princeton Univ. Press, Princeton, NJ, 1986.

G. Kato

1. *The Heart of Cohomology.* Springer, Dordrecht, 2006.
2. *u-Singularity and t-topos theoretic entropy.* Int. J. Theor. Phys. 49 (2010), 1952-1960, 265-298.
3. *Elements of Temporal Topos.* Abramis Sci. Publ., 2013.

S. Kobayashi

1. *Differential Geometry of Complex Vector Bundles.* Princeton Univ. Press, Princeton, 1987.

E. Kunz

1. *Kähler Differentials.* Vieweg, Braunschweig, 1986.

G. Lassner–A. Uhlmann

1. *On Op*-algebras of unbounded operators.* Trudy Mat. Inst. Steklov 135 (1978), 171-176.

H. B. Lawson, Jr.–M.-L. Michelsohn

1. *Spin Geometry.* Princeton Univ. Press, Princeton, NJ, 1989.

G. Ludwig

1. *Deutung des Begriffs "physikalische Theorie" und axiomatische Grunglegung der Hilbertraumstruktur des Quantenmechanik durch Hauptsätze des Messens.* Lect. Notes Phys., No. 4. Springer-Verlag, Berlin, 1970.

2. *Foundations of Quantum Mechanics I.* Springer-Verlag, Berlin, 1983.

S. Mac Lane

1. *Categories for the Working Mathematician.* Springer-Verlag, New York, 1971. [Referred also just by, CWM].

2. *Homology* (3rd Cor. Pr.). Springer-Verlag, Berlin, 1975.

3. *Mathematics: Form and Function.* Springer-Verlag, Berlin, 1986.

S. Mac Lane–I. Moerdijk

1. *Sheaves in Geometry and Logic. A First Introduction to Topos Theory.* Springer-Verlag, New York, 1992. [For convenience, we occasionally refer to this item simply by MM].

Sh. Magid (Ed.)

1. *On Space and Time.* Cambridge Univ. Press, Cambridge, 2008.

A. Mallios

1. *Topological Algebras. Selected Topics.* North-Holland, Amsterdam, 1986. [This item will be also indicated, throughout the text, for convenience, by TA].

2. *Lectures on Differential Geometry. An Introduction. Theory of Differential Manifolds and of Lie Groups.* Kardamitsa Publ., Athens, 1992. [Greek]

3. *On an abstract form of Weil's integrality theorem.* Note Mat. 12(1992), 167-202.

4. *On geometric topological algebras.* J. Math. Anal. Appl. 172(1993), 301-322.

5. *On an Axiomatic Treatment of Differential Geometry via Vector Sheaves. Applications.* Math. Japonica (Intern. Plaza) 48(1998), 93-180.

6. *Geometry of Vector Sheaves. An Axiomatic Approach to Differential Geometry, Volume I: Vector Sheaves, General Theory.* Kluwer, Dordrecht, 1998. [Referred occasionally, just, by VS, I].

7. *Geometry of Vector Sheaves. An Axiomatic Approach to Differential Geometry, Volume II: Geometry Examples and Applications.* Kluwer, Dordrecht, 1998. [Referred also, just, by VS, II].

8. *On an axiomatic approach to geometric prequantization: a classification scheme à la Kostant–Souriau–Kirillov.* J. Math. Sci. (New York) 95(1999), 2648-2668.

9. *Abstract differential geometry, singularities, and physical applications.* in "*Topological Algebras with Applications to Differential Geometry and Mathematical Physics*". Proc. Fest-Coll. in Honour of Professor Anastasios Mallios (16-18 Sept. 1999). P. Strantzalos and M. Fragoulopoulou (Eds.). Dept. Math., Univ. of Athens, Athens, 2002, pp. 11-18.

10. *Abstract differential geometry, general relativity, and singularities.* in "*Unsolved Problems in Mathematics for the 21st Century: A Tribute to Kiyoshi Iséki's 80th Birthday*", J. M. Abe and S. Tanaka (Eds.), IOS Press, Amsterdam, 2001, pp. 77-100.

11. *K-Theory of topological algebras and second quantization.* Acta Univ. Oulu (Sci. Rer. Natur.) A 408. J. Archippainen and M. Filali (Eds.). Oulu 2004: pp. 145-160.

12. *Remarks on "singularities".* Preprint (2002), arXiv: gr-qc/0202028.

13. *On localizing topological algebras.* Contemp. Math. 341(2004), 79-95.

14. *Quantum gravity and "singularities".* Note Mat. 25(2006), 57-76.

15. *Modern Differential Geometry in Gauge Theories. Maxwell Fields, Volume I.* Birkhaüser, Boston, 2006. [Referred occasionally still, as GT, II]

16. *Geometry and physics today.* Int. J. Theor. Phys. 45(2006), 1557-1593.

17. *On algebra spaces.* Contemp. Math. 427(2007), 263-283.

18. *A-invariance: An axiomatic approach to quantum relativity.* Int. J. Theor. Phys. 47(2008), 1929-1948.

19. *On topological algebra schemes.* Math. Studies 4(2008). Est. Math. Soc., Tartu: pp. 65-72.

20. *Relational mathematics: A response to quantum relativity.* Publ. École Norm. Sup. Takaddoum, Rabat, 2007/2010: pp. 61-68.

21. *Modern Differential Geometry in Gauge Theories. Yang-Mills Fields, Volume II.* Birkhaüser, Boston, 2010. [Referred also, just, by GT, II]

22. *On Utiyama's theme through "A-invariance".* Complex Anal. Oper. Theory 6 (2012), 775-780.

23. *"A-invariance", field interactions, and all that: invariant determination/"interpretation" of interactions.* (to appear)

A. Mallios–A. Conte-Thrasyvoulidou–Z. Daoultzi-Malamou

1. *Geometry of an A-bilinear form, Darboux theorem: A Lagrangian perspective.* J. Math. Anal. 5 (2014), 20-35.

A. Mallios–P. P. Ntumba

1. *Pairings of sheaves of A-modules.* Quest. Math. 31(2008), 397-414.

2. *On a sheaf-theoretic version of the Witt's decomposition theorem.* Rend. Circ. Mat. Palermo 58(2009), 155-168.

3. *Fundamentals for symplectic A-modules. Affine Darboux theorem.* Rend. Circ. Mat. Palermo 58(2009), 169-198.

4. *On extending A-modules through the coefficients.* Mediterr. J. Math. 10(2013), 73-89.

A. Mallios–I. Raptis

1. *Finitary spacetime sheaves of quantum causal sets: curving quantum causality.* Int. J. Theor. Phys. 40(2001), 1885-1928.

2. *Finitary Čech–de Rham cohomology: much ado without C∞-smoothness.* Int. J. Theor. Phys. 41(2002), 1857-1902.

3. *Finitary causal and quantal vacuum Einstein gravity.* Int. J. Theor. Phys. 42(2003), 1469-1509.

4. *Abstract Differential Geometry in Quantum Gravity.* (to appear)

A. Mallios–E. E. Rosinger

1. *Abstract differential geometry, differential algebras of generalized functions, and de Rham cohomology.* Acta Appl. Math. 55(1999), 231-250.

2. *Space-time foam dense singularities and de Rham cohomology.* Acta Appl. Math. 67(2001), 59-89.

A. Mallios–E. Zafiris

1. *The homological Kähler-de Rham differential mechanism, part I: Application in general theory of relativity.* Adv. Math. Phys. (2011), 191083.

2. *The homological Kähler - de Rham differential mechanism: II. Sheaf-theoretic localization of quantum dynamics.* Adv. Math. Phys. (2011), 189801.

3. *A functorial approach to quantum gravity via differential sheaf theory.* Adv. Develop. Mod. Phys. 1 (1) (2012), 1-25.

Yu. I. Manin

1. *Mathematics and Physics.* Birkhäuser, Boston, 1981.

J. E. Marsden–T. S. Ratiu

1. *Introduction to Mechanics and Symmetry. A Basic Exposition of Classical*

Mechanical Systems (2nd ed.). Springer-Verlag, New York, 1999.

M. E. Mayer

1. *Introduction to the Fiber-Bundle Approach to Gauge Theories*, in *"Fiber Bundle Techniques in Gauge Theories"*. Lect. Notes in Phys., No. 67, W. Drechsler and M. E. Mayer (Eds.). Part I, Springer-Verlag, Berlin, 1977.

2. *Gauge fields and characteristic classes.* (to appear).

J. S. Milne

1. *Étale Cohomology.* Princeton Univ. Press, Princeton N. J., 1980.

G. L. Naber

1. *Topology, Geometry, and Gauge Fields. Foundations* (2nd ed.). Springer, New York, 2011.

2. *Topology, Geometry and Gauge Fields. Interactions* (2nd ed.). Springer, New York, 2011.

J. A. Navarro González–J. B. Sancho de Salas

1. C^∞-*Differentiable Spaces.* Lecture Notes Math. No 1824. Springer-Verlag, Berlin, 2003.

J. Nestruev

1. *Smooth Manifolds and Observables.* Springer-Verlag, New York, 2003.

P. P. Ntumba

1. *Cartan - Dieudonné theorem for $\mathcal{A}$-modules.* Mediterr. J. Math. 7 (2010), 445-454.

2. *The symplectic Gram-Schmidt theorem and fundamental geometries for $\mathcal{A}$-modules.* Czech. Math. J. 62 (2012), 265-278.

3. *Clifford $\mathcal{A}$-algebras of quadratic $\mathcal{A}$-modules.* Adv. App. Clifford Alg. 22 (2012), 1093-1107.

P. P. Ntumba–A. C. Anyaegbunam

1. $\mathcal{A}$-*transvections and Witt's theorem in symplectic $\mathcal{A}$-modules.* Mediterr. J. Math. 8 (2011), 509-524.

P. P. Ntumba–B. Y. Yizengaw

1. *On the commutativity of the Clifford and scalar extension functors.* (to appear)

L. O' Raifeartaigh

1. *The Dawning of Gauge Theory.* Princeton Univ. Press, Princeton, NJ, 1997.

M. Papatriantafillou

1. *The category of differential triads.* Bull. Greek Math. Soc. 44(2000), 129-141.

2. *Initial and final differential structures.* in *Proc. Intern. Conf. on "Topological Algebras and Applications"*, Rabat 2000. École Normale Supérieure, Takaddoum, Rabat, 2004, pp. 115-123.

3. *Projective and inductive limits of differential triads.* in *"Steps in Differential Geometry"* (Debrecen 2000). Inst. Math. Inform., Debrecen, 2001, pp. 251-262.

4. *Pre-Lie groups in abstract differential geometry.* Mediterr. J. Math. (2014), 1-14.

5. *Abstract Differential Calculus.* (to appear)

6. *Abstract Differential Structures. A Functorial Approach.* (Book, in preparation)

M. J. Pflaum

1. *Analytic and Geometric Study of Stratified Spaces.* Lect. Notes in Math., No. 1768. Springer-Verlag, Berlin, 2001.

E. Prugovečki

1. *Quantum Mechanics in Hilbert Space* (2nd ed.). Academic Press, New York, 1981.

M. Puta

1. *Hamiltonian Mechanical Systems and Geometric Quantization.* Kluwer, Dordrecht, 1993.

I. Raptis

1. *Finitary spacetime sheaves.* Int. J. Theor. Phys. 39(2000), 1703-1716.

2. *Recollections of interactions with Anastasios Mallios on the problem of Quantum Gravity and life in general.* in *"Topological Algebras with Applications to Differential Geometry and Mathematical Physics".* Proc. Fest-Coll. in honour of Prof. Anastasios Mallios, Sept. 16-18, 1999. P. Strantzalos and M. Fragoulopoulou (Eds.). Dept. Math. Univ. of Athens, Athens, 2002, pp. 86-89.

3. *A sheaf and topos-theoretic perspective on finitary spacetime and gravity.* in *"Sheaves and Topoi in Theoretical Physics".* Workshop, Imperial College, London, 11-12 July 2002.

4. *Finitary-algebraic "resolutions" of the inner Schwartzschild singularity.* Int. J. Theor. Phys. 45(2006), 79-128.

5. *Glafka-2004: Categorical quantum gravity.* Intern. J. Theor. Phys. 45(2006), 1499-1527.

6. *Finitary topos for locally finite, causal and quantal vacuum Einstein gravity.* Int. J. Theor. Phys. 46(2007), 688-739.

7. *A dodecalogue of basic didactics from applications of abstract differential geometry to quantum gravity.* Int. J. Theor. Phys. 46(2007), 3009-3021.

8. (manuscript)

E. E. Rosinger

1. *Non-Linear Partial Differential Equations. An Algebraic View of Generalized Solutions.* North-Holland, Amsterdam, 1990.

2. *Parametric Lie Group Actions on Global Generalized Solutions on Nonlinear PDEs. Including a Solution to Hilbert's Fifth Problem.* Kluwer, Dordrecht, 1998.

H.-J. Schmidt

1. *Axiomatic Characterization of Physical Geometry.* Lect. Notes Phys., No. 111. Springer-Verlag, Berlin, 1979.

A. S. Schwarz

1. *Topology for Physicists.* Springer-Verlag, Berlin, 1994.

S. A. Selesnick

1. *Correspondence principle for the quantum net.* Int. J. Theor. Phys. 30(1991), 1273-1292.

2. *Quanta Logic and Spacetime* (2nd ed.). World Scientific, Singapore, 2003.

3. *Some quantum symmetries and their breaking I.* Int. J. Theor. Phys. 51(2012), 871-900.

4. *Some quantum symmetries and their breaking II.* Int. J. Theor. Phys. 52(2013), 1088-1121.

5. *Watts cohomology for a class of Banach algebras and the duality of compact abelian groups.* Math. Z. 130 (1973), 313-323.

6. *Line bundles and harmonic analysis on compact groups.* Math. Z. 146 (1976), 53-67.

S. B. Simon

1. *Holonomy, the quantum adiabatic theorem, and Berry's phase.* Phys. Rev. 51(1983), 2167-2170.

R. D. Sorkin

1. *Finitary substitute for continuous topology.* Intern. J. Theor. Phys. 30(1991), 923-947.

L. Smolin

1. *Three Roads to Quantum Gravity.* Basic Books, New York, 2001.

J.-M. Souriau

1. *Structure of Dynamical Systems. A Symplectic View of Physics.* Birkhäuser, Boston, 1997.

S. Sternberg

1. *On the role of field theories in our physical conception of geometry.* in *"Differential Geometrical Methods in Mathematical Physics II".* Proc., Bonn 1977. K. Bleuler et al. (Eds): Lect. Notes in Math., No. 676. Springer-Verlag, Berlin, 1978, pp. 1-80.

F. Strocchi

1. *Elements of Quantum Mechanics of Infinite Systems.* World Scientific, Singapore, 1985.

B. R. Tennison

1. *Sheaf Theory.* Cambridge Univ. Press, Cambridge, 1975.

P. Tondeur

1. *Introduction to Lie Groups and Transformations Groups.* Lect. Notes Math., No. 7. Springer-Verlag, Berlin, 1969.

R. Torretti

1. *Relativity and Geometry.* Dover, New York, 1996.

L. W. Tu

1. *An Introduction to Manifolds* (2nd ed.). Springer, New York, 2011.

R. Utiyama

1. *Invariant theoretical interpretation of interactions.* Phys. Rev. 101(1956), 1597-1607.

V. S. Varadarajan

1. *Lie Groups, Lie Algebras, and Their Representations.* Springer-Verlag, New York, 1984.

E. Vassiliou

1. *Geometry of Principal Sheaves*, Mathematics and Its Applications (New York), No. 578. Springer, Dordrecht, 2005.

R. Vargas Le–Bert

1. *On complex manifolds and observable schemes.* arXiv: 1203.5806v1 [math.CV] 26 Mar.

A. Weil

1. *Basic Number Theory* (3rd ed.). Springer-Verlag, Berlin, 1974.

C. von Westenholz

1. *Differential Forms in Mathematical Physics* (rev. ed.). North-Holland, Amsterdam, 1981.

H. Weyl

1. *Space-Time-Matter.* Dover Publ., 1922/1950.

2. *Classical Groups.* Princeton Univ. Press, Princeton, NJ, 1946/1953.

3. *Philosophy of Mathematics and Natural Science.* Princeton Univ. Press, Princeton, NJ, 2009.

F. Wilczek–A. Shapere

1. *Geometric Phases in Physics*, World Scientific, 1989.

E. Witten

1. *Topological quantum field theory.* Comm. Math. Phys. 117 (1988), 353-386.

L. Wittgenstein

1. *Culture and Value* (rev. ed.). G. H. von Wright (Ed.). Blackwell, 1998.

2. *Remarks on Aesthetics, Psychology and Religious Belief.* Blackwell, London, 2003.

N. M. J. Woodhouse

1. *Geometric Quantization* (2nd ed.). Oxford Univ. Press, Oxford, 1997.

C. N. Yang

1. *Magnetic monopoles, fiber bundles, and gauge fields.* Ann. N.Y. Acad. Sci. 294(1977), 86-97.

C. N. Yang–R. L. Mills

1. *Conservation of isotopic spin and isotopic gauge invariance.* Phy. Rev. 96(1954), 191-195.

E. Zafiris

1. *Probing quantum structure through Boolean localization systems.* Int. J. Theor. Phys. 39 (2000), 2761-2778.

2. *Quantum event structures from the perspective of /Grothendieck topoi.* Found. Phys. 34 (2004), 1063-1090.

3. *Interpreting observables in a quantum world from the categorial standpoint.* Int. J. Theor. Phys. 43 (2004), 265-298.

4. *Boolean coverings of quantum observable structure: A setting for an abstract differential geometric mechanism.* J. Geom. Phys. 50(2004), 99-114.

5. *Generalized topological covering systems on quantum events' structures.* J. Phys. A: Math. Gen. 39(2006), 1485-1505.

6. *Topos-theoretic classification of quantum events structures in terms of Boolean reference frames.* Intern. J. Geom. Meth. Mod. Phys. 3(2006), 1501-1527.

7. *Quantum observables algebras and abstract differential geometry: The topos-theoretic dynamics of diagrams of commutative algebraic localizations.* Int. J. Theor. Phys. 46(2007), 319-382.

8. *Physical Principles of Functorial Gauge Localization and Dynamics with a View Toward Quantum Gravity.* (Book, in preparation)

9. *Sheaf-theoretic representation of quantum measure algebras.* J. Math. Phys. 47(2006), 092103-092103.22.

10. *Boolean information sieves: a local-to-global approach to quantum information.* Int. J. Gen. Sys. 39 (2010), 873-895.

E. Zafiris–V. Karakostas

  1. *A categorial semantic representation of quantum event structures.* Found. Phys. 43(2013), 1090-1123.

D. Zhang

  1. *Young and contemporary mathematics.* Math. Intelligencer 15(1993), 13-21.

J. M. Ziman

  1. *Elements of Advanced Quantum Theory.* Cambridge Univ. Press, Cambridge, 1969.

Index

Printed in the United States
By Bookmasters